NOVEL DRUG DELIVERY SYSTEMS IN THE MANAGEMENT OF CHRONIC DISEASES

NOVEL DRUG DELIVERY SYSTEMS IN THE MANAGEMENT OF CHRONIC DISEASES

Edited by

Raj K. Keservani, PhD

Umesh D. Laddha, PhD

Eknath D. Ahire, MS, PhD

Sanjay J. Kshirsagar, PhD

AAP | APPLE ACADEMIC PRESS

First edition published 2025

Apple Academic Press Inc.
1265 Goldenrod Circle, NE,
Palm Bay, FL 32905 USA

760 Laurentian Drive, Unit 19,
Burlington, ON L7N 0A4, CANADA

CRC Press
2385 NW Executive Center Drive,
Suite 320, Boca Raton FL 33431

4 Park Square, Milton Park,
Abingdon, Oxon, OX14 4RN UK

Library and Archives Canada Cataloguing in Publication

CIP data on file with Canada Library and Archives

Library of Congress Cataloging-in-Publication Data

CIP data on file with US Library of Congress

ISBN: 978-1-77491-586-8 (hbk)
ISBN: 978-1-77491-587-5 (pbk)
ISBN: 978-1-03265-488-1 (ebk)

About the Editors

Raj K. Keservani, PhD
Associate Professor, Faculty of B. Pharmacy, CSM Group of Institutions, Prayagraj, Uttar Pradesh, India

Raj K. Keservani, PhD, MPharm, is currently working as an Associate Professor in the Faculty of B. Pharmacy at CSM Group of Institutions, Prayagraj, India. He has more than 12 years of academic (teaching) experience from various institutes in India in pharmaceutical education. He has supervised more than 25 UG students for their research work, has published 31 peer-reviewed papers in the field of pharmaceutical sciences in national and international journals, 1 patent, 60 book chapters, 3 co-authored books, and 30 edited books. He is also active as a reviewer for several international scientific journals. Dr. Keservani graduated with a pharmacy degree from the Department of Pharmacy, Kumaun University, Nainital (Uttarakhand), India. He received his Master of Pharmacy (MPharm) (specialization in pharmaceutics) from the School of Pharmaceutical Sciences, Rajiv Gandhi Proudyogiki Vishwavidyalaya, Bhopal, India. He received his PhD degree from IKG Punjab Technical University, Jalandhar, Punjab, India. Dr. Keservani is a life member of the Society of Pharmaceutical Education and Research (SPER). His research interests include nutraceutical and functional foods, novel drug delivery systems (NDDS), transdermal drug delivery/drug delivery, health science, cancer biology, and neurobiology.

Umesh D. Laddha
Assistant Professor, Department of Pharmaceutics, MET's, Institute of Pharmacy, Bhujbal Knowledge City, Adgaon, Nashik, Maharashtra, India

Umesh Dilip Laddha, PhD, MPharm, is currently working as an Assistant Professor in the Department of Pharmaceutics at MET's, Institute of Pharmacy, Bhujbal Knowledge City, Adgoan, Nashik, affiliated to Savitribai Phule Pune University, Pune. He has obtained his graduation (BPharm) and master's degree from Savitribai Phule Pune University, Pune. He is pursuing a PhD from Savitribai Phule, Pune University, Pune. He has a total of eight years of academic experience. He is an editorial board member of the *International*

Journal of Ophthalmology and Visual Science (USA) and the *Asian Journal of Pharmaceutical Research* (India). He is also working as a reviewer for various journals of Springer and Elsevier. He is the author of eight books and four chapters. He has published 16 research and review articles in well-reputed national and international journals. He has one Indian patent to his credit, and two others are in progress. He has received 11 awards, including Young Scientist Award – Bronze medal, Researcher of the Year, Best Teacher Award, and many more. He has presented his research work at various international and national conferences. He has worked on nanotechnology-based drug delivery systems and novel drug delivery systems.

Eknath D. Ahire
Assistant Professor, Department of Pharmaceutics, MET's, Institute of Pharmacy, Bhujbal Knowledge City, Adgaon, Nashik, Maharashtra, India

Eknath D. Ahire, PhD, MSPharm, is currently working as an Assistant Professor in the Department of Pharmaceutics at MET's, Institute of Pharmacy, Bhujbal Knowledge City, Adgoan, Nashik, affiliated to Savitribai Phule Pune University, Pune. Currently, he is admitted as a PhD scholar with Savitribai Phule, Pune University, Maharashtra, India. He was awarded with a NFST (RGNF) PhD Research Scholar fellowship by the Government of India. He is working for QCS Labs LLP, Mumbai, as an inspector for Nigeria regulatory market pre-shipment inspection under the Earn and Learn Scheme. He has a total of five years of experience, working as a Formulation Scientist II at Torrent Research Centre – Gandhinagar, Gujarat, as well as four years of academic experience. He is working as a reviewer for several reputed international journals, including *Pharmaceutical Development and Technology* (Taylor and Francis), *International Journal of Applied Nanotechnology Research (IJANR)* (IGI Global), and *Exploration of Immunology Journal*. He is an editorial board member of *Acta Scientific and Community Acquired Infection*, an open access journal. He is a life-member of the Association of Pharmaceutical Teachers of India (APTI). He has more than 30 publications (research, review, and book chapters) published in different national and international well-reputed and indexed journals with six textbooks. He has participated in and attended several national and international conferences and workshops. He has worked on nanotechnology-based novel drug delivery systems, nutraceuticals, and immunology. He obtained his graduation (BPharm) degree from Savitribai Phule Pune University and his master's degree (MSPharm) in Pharmaceutics from the National Institute of Pharmaceutical Education and Research-Ahmedabad (NIPER-A).

Sanjay J. Kshirsagar, PhD

Professor and Principal at MET's Institute of Pharmacy, Bhujbal Knowledge City, Adgoan, Nashik, affiliated to Savitribai Phule Pune University, Pune, Maharashtra, India

Sanjay J. Kshirsagar, PhD, is working as a Professor and Principal at MET's Institute of Pharmacy, Bhujbal Knowledge City, Adgoan, Nashik, affiliated with Savitribai Phule Pune University, Pune, India. He earned his PhD from JNTU Hyderabad. He has over 19 years of experience in the pharmacy field. He has guided 50 students for MPharm degrees, and eight students are pursuing their PhDs under his guidance. He has published over 160 research and review articles in well-reputed national/international journals and has a cumulative impact factor of 70. He has been awarded several awards, such as Best Research Guide by the national level Pharma Innova, Best Original Paper Award, Pharma Excipients Poster Award from International AG, SCHWEIZ, and many more. He is a member of the Board of Studies in Pharmaceutics of Savitribai Phule Pune University, Pune, and an inspector of the Pharmacy Council of India, India. He has been granted one patent by the Indian Patent Office, and four more are in progress. He is also the author of five books and three chapters. He has presented over 20 papers in national and international conferences. He is a well-known personality in the field of pharmacy. He is an expert in colon targeted drug delivery system.

Contents

Contributors

Eknath D. Ahire
Department of Pharmaceutics, MET's Institute of Pharmacy, Bhujbal Knowledge City, Adgaon, Nashik, Maharashtra, India

Sapana P. Ahirrao
MET's Institute of Pharmacy, Bhujbal Knowledge City, Adgaon, Nashik, Maharashtra, India

Vishal S. Bagul
Department of Pharmacognosy, H. R. Patel Institute of Pharmaceutical Education and Research, Karwand Naka, Shirpur, Maharashtra, India

Deepti D. Bandawane
Department of Pharmacology, PES Modern College of Pharmacy, Pune, Maharashtra, India

Rohan K. Barse
Department of Pharmaceutics, Yashwantrao Bhonsale College of Pharmacy, Sawantwadi, Sindhudurg, Maharashtra, India

Deepak S. Bhambere
MET's Institute of Pharmacy, Bhujbal Knowledge City, Adgaon, Nashik, Maharashtra, India

Bhavesh V. Bhavsar
Vice President, KriGen Pharmaceuticals LLC, 800 Edwards Brothers Dr, Lillington, North Carolina, USA

Somnath D. Bhinge
Department of Pharmaceutical Chemistry, Rajarambapu College of Pharmacy, Kasegaon, Maharashtra, India

Mangesh A. Bhutkar
Department of Pharmaceutics, Rajarambapu College of Pharmacy, Kasegaon, Maharashtra, India

Varsha Chaudhari
Department of Pharmacognosy, Matoshri College of Pharmacy, Eklahare, Maharashtra, India

Rohit Choure
Department of Pharmaceutics, Sanjivani College of Pharmaceutical Education and Research, Kopargaon, Maharashtra, India

Neelam L. Dashputre
Department of Pharmacology, MET's Institute of Pharmacy, Bhujbal Knowledge City, Adgaon, Nashik, Maharashtra, India

Sumit Deore
School of Pharmaceutical Sciences, Sandip University, Nashik, Maharashtra, India

Prashant K. Deshmukh
Department of Pharmaceutics, Dr. Rajendra Gode College of Pharmacy, Malkapur, Maharashtra, India

Ramdas Dolas
School of Pharmaceutical Sciences, Sandip University, Nashik, Maharashtra, India

Sachin Gaikwad
Department of Pharmaceutics, Sanjivani College of Pharmaceutical Education and Research, Kopargaon, Maharashtra, India

Bhadane Roshani Giridhar
Pharmaceutics Department, MET's Institute of Pharmacy, Bhujbal Knowledge City, Affiliated to Savitribai Phule Pune University, Nashik, Maharashtra, India

Vishal Gulecha
School of Pharmaceutical Sciences, Sandip University, Nashik, Maharashtra, India

Pravin Jha
School of Pharmaceutical Sciences, Sandip University, Nashik, Maharashtra, India

Jayesh D. Kadam
Department of Pharmacology, MET's Institute of Pharmacy, Bhujbal Knowledge City, Adgaon, Nashik, Maharashtra, India

Amit Vitthal Kakad
Pharmaceutics Department, MET's Institute of Pharmacy, Bhujbal Knowledge City, Affiliated to Savitribai Phule Pune University, Nashik, Maharashtra, India

Smita P. Kakad
Department of Pharmaceutics, MET's Institute of Pharmacy, Nashik, Maharashtra, India

Girish A. Kashid
Department of Pharmaceutical Chemistry, Sanjivani College of Pharmaceutical Education and Research, Kopargaon, Maharashtra, India

Akshada Kothule
Department of Pharmaceutics, Sanjivani College of Pharmaceutical Education and Research, Kopargaon, Maharashtra, India

Sanjay J. Kshirsagar
Department of Pharmaceutics, MET's Institute of Pharmacy, Bhujbal Knowledge City, Affiliated to Savitribai Phule Pune University, Adgaon, Nashik, Maharashtra, India

Abhijeet Kulkarni
School of Pharmaceutical Sciences, Sandip University, Nashik, Maharashtra, India

Nikita S. Kumawat
R. C. Patel Institute of Pharmaceutical Education and Research, Shirpur, Dhule, Maharashtra, India

Vivek S. Kumawat
R. C. Patel Institute of Pharmaceutical Education and Research, Shirpur, Dhule, Maharashtra, India

Umesh Dilip Laddha
Department of Pharmaceutics, MET's Institute of Pharmacy, Bhujbal Knowledge City, Affiliated to Savitribai Phule Pune University, Nashik, Maharashtra, India

Kamlesh D. Mali
R. C. Patel Institute of Pharmaceutical Education and Research, Shirpur, Dhule, Maharashtra, India

Gowtham Menon
Department of Pharmaceutics, Sanjivani College of Pharmaceutical Education and Research, Kopargaon, Maharashtra, India

Kailas K. Moravkar
Principal Research Engineer, Regeron Inc., Chuncheon, Gangwon-Do, South Korea

Mahesh P. More
Department of Pharmaceutics, Dr. Rajendra Gode College of Pharmacy, Malkapur, Maharashtra, India

Mayur More
Department of Pharmaceutics, Sanjivani College of Pharmaceutical Education and Research, Kopargaon, Maharashtra, India

Anuja Shivaji Nalkar
Bachelor of Pharmacy, MET's Institute of Pharmacy, Bhujbal Knowledge City, Affiliated to Savitribai Phule Pune University, Nashik, Maharashtra, India

Sushil Narkhede
School of Pharmaceutical Sciences, Sandip University, Nashik, Maharashtra, India

Pooja Shankar Niphade
Pharmaceutics Department, MET's Institute of Pharmacy, Bhujbal Knowledge City, Affiliated to Savitribai Phule Pune University, Nashik, Maharashtra, India

Vishal Pande
N. N. Sattha College of Pharmacy, Ahmednagar, Maharashtra, India

Chandrakantsing V. Pardeshi
Department of Pharmaceutics, R. C. Patel Institute of Pharmaceutical Education and Research, Shirpur, Maharashtra, India

Jayvadan K. Patel
Nootan Pharmacy College, Faculty of Pharmacy, Sankalchand Patel University, Visnagar, Gujarat, India

Prashant B. Patil
Nootan Pharmacy College, Faculty of Pharmacy, Sankalchand Patel University, Visnagar, Gujarat, India; Department of Pharmaceutical Chemistry, H. R. Patel Institute of Pharmaceutical Education and Research, Karwand Naka, Shirpur, Maharashtra, India

Surabhi B. Patil
Department of Pharmacology, MET's Institute of Pharmacy, Bhujbal Knowledge City, Adgaon, Nashik, Maharashtra, India

Deepak Pisote
Department of Pharmaceutical Biotechnology, Sanjivani College of Pharmaceutical Education and Research, Kopargaon, Maharashtra, India

Dheeraj S. Randive
Department of Pharmaceutics, Rajarambapu College of Pharmacy, Kasegaon, Maharashtra, India

Layba Sajjadali Sayyad
Pharmaceutics Department, MET's Institute of Pharmacy, Bhujbal Knowledge City, Affiliated to Savitribai Phule Pune University, Nashik, Maharashtra, India

Raosaheb S. Shendge
Department of Pharmaceutics, Sanjivani College of Pharmaceutical Education and Research, Kopargaon, Maharashtra, India

Milind Shewale
Department of Pharmaceutics, MET's Institute of Pharmacy, Nashik, Maharashtra, India

Rushikesh Shinde
Department of Pharmaceutics, MET's Institute of Pharmacy, Nashik, Maharashtra, India

Nitin R. Shirsath
Department of Pharmaceutics, H. R. Patel Institute of Pharmaceutical Education and Research, Karwand Naka, Shirpur, Maharashtra, India

Chetan Sonawane
Department of Pharmaceutics, MET's Institute of Pharmacy, Nashik, Maharashtra, India

Ganesh H. Wadkar
Department of Pharmacognosy, Rajarambapu College of Pharmacy, Kasegaon, Maharashtra, India

Vivek Wagh
Department of Pharmaceutics, Sanjivani College of Pharmaceutical Education and Research, Kopargaon, Maharashtra, India

Amar Zalte
School of Pharmaceutical Sciences, Sandip University, Nashik, Maharashtra, India

Abbreviations

AATD	alpha-1 antitrypsin deficiency
ACUPA	(S)-2-(3-((S)-5-amino-1-carboxypentyl) ureido) pentanedioic acid
AD	Alzheimer's disease
ADA	American Diabetes Association
ADCs	antibody-drug conjugates
AETs	active efflux transporters
AFP	alpha fetoprotein
AGE	advanced glycation end
AI	artificial intelligence
AIDS	acquired immune deficiency syndrome
ALK	anaplastic lymphoma kinase
AML	acute myelocytic leukemia
ANFIS	adaptive neural-fuzzy inference system
ANNs	artificial neural networks
APCs	antigen presenting cells
APOE	apolipoprotein E
Apt	aptamer
ARDS	acute respiratory distress syndrome
ARV	antiretroviral
ASGPR	asialoglycoprotein receptor
ATRAM	acidity-triggered rational membrane
AuNPs	gold nanoparticles
AZT	azacytidine
BBB	blood-brain-barrier
BC	breast cancer
BCG	bacillus Calmette–Guérin
Bcl-2	B-cell lymphoma 2
BDDS	bioadhesive drug delivery system
BSA	bovine serum albumin
CAD	coronary artery disease
CART	combination antiretroviral therapy

CB	chronic bronchitis
CD	cellulose derivative
CDC	centers for disease control and prevention
CES2	carboxylesterase 2
CF	cystic fibrosis
CGNCs	chitosan-gellan nanocapsules
CHD	coronary heart disease
CHF	congestive heart failure
CMC	carboxymethyl cellulose
CMC	critical micelle concentration
CMT	carrier-mediated transport
CNS	central nervous system
CO2H	carboxylic acid
COPD	chronic obstructive pulmonary disease
CPP	cell penetrating peptide
CRC	colorectal cancer
CSII	continuous subcutaneous insulin infusion
CT	computer tomography
DAN	Daunorubicin
DBCO	di-benzocyclooctyne
DDS	drug delivery system
DED	diabetic eye disease
DHA	docosahexaenoic acid
DLS	dynamic light scattering
DM	diabetes mellitus
DMARDs	disease-modifying anti-rheumatic drugs
DME	diabetic macular edema
DOCA	deoxycholic acid
DOPA	3,4-dihydroxyphenyl-l-alanine
DOX	doxorubicin
DP	diploma of polymerization
DPI	dry powder inhalers
DR	diabetic retinopathy
DSPE	1,2-distearoyl-sn-glycero-3-phosphatidylethanolamine
ECG	electrocardiogram
ECSs	extracellular spaces
EGFR	epidermal growth factor receptor

EPA	Environmental Protection Agency
ESCs	embryonic stem cells
EUS	endoscopic ultrasonography
FACS	flow cytometry analysis
FBS	fetal bovine serum
FDA	Food and Drug Administration
FIS	fuzzy inference system
FLX	fluoxetine
FPF	fine particle fraction
GA	genetic algorithm
GAD	glutamic acid decarboxylase
GAG	glycosaminoglycans
GBA	galbanic acid
GBM	glioblastoma multiforme
GC	gastric cancer
GDNF	glial cell-derived neurotrophic factor
GI	gastrointestinal
GM-CSF	granulocyte-macrophage colony-stimulating factor
GNPs	gelatin nanoparticles
HA	hydroxyapatite
HAART	highly active antiretroviral therapy
HCC	hepatocellular carcinoma
HFA	hydrofluoroalkane
HGs	hydrogels
HIV	human immunodeficiency virus
HLB	hydrophilic-lipophilic balance
HMCCs	hematological cancer cells
HMPWCs	hollow manganese Prussian white nanocapsules
HPMC	hydroxypropyl methylcellulose
HRG	heregulin
HRQoL	health-related quality of life
IB	ideal ibuprofen
IBD	inflammatory bowel disease
IDDM	insulin dependent diabetes mellitus
IRMA	intra-retinal microvascular abnormalities
LC	lung cancer
MAB	monoclonal antibody

MBs	microbubbles
MC	methylcellulose
MCF-7	Michigan Cancer Foundation-7
MCI	mild cognitive impairment
MDI	multiple daily injections
MDIs	metered dose inhalers
MDR-TB	multidrug-resistant tuberculosis
MI	myocardial infarction
MLP	multilayer perceptron
MLT	macular laser therapy
MMAD	mass median aerodynamic diameter
MPEG	methoxy polyethylene glycol
MPs	microparticles
MPS	mononuclear phagocytic system
MS	multiple sclerosis
MT	montmorillonite
MTX	methotrexate
NAC	N-acetyl cysteine
NDDS	novel drug delivery system
NDs	neurodegenerative diseases
NE	nanoemulsion
NEG	nano-emulsion based gel
NF-κB	nuclear factor-κB
NHL	non-Hodgkin lymphoma
NIDDM	non-insulin dependent diabetes mellitus
NK	natural killer
NLCs	nanostructured lipid carriers
NMDA	N-methyl-D-aspartate
NN	neural network
NPDR	non-proliferative diabetic retinopathy
NPs	nanoparticles
NSAIDs	nonsteroidal anti-inflammatory drugs
NTG	normal-tension glaucoma
NVD	new vessels on the disk
NVE	new vessel elsewhere
ODFs	Oro-dispersible films
PAC	pancreatic cancer

PAUF	pancreatic adenocarcinoma up-regulated factor
PC	phosphatidylcholine
PC	prostate cancer
PCL	polycaprolactone
PCLS	precision-cut lung slices
PD	Parkinson's disease
PDA	olydopamine
PDMAEMA	poly(2-dimethylaminoethyl methacrylate)
PDR	proliferative diabetic retinopathy
PE	phosphatidyl ethanolamine
PEG	polyethylene glycol
PEG-PLA	poly(ethylene glycol)-block-poly(lactic acid)
PEI	polyethyleneimine
PEO	polyethylene oxide
PEP	post-exposure prophylaxis
PET	positron emission tomography
PGA	polyglycolides
PIC	pancreatic islet cancer
PLA	polylactic acid
PLGA	poly(D,L-lactide-co-glycolide)
PLL	poly-L-lysine
PLT	panretinal laser therapy
PMMA	poly(methyl methacrylate)
PNPs	polymeric nanoparticles
PPAR	peroxisome proliferator-activator receptor
PPE	poly(propylene imine)
PSA	pressure-sensitive adhesives
PU	polyurethane
PVA	polyvinyl alcohol
PVP	polyvinylpyrrolidone
RA	rheumatoid arthritis
RC	renal cancer
RES	reticuloendothelial system
RISC	RNA-induced silencing complex
RMT	receptor-mediated transcytosis
RNAi	RNA interference
ROP	ring-opening polymerization

RV	residual volume
SAP	super adsorbent polymer
SD	solid dispersion
SEVI	semen-derived enhancer of viral infection
SGF	simulated gastric fluid
SIF	simulated intestinal fluid
siRNA	slight interfering RNA
SLMs	solid lipid microparticles
SLNs	solid lipid nanoparticles
SOB	shortness of breath
SR	sustained release
SRT	sertraline
ssRNA	single-stranded RNA
TB	tuberculosis
TEM	transmission electron microscopy
Tg	glass transition temperature
TGA	thermogravimetric analysis
TRP2	tyrosinase related protein 2
TS	terbutaline sulfate
TZDs	thiazolidinedione
UC	ulcerative colitis
VTDR	vision threatening DR
W/O	water-in-oil
WHO	World Health Organization
ZnONPs	zinc oxide nanoparticles

Preface

Chronic diseases are those that persist for an extended period of time and progress slowly. They are the leading cause of mortality among adults in nearly every country, and it is projected that this number will increase by 17% over the next decade. Approximately one-third of all adults worldwide suffer from multiple chronic illnesses. These diseases are recognized as the primary cause of death globally, accounting for over 60% of annual deaths worldwide. However, managing chronic diseases using conventional methods poses a significant challenge for healthcare providers.

Conventional systems face several limitations, such as frequent administration, high drug concentration, non-specific drug delivery, and rapid metabolism. These limitations ultimately result in poor patient compliance and adherence to treatment. Overcoming these limitations is urgently needed to improve the health status of patients with chronic diseases.

This book covers recent trends of novel drug delivery systems (NDDS) in the context of the most frequently occurring chronic diseases. To the best of our knowledge, all chapters contain the latest updates on NDDS with respect to individual chronic diseases. The authors of all chapters possess a good knowledge of the assigned topics. Moreover, all these authors are currently engaged in the development of NDDS to address different diseased conditions. Recent developments in research, specifically in drug delivery systems, have aimed to address almost all problems associated with the conventional approach. This makes drug delivery more site-specific and reduces the frequency of administration.

The book, titled *Novel Drug Delivery Systems in Management of Chronic Diseases*, comprises 12 chapters divided into 4 parts. These parts discuss the fundamentals of chronic diseases, novel drug delivery systems in CNS diseases, novel drug delivery systems in fatal diseases, and novel drug delivery systems and future aspects.

Part I, Fundamentals of Chronic Diseases, consists of two chapters. Chapter 1: Overview on Chronic Diseases, written by Surabhi B. Patil and her coworker, discusses chronic illnesses, risk factors, and their treatments.

Chapter 2: Overview on NDDS for Betterment of Chronic Diseases, written by Kamlesh et al., discusses the treatment of various ailments.

Carrier-based medication delivery has proven to be a successful drug delivery technique that also improves patient compliance.

Chapter 3: Insight into Suitable Polymers and Materials for the Preparation of NDDS, written by Amit Vitthal Kakad and his colleagues, describes the versatility of polymers in novel dosage forms. It offers significant advantages over the barriers that arise in traditional dosage forms.

Part II, Novel Drug Delivery System in Central Nervous System Diseases, consists of three chapters. In Chapter 4: Novel Drug Delivery Systems for the Treatment of Chronic Alzheimer's, Kulkarni et al. comprehensively cover the neuropathogenesis of Alzheimer's disease (AD), current treatment modalities, and drug delivery approaches, including nanotechnology applications.

Chapter 5: Parkinson's Targeting: A Novel Drug Delivery Approach, written by Sapana P. Ahirrao and her coworker, discusses an effective therapeutic approach in the treatment of chronic CNS diseases using these novel drug delivery approaches.

Chapter 6: Nose to Brain Delivery as Novel Drug Delivery System, written by Smita P. Kakad and her associates, discusses an effective approach for CNS diseases such as migraine, Alzheimer's disease, Neuro-AIDS, and brain tumors for effective targeting.

Part III, Novel Drug Delivery System in Fatal Disease, consists of four chapters. Chapter 7: Novel Drug Delivery Systems in Treatment of Cancer: A Better Future Hope, written by Prashant Bhagwan Patil and his associates, discusses novel drug delivery-based strategies like ligand/receptor-based targeting, penetrating peptide, and chemical conjugation of cancer chemotherapeutics.

Chapter 8: Novel Drug Delivery Systems in HIV/AIDS and Associated Complications, written by Deepak et al., reports on nano-based antiretroviral drugs that are superior due to their reduced adverse effects and high efficiency in controlling AIDS.

Chapter 9: Novel Drug Delivery Systems for Diabetes: An Immense Hope, written by Gowtham Menon and his colleagues, provides information on innovative drug delivery methods for antidiabetics, as well as potential problems.

Chapter 10: Novel Drug Delivery Systems in Management of Asthma and Pulmonary Diseases, written by Mangesh et al., focuses on significant contributions to enhancing the efficiency of drug delivery and targeted delivery to the lungs for the effective treatment of several respiratory diseases.

Chapter 11: New Frontiers for Management of Diabetic Retinopathy, written by Umesh Dilip Laddha and his associates, provides insight into the development of diabetic retinopathy, limitations of current treatments, and applications of novel delivery systems for better care.

Part VII, Novel Drug Delivery System and Future Aspects, consists of one chapter. In Chapter 12: Artificial Intelligence: A New Era of NDDS, written by Rohan K. Barse, the field of artificial intelligence (AI) is explored for its tremendous growth potential in research and development.

PART I

Fundamentals of Chronic Diseases

CHAPTER 1

Overview of Chronic Diseases

NEELAM L. DASHPUTRE,[1] UMESH DILIP LADDHA,[2] EKNATH D. AHIRE,[2] DEEPTI D. BANDAWANE,[3] SURABHI B. PATIL,[1] and JAYESH D. KADAM[1]

[1]*Department of Pharmacology, MET's Institute of Pharmacy, Bhujbal Knowledge City, Adgaon, Nashik, Maharashtra, India*

[2]*Department of Pharmaceutics, MET's Institute of Pharmacy, Bhujbal Knowledge City, Adgaon, Nashik, Maharashtra, India*

[3]*Department of Pharmacology, PES Modern College of Pharmacy, Pune, Maharashtra, India*

ABSTRACT

Chronic diseases are those that last for a long period of time and advance slowly. The four primary forms of chronic diseases, according to the World Health Organization (WHO), are cardiovascular diseases (such as heart attacks and strokes), chronic respiratory diseases (such as COPD and asthma), cancer, and diabetes. Chronic illnesses are the main cause of death worldwide, accounting for more than 60% of all deaths each year. Age-related changes, combined with plenty of increasing physical, cognitive, and psychological health issues, accelerate functional decline, lower quality of life, and lower survival rates. Chronic illness management has proven to be difficult due to a number of factors, including a lack of information technologies in outpatient settings, many sources of non-integrated data, restricted access to and usage of specialists, including educational facilities, and time constraints.

1.1 INTRODUCTION

Chronic diseases are the most common reason for mortality among adults in practically every country, and the number is expected to rise by another

Novel Drug Delivery Systems in the Management of Chronic Diseases.
Raj K. Keservani, Umesh D. Laddha, Eknath D. Ahire, and Sanjay J. Kshirsagar (Eds.)

17% within the next 10 years. Approximately one-third of all adults in the world suffer from numerous chronic illnesses. In the United States, 6 out of 10 adults have a chronic disease, and 4 out of 10 adults have 2 or more. Chronic diseases are expected to account for about 35 million out of the 58 million deaths in 2005, according to estimates. Tobacco use, insufficient physical activity, bad dietary habits, and alcoholism are all key contributors to the most common chronic diseases. Cardiovascular disease, stroke, malignancy, diabetes, respiratory diseases, and arthritis are examples of chronic diseases. Chronic illnesses have long been a key death and disability factor in high-income countries. Chronic illnesses are responsible for more than 70% of global deaths and also more than 87% in the United States of America. A chronic illness affects about one out of every two Americans. Aside from the human welfare cost, treatment of chronic disease accounts for around three-fourths of all healthcare expenditures in the United States [Centers for Disease Control and Prevention (CDC)]. Chronic diseases have a direct impact on healthcare spending, labor productivity, and the economy. Noncommunicable illnesses represent two-thirds of the burden of global disease in the mid-nineties, and they are anticipated to increase by three-quarters by 2030, often in tandem with economic growth. Risks to public health take place on two fronts simultaneously, which is of significant concern in many countries with low- and middle-income: In the slums of today's large cities, we are seeing non-communicable diseases occur due to poor diets and practices, alongside undernutrition (Pinquart & Sorensen, 2001) (Figure 1.1).

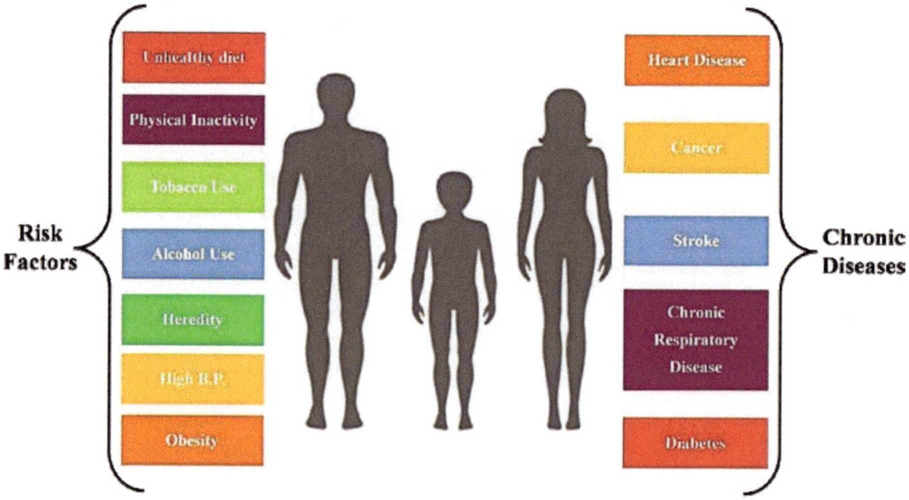

FIGURE 1.1 Factors determining health and chronic diseases.

1.2 CHRONIC DISEASES OF THE CENTRAL NERVOUS SYSTEM (CNS)

1.2.1 ALZHEIMER'S DISEASE (AD)

Alzheimer's disease (AD) is a progressive neurodegenerative illness that begins gradually and continues to worsen over time. It is considered to be the cause of 60–70% of the total dementia cases. The trouble recalling recent events is the most prevalent early symptom. As the illness advances, language issues, disorientation (including getting lost easily), losing interest, mood changes, self-neglect, and behavioral concerns are all possible indicators. When a person's health begins to worsen, they often withdraw from their family and community. The body's functioning diminishes with time, finally leading to death. Despite the fact that the rate of progression varies, the average lifespan following detection is three to nine decades. AD is the most prevalent type of dementia. It was named after the psychiatrist Alois Alzheimer. He was the first to characterize the condition in the early twentieth century. Alzheimer's patients lack their memory and ability to focus with time. Orientation in space and time becomes more challenging, and they find it more difficult to operate on their own in daily life. As the condition worsens, those who are affected will require greater assistance. A variety of treatments, including non-drug and drug treatments, can help slow the progression of AD, but it is impossible to cure the condition or stop it from progressing. This emphasizes the importance of proper care and support for people with AD: they require caring and solid relationships as well as a tolerant environment. Many family members require assistance as well (Azermai et al., 2012).

1.2.1.1 SYMPTOMS

Alzheimer's illness is commonly associated with forgetfulness. However, it can cause a wide range of symptoms and manifest itself in a variety of ways. How someone responds to and copes with AD is influenced not just by the alterations in their brain but also by their character, experiences, present conditions, and interpersonal interactions.

1.2.1.1.1 Memory and Cognitive Ability

As people get older, their memory, as well as other cognitive functions, deteriorate. It is a natural component of aging to lose the ability to react swiftly and freely to new events. In unfamiliar regions, it could be difficult to notice and handle problems. However, it is still easy to draw on the information accumulated over time in order to remain oriented, self-sufficient, and capable of making sensible decisions (Ahire et al., 2020).

This is not the case for those with AD. Their memory deteriorates over time. Short-term memory is initially more damaged. This indicates that individuals forget about recent events yet recall those from the past. Long-term memory, on the other hand, degrades over time.

1.2.1.1.2 Speech and Language

Every now and then, we all have to look for the proper word or feel tongue-tied. Individual words, on the other hand, are becoming increasingly difficult to remember. As dementia worsens, it becomes increasingly hard to remember the correct terms. Thus, people resort to using words that are inappropriate for the situation. As a consequence of this, others will have a hard time understanding them. Dementia patients also lose track of what words imply, making it difficult to follow conversations. As a consequence of this, it becomes progressively difficult to communicate vocally.

1.2.1.1.3 Mental Health and Changes in Behavior

Many patients with Alzheimer's have notable changes in behavior. Their personality may change dramatically in the future. They may become unusually afraid, suspicious, or passive, as well as hostile. These changes might occur suddenly, resulting in wrath, or they can be a relatively slow process, as with listlessness. This type of abnormal conduct could be linked to the condition. Fear, apathy, or hostility, on the other hand, can be entirely natural reactions to the disease's circumstances. For instance, a person with AD is constantly put in perplexing situations in which they act "incorrectly" (Azermai et al., 2012).

1.2.1.2 CAUSES

As time passes by with AD, more brain cells are lost. It's unclear why this occurs. One aspect that is understood is that persons with AD have insufficient amounts of acetylcholine, a key chemical messenger in the brain.

Protein particles have also been observed to accumulate in the brain. The nerve cells may perish as a consequence of this. However, the exact causation of AD remains unknown. Several elements are most likely at play.

1.2.1.3 RISK FACTORS

AD is more prone to develop as people get older, starting around the age of 65. Many researchers have looked into whether certain life events, diseases, or activities enhance or decrease the chance of AD. Although no definitive solutions have yet been found, research suggests that the following elements may play a role:

- diabetes;
- cardiovascular diseases;
- obesity; and
- tobacco.

1.2.1.4 DIAGNOSIS

Whereas the person is still living, it is not feasible to identify AD with 100% certainty with the currently available diagnostics. If the symptoms are present and no other reason can be determined, the condition is diagnosed. Imaging techniques like CT (Computed tomography) or MRI (Magnetic Resonance Imaging) are insufficient to determine whether or not someone has AD.

It's crucial not to jump to a diagnosis of Alzheimer's, considering symptoms like dementia, changes in behavior, and issues with orientation can be because of a number of things.

1.2.1.5 TREATMENT

Alzheimer's patients require long-term assistance and therapy. An entire team of individuals may be involved in delivering care, depending on the requirements, their life situation, and the stage of the disease. Volunteers, as well as skilled doctors, psychiatrists, nurses, and social workers, could be among them. Both medication-assisted and non-drug interventions aim both medication-assisted and non-drug interventions aim to:

- Enhance the quality of life;
- Maintain the ability to perform daily duties and remain independent for as long as feasible;

- Lessen mental health problems; and
- Assist family members in coping better.

Cholinesterase inhibitors, memantine, and extracts from the leaves of the Ginkgo biloba tree are among the most frequently used drugs to treat AD.

Memory and orientation training, completing everyday tasks as a team, art therapy, aromatherapy, and animal-assisted or music therapy are all examples of non-drug treatments. Massages and physical activities can also assist. It's also crucial for family members to receive caregiver training (https://www.ncbi.nlm.nih.gov/books/NBK279360/).

1.2.2 PARKINSON'S DISEASE (PD)

Parkinson's disease (PD) is a neurological illness affecting the brain. It usually develops into old age and inhibits a person's mobility. The actual reason for this is unknown.

Despite the fact that there is no cure at this time, there are effective therapies available to alleviate the symptoms. The most common treatment for PD is medication. It usually takes a long time to progress. Many people can live an almost independent life for so long after they've been diagnosed.

1.2.2.1 SYMPTOMS

The symptoms of PD can be rather varied. Typical signs and symptoms are:

1. **Akinesia (Unusual Movement):** Those who are impacted can only slowly move. They may move shakily, taking only brief, increasingly little steps. It's tougher for them to get moving, and taking the initial step is frequently the most difficult. Once they're moving, it's much more difficult to get them to stop, such as by commanding them to remain still. They quit raising their hands when walking as the condition worsens. Being able to hold items with their hands is becoming more difficult, as does performing jobs that need dexterity, such as tying shoelaces. The face begins to look mask-like over time due to decreased facial muscle activation. Speech gets quieter and more monotonous, and swallowing becomes more difficult.
2. **Muscle Stiffness (Rigidity):** The arms, legs, and neck, in particular, can become stiff and strained over time. Even if someone else

attempts to move the person's legs and arms, it's difficult, if not impossible, and they stiffen up even more as a result. Aches are frequently caused by excessive muscle tension.

3. **Tremor at Rest:** Most persons with PD experience a tremor at rest (involuntary shaking). It gets worse as they move, and it's especially bad in the hands. People's handwriting becomes small and less legible as a consequence of this.
   Symptoms are usually more noticeable in one body part. Other symptoms of PD include urinary and digestive problems, dizziness and faintness, rising oil secretion on the skin, concentration problems, and depression (Banerjee et al., 2011).

1.2.2.2 CAUSES

Different types of the disease exist, depending on what causes it. The most frequent type of PD is idiopathic PD. The term "idiopathic" describes a condition for which there is no known cause. PD affects the nerves that produce dopamine, a chemical messenger. One of dopamine's functions is to convey electrical signals from the brain to the muscles via nerves. Our movements are controlled by these signals. The capacity to initiate or coordinate motions is harmed when these cells are damaged. This can make it hard to maintain your balance, putting you at risk for falls and bone fractures.

However, other nervous system illnesses, such as dementia, can elicit Parkinson's-like symptoms. These symptoms can also be caused by infections, tumors, and medications such as psychiatric drugs.

1.2.2.3 DIAGNOSIS

Testing and a thorough discussion of patient history are usually used to obtain a diagnosis. Going to the doctor alongside family or friends can be beneficial because they may have observed signs that you haven't seen. In addition to a routine physical examination, the doctor will assess your responses, responsiveness (to pain or pressure, for example), and quickness (if your joints can move freely or whether your muscles oppose movement).

L-Dopa test is occasionally used to ensure that the diagnosis is correct. This test entails taking the Dopa (Levodopa) drug to evaluate if the symptoms improve. PD is characterized by quick improvement.

It's difficult to tell the difference between Parkinson's and other disorders, especially in the beginning. After that, it's a good thought to check and see how symptoms progress. Scans such as CT (Computed tomography) or MRI (Magnetic Resonance Imaging) are occasionally used to eliminate other disorders (Banerjee et al., 2011).

1.2.2.4 TREATMENT

Treatment isn't always necessary in the initial stages of the condition when the signs aren't causing problems. If the symptoms worsen, a medication that compensates for the absence of dopamine can help. This can often help to alleviate the symptoms. However, because these drugs can't prevent the disease from getting worse, they lose their effectiveness over time. Doctors must then continue to alter the medication's type and dosage. A customized pump can be employed to ensure that the medication has a continual impact. This allows the medicine to enter the small intestine through the skin or straight into the intestine.

Occupational therapy is a therapy that involves performing daily motions and activities. Muscles get weaker when persons with PD are unable to move their bodies as much. Special activities and sports are designed to help offset this by improving mobility and coordination. If your tone becomes weaker and you are unable to communicate, speech therapy may be an option.

Deep brain activation is occasionally indicated if the medication does not adequately treat the symptoms. Electrodes are implanted in specific parts of the brain during surgery. They emit electrical impulses that influence muscle function on a constant basis (https://www.ncbi.nlm.nih.gov/books/NBK293713/).

1.3 CHRONIC DISEASES OF THE OCULAR SYSTEM

1.3.1 GLAUCOMA

With age, our vision often deteriorates. Apart from the natural aging process, people can develop eye diseases that impair their vision or even cause blindness. The most frequent eye illness is glaucoma. Glaucoma affects around 2 out of every 100 adults over 40 years of age in the industrialized world.

"Glaucoma" is a term used to describe a variety of eye disorders that all include damage to the optic nerve. This damage causes ever-widening gaps in the field of vision, which are usually undetectable at first. When you look straight ahead, without moving your eyes, your field of vision is what you can see. Your ability to see things clearly (visual acuity) deteriorates as the disease progresses. Open-angle glaucoma is the most common type of glaucoma., which affects over 90% of people. This type of cancer grows slowly, and damage to the optic nerve does not produce vision issues for many years. The most popular treatments are eye drops and surgery. Angle-closure glaucoma is a less frequent type of glaucoma that can go untreated for years before causing serious vision loss and other issues.

1.3.1.1 SYMPTOMS

Glaucoma patients can no longer see, or may only see to a limited extent, specific areas within their range of vision. Blind spots form at the borders of the field of vision, frequently at the point of clearest vision (macula). Although central vision is not impacted at first, these issues are generally undetectable first.

However, the blind spots might make getting around in normal life more challenging. When driving, for example, looking straight ahead allows you to see the street ahead of you but not the things on the walkways to the right and left. It's also difficult to adapt to differences in light, such as when you walk into a dark hallway outside. And impediments such as steps or the curb can become increasingly difficult to perceive. Falls are more likely as a consequence of this.

Angle-closure glaucoma can cause an acute glaucoma attack. This occurs when intraocular pressure rises suddenly and dramatically. Acute visual issues, reddening of the eye, strong headache and eye irritation, and nausea are all common symptoms (Alexander & Berson, 1988).

1.3.1.2 CAUSES

An increase in intraocular pressure is a common cause of glaucoma (pressure in the eye). Intraocular pressure develops between both the cornea and the lens in the compartments of the eye. The fluid in the chambers is called aqueous humor, and it is created by the eyes themselves. It drains out via a very narrow tube (Schlemm's canal) after flowing constantly from the rear posterior and anterior chamber. This cycle aids in maintaining a constant level of healthy ocular pressure. The cornea, iris, and lens all receive nourishment from the aqueous fluid.

If the aqueous humor does not drain properly, it builds up, increasing the pressure within the eye. Nerve cells may perish as a consequence of the "squashing" of the optic nerve. The strength of the optic nerve will determine if high intraocular pressure produces injury; only certain persons will be sensitive to optic nerve damage. The aqueous humor outflow can be totally obstructed in angle-closure glaucoma. If this happens, no aqueous humor can exit the eye, and the eye immediately begins to enlarge. The same unit as blood pressure, mmHg (millimeters of mercury), is used to measure intraocular pressure. Normal blood pressure readings are around 10 and 21 mmHg. However, increased intraocular pressure isn't necessarily the cause of glaucoma. The optic nerve is damaged in around half of all persons who develop glaucoma despite having a very high intraocular pressure. Normal pressure is assumed to have already harmed these individuals. Because of this, glaucoma of this type is also called "normal-tension glaucoma" (NTG).

1.3.1.3 RISK FACTORS

High intraocular pressure affects about 4 out of every 100 adults over 40 years of age. However, only a small proportion of them will acquire glaucoma: Within five years, around 10 out of every 100 individuals with higher intraocular pressure will lose their vision. Several variables influence your risks, such as the thickness of the cornea and the pressure inside your eye. People who have very high intraocular pressure are far more likely to get glaucoma than those who have slightly higher intraocular pressure.

Old age, a genetic history of glaucoma, extreme nearsightedness, and diabetes are all risk factors. Dark-skinned folks are often more likely to get glaucoma.

1.3.1.4 DIAGNOSIS

Glaucoma can be diagnosed by an eye doctor well before the first symptoms begin. They will initially inquire about any symptoms you may be experiencing before examining your eye. An ophthalmoscope is a specific piece of equipment that is used to check inside the eye and see if the optic nerve is injured. Intraocular pressure is also evaluated, commonly using an applanation tonometry approach. A drop of liquid is used to numb the eye, and then the small monitoring gadget's tip is placed against the cornea.

Your eye doctor will also evaluate your field of vision if you are suspected of having glaucoma. This is done to examine if any areas of the field of vision have been damaged and if any blind spots have arisen (Alexander & Berson, 1988).

1.3.1.5 TREATMENT

The gradual loss of eyesight can be stopped or slowed by lowering intraocular pressure. Glaucoma, on the other hand, cannot be healed since the damage to the visual nerve is irreversible. If the intraocular pressure is high, Glaucoma is commonly treated with eye drops. The eye drops are used once or multiple times each day, depending on the formulation. Regular eye examinations are also necessary in this case. The examinations are scheduled one to three times a year, depending on how advanced the glaucoma is.

If the drug does not work, stops working after a little while, or isn't well-tolerated, surgery or laser therapy may be explored.

Acute glaucoma attacks must be treated as soon as possible to avoid optic nerve damage. If you have any of the normal signs, such as acute vision issues or eye pain, you should consult a physician or go to the hospital right away. They may employ laser therapy to remove the fluid that has accumulated there (https://www.ncbi.nlm.nih.gov/books/NBK367579/).

1.3.2 DIABETIC RETINOPATHY (DR)

Diabetic retinopathy (DR) is a vision-related consequence of diabetes. Injury to the arteries of the light-sensitive cells in the rear of the eye causes this condition (retina). DR may cause no symptoms or just minor vision abnormalities at first. However, it has the potential to cause blindness. Anyone with type 1 diabetes or type 2 diabetes can develop the illness. This eye issue is more prone to develop the more you have diabetes and the less well your blood glucose is controlled.

DR is classified into two types:

1. **Non-Proliferative Diabetic Retinopathy (NPDR):** It is a kind of DR in which blood vessels in the retina leak. Fluid, hemorrhage, or lipid in the retina are all signs of this. These blood vessels eventually close, resulting in ischemia or insufficient flow of blood.

2. **Proliferative Diabetic Retinopathy (PDR):** It is a condition in which aberrant blood vessels form in response to ischemia as the illness advances. These aberrant veins can bleed blood into the jelly material (vitreous) that covers your eye, causing tractional alterations to the retina's surface, detaching it, and, in the worst-case scenario, causing serious vision loss (Deeks, 2001).

1.3.2.1 CAUSES

Diabetic retinopathy (DR), which is described as damage to the retinal vessels of the eye, can cause cataracts, glaucoma, and DR. If not treated, these damaged vessels can cause inadequate blood flow (ischemia), swelling, and eventually legal blindness.

1.3.2.2 SYMPTOMS

In the initial phases of DR, you may not notice any symptoms. As the illness worsens, you may experience:

- In your eyesight, there are dark or empty places;
- Spots or black strings appear in your field of view (floaters);
- A vision that shifts from one moment to the next;
- Vision is hazy;
- Loss of vision.

1.3.2.3 RISK FACTORS

Diabetic retinopathy can affect everyone who has diabetes. The following factors can enhance your chances of having the condition:

- Diabetes has been a part of my life for a long time;
- Your blood glucose levels are out of control;
- Blood pressure that is too high;
- Tobacco used during pregnancy causes high cholesterol;
- Being black, Hispanic, or Native American has its advantages and disadvantages.

1.3.2.4 DIAGNOSIS

1. **Diabetic Retinopathy:** It can be detected during a routine eye checkup by an ophthalmologist.
2. **Acuity:** It relates to the clarity with which you can see.
3. **Intraocular Pressure:** It is measured to ensure that no indications of glaucoma are present.
4. **Muscle Function in the Eyes:** It refers to your ability to move your eyes.
5. **Peripheral Vision:** It is the ability to see out of the corners of your eyes.
6. **Pupil Response:** This test examines your pupils' reactions to light. Your eye specialist will then place drops in your eyes. Your pupils dilate as a consequence of the drops. During this examination, the doctor will look for the following:
   - Blood vessels that are abnormal;
   - You have a blemish in the center of your eye;
   - The formation of angiogenesis;
   - Swelling of the retina.

1.3.2.5 TREATMENT

When developing a treatment plan, your healthcare professional will examine the following factors:

- Age;
- Medical background;
- The amount of the injury to the retina;
- Visual acuity is the ability to see clearly;
- HgbA1c.

Your health professional may take a wait-and-see attitude in the initial stages of the condition, especially if you have decent eyesight. You'll get regular eye tests during this time, but no more treatment is required. Every 2 to 4 months, some people require an eye exam (Deeks, 2001).

Other alternatives for treatment include:

1. **Injections:** Drugs, such as anti-vascular endothelial growth factor medicines or corticosteroids, are injected into your eye by your healthcare professional. These medications aid in the slowing of disease progression and the improvement of eyesight.
2. **Laser Surgery:** Your healthcare practitioner will use a laser to minimize the inflammation in your eye and the formation of new blood vessels. The lasers can either constrict capillaries or stop them from leaking.
3. **Vitrectomy:** If you have foggy vision owing to leaking blood vessels, your healthcare physician may consider this outpatient eye surgery. The eye doctor creates a small cut in the eye during a vitrectomy. The blood vessels can be repaired, and scar tissue can be removed by the physician (https://www.mayoclinic.org/diseases-conditions/diabetic-retinopathy/symptoms-causes/syc-20371611#:~:text=Diabetic%20retinopathy%20(die%2Duh%2D,or%20only%20mild%20vision%20problems).

1.4 CHRONIC DISEASES OF THE HEART AND RESPIRATORY SYSTEM

1.4.1 HEART FAILURE

Heart failure occurs when the heart is unable to pump a sufficient amount of blood throughout the body. Their organs, muscles, as well as other forms of tissue, don't get enough oxygen as a result. Congestive heart failure (CHF) or cardiac insufficiency are other terms for heart failure. Heart failure can occur suddenly, such as after a cardiac arrest, or it can develop gradually as a consequence of chronic hypertension or coronary heart disease (CHD). Depending on the severity of your heart failure, it may go undetected, create only minimal symptoms, or have a significant impact on your physical fitness. It is difficult for those with advanced heart failure to carry out typical daily activities. Various treatments can help you live longer by alleviating symptoms.

1.4.1.1 SYMPTOMS

Heart failure can manifest itself in a variety of ways. Chest pain and trouble breathing are the most typical symptoms, as are weariness and exhaustion.

Fluid accumulation in the toes, ankles, or legs, or – less commonly – the vaginal area or stomach (belly).

A quick heartbeat, frequent urination at night, trouble concentrating, and a common cold are all possible symptoms. People may gain weight as a consequence of fluid build-up. However, there are a variety of alternative reasons for these symptoms. A number of patients with heart failure also have other medical problems. As a result, early indicators of heart failure can be difficult to spot.

1.4.1.2 CAUSES

Another health condition that destroys the heart tissue or blood vessels is frequently the cause of heart failure. The following are some common reasons:

- coronary artery disease (CAD);
- angina pectoris;
- blood pressure that is too high.

The arteries that deliver oxygen-rich blood to the heart are excessively narrow in coronary artery disease (CAD, commonly known as coronary heart disease or CHD). This can lead to a lack of blood supply to the heart, resulting in heart problems. During a cardiac arrest, a few of these blood channels are suddenly stopped, preventing blood from reaching the portion of the heart and causing muscle tissue to die. Consistently high blood pressure might cause the flexibility of blood vessels to deteriorate. The heart must thus constantly work against greater resistance. Various modifications occur in the body to combat this problem and ensure that all regions of the body are provided with blood. For example, the heart muscle thickens and gets larger, increasing the heart's circulation of blood at first. However, over time, the tissue may harden, and the heart valves may expand. This can weaken the heart's pumping of blood in the long run, ultimately leading to heart problems.

Conditions affecting the heart muscle, heart valves, or the sac around the heart can also cause heart failure (pericardium). Inflammations, congenital heart muscle defects, heart valve anomalies, and irregular heart rhythms like atrial fibrillation are examples of such disorders. Abuse of alcohol, drugs, or certain medications can potentially harm heart muscle tissue (Mottram et al., 2002).

1.4.1.3 RISK FACTORS

The following are the primary risk factors for heart failure:

- Diabetes type 2;
- Smoking;
- Consuming a large amount of alcohol;
- Having a major weight problem (obesity);
- Medications used in cancer therapy.

People who have a genetic tendency for less uncommon heart disorders, such as certain forms of heart muscle dysfunction, are more likely to develop heart failure.

1.4.1.4 DIAGNOSIS

Trouble breathing, tiredness, swelling owing to water retention, irregular heart sounds (murmurs), and crackling sounds when inhalation are all indicators of heart failure. However, in order to effectively identify heart failure, doctors must do a number of tests, including:

- An ECG (electrocardiogram) to check your heart's electrical activity and beat;
- A blood test was performed to determine the concentrations of several substances in your bloodstream; and
- An echocardiogram (sometimes known as an "echo test") is an ultrasound study of the heart.

Doctors use echo testing to determine how well the heart pumps blood and also how the atrioventricular valves are functioning.

It's also critical to figure out what's causing heart problems so that any underlying illnesses may be addressed.

1.4.1.5 TREATMENT

Heart failure treatment consists of numerous components:

1. **Treating the Underlying Condition:** If heart failure occurred as a result of some other form of cardiovascular diseases, such as excessive blood pressure, the underlying disease must also be addressed.

2. **Medication:** Beta-blockers, ACE inhibitors, and diuretics can help ease symptoms and lower the load on the heart. The symptoms of heart failure will determine the drug used.
3. **Physical Activity:** For persons with heart failure, there are special cardiac training programs that focus on individually tailored training to increase stamina and muscle mass. These programs help people improve their physical fitness and also their overall quality of life (https://www.ncbi.nlm.nih.gov/books/NBK279539/).

1.4.2 ASTHMA AND COPD

People with asthma have their airways narrowed (temporarily or permanently), making it harder to breathe. Asthma generally manifests itself as wheezing, coughing, and breathlessness in bouts or "attacks." The lungs may slowly take in less oxygen if not treated properly, compromising your overall physical fitness. Chronic obstructive pulmonary disease (COPD) is a lung illness that causes persistent damage and narrows the airways (bronchi). Once the condition worsens to this level, breathing becomes difficult. Even simple activities like walking upstairs, gardening, or taking walks can leave you out of breath at that point. COPD does not appear anywhere. Rather, it grows gradually over a longer period of time. Symptoms such as a chronic cough are sometimes misdiagnosed as a "typical" smoker's cough or asthma. When people's symptoms worsen, they're more likely to be diagnosed with COPD.

1.4.2.1 SYMPTOMS OF ASTHMA AND COPD

The following symptoms are associated with asthma:

- Breathing problems;
- When exercising, you may experience a lack of breath;
- Wheezing, whistling, and rattling are examples of sounds made when breathing (out);
- Tightness in the chest;
- Coughing and/or the desire to cough are both symptoms of asthma.

The following are some of the common COPD symptoms:

- In severe COPD, breathing problems occur during physical activity as well as at rest;
- Coughing regularly or for a long time;
- Phlegm (coughed-up phlegm);
- Wheezing and other breathing noises;
- If you have a cold or flu, your symptoms will get worse.

1.4.2.2 RISK FACTORS

Asthma runs in some families but not in others, suggesting that genes may play a role. Low birth weight and having other allergy-related illnesses, including hay fever or eczema, either oneself or in your family, are also risk factors. Boys are more likely than girls to suffer from asthma. If children are exposed to cigarette smoke, they are more prone to develop asthma. Smoking parents can lower their risk by quitting the habit.

The association between smoke and COPD is well established. While not all smokers acquire COPD, the majority of those who do have it smoke or have smoked in the past. Other sorts of regular respiratory irritation, such as occupational exposure to specific types of dust particles, can also raise your risk of acquiring the condition. Certain variables that are present at birth are thought to make some individuals more inclined to acquire COPD later in life. Problems with lung prenatal development, for example, include the unusual "alpha-1 antitrypsin deficiency (AATD)," which affects 1 to 2 people with COPD out of 100 (Mottram et al., 2002).

1.4.2.3 DIAGNOSIS

Doctors can only tell if their patient's symptoms are due to asthma or another condition if they take all of their symptoms and test results into account. Asthma is frequently diagnosed by looking at:

1. **Anamnesis (In-Depth Discussion with the Doctor):** This will include information such as the patient's symptoms, health information, and lifestyle choices.
2. **Physical Examination:** Your lungs, cardiac system, circulation, and overall health are all examined.

A pulmonary function test (peak flow monitoring and/or spirometry) can be used to assess lung function. The peak flow meter determines how quickly you can expel air from your lungs. Both the quantity you blow out and the speed at which you blast it out are assessed in spirometry.

COPD is frequently misdiagnosed at first. It's a common misconception among smokers that it's just a harmless "smoker's cough." However, indicators of COPD include a chronic cough and frequent breathlessness.

A family doctor can often do the initial testing. Any additional tests must be performed by a pulmonologist (a lung expert). Your physician will first perform a physical examination and take a blood test, as well as to inquire about any other problems you may have. Then, there are tests to see how well your lungs are working (lung function tests, spirometry). Further tests may be ordered based on the results to eliminate other illnesses such as asthma, cardiac arrest, or lung cancer (LC).

1.4.2.4 TREATMENT

The main objective of asthma therapy is to lessen the severity and frequency of problems as much as feasible. People with asthma should be able to live as ordinary a life as feasible with the therapy. It's crucial to understand that medication-assisted asthma treatment has extremely few negative effects. This is also true for asthmatic women who are expecting a child. A doctor can assist you in determining which medications are appropriate for you.

Asthma is treated with two types of medications:

- Medication that acts quickly and is given only when it is needed (reliever or "rescue" drug); and
- A long-acting drug that is administered on a regular basis.

  COPD treatment aims to alleviate symptoms, making daily life easier and enhancing the quality of life. The goal is also to limit the disease's progression and prevent exacerbations.

  Medication is a crucial aspect of COPD treatment on a regular basis. It could either be aspirated (breathed in) or consumed in tablet form. Combinations of various medications are routinely utilized based on how far the sickness has progressed. There are a variety of drugs available, including:

- To make it simpler to breathe, medications that open up the airways are often inhaled as a powder. Beta-2 agonists, anticholinergics, and methylxanthines are among them.
- Steroid sprays or pills can help to decrease airway inflammation.
- PDE4 inhibitors: These medications have anti-inflammatory properties as well (https://www.ncbi.nlm.nih.gov/books/NBK279520/; https://www.ncbi.nlm.nih.gov/books/NBK315789/).

1.5 CHRONIC DISEASES OF THE DIGESTIVE SYSTEM

1.5.1 INFLAMMATORY BOWEL DISEASES (IBDS)

Inflammatory bowel disease (IBD) is a set of diseases that affect the intestines and cause chronic inflammatory responses (pain and swelling). Crohn's disease and ulcerative colitis (UC) are examples of IBD. The digestive system is affected by both types. Treatments can aid in the management of this long-term illness.

IBD is divided into two types: Crohn's disease and UC. The following are examples of different types: Crohn's disease is a digestive system disease that causes discomfort and swelling. It can affect the anus on any side of the mouth. The small intestine and top portion of the large intestine are the most usually affected areas. UC is a condition that produces swelling and ulceration within the large intestine. Microscopic colitis is a type of colitis that causes inflammatory responses that can only be seen under a microscope.

1.5.1.1 SYMPTOMS

Symptoms of IBD might come and go. They might be moderate or severe, and they can manifest themselves immediately or gradually. IBD flares are periods of increased IBD symptoms. You're in remission when you don't have any symptoms.

Symptoms of IBD include:

- Pain in the abdomen (belly);
- Constipation or a strong desire to poop (bowel urgency);
- Bloating and gas;
- Appetite loss or decreased appetite;
- In the stool, there is mucus or blood;
- Stomach ache.

IBD can also lead to:

- Fatigue;
- Fever;
- Eyes that are itchy, red, and uncomfortable;
- Joint discomfort;
- Nausea and vomiting are common side effects;
- Rashes and sores on the skin (ulcers);
- Vision issues.

1.5.1.2 CAUSES

Researchers are still attempting to figure out why some people acquire IBD. Three things appear to be at play:

1. **Genetics:** One in every four patients with IBD has a family history of the disorder.
2. **Immune System Response:** Normally, the immune system fights infections. The immune system of persons with IBD misidentifies foods as alien things. It produces antibodies (proteins) to combat the threat, resulting in IBD symptoms.
3. **Environmental Triggers:** Personal or family history of IBD may get the disease after being exposed to an environmental trigger. Smoking, stress, pharmaceutical use, and depression are examples of these causes.

1.5.1.3 DIAGNOSIS

Symptoms of Crohn's disease and UC are similar. Neither condition can be diagnosed with a single test. Your health professional will inquire about your symptoms in order to make a diagnosis. A platelet count (CBC) and stool test may be ordered to look for symptoms of intestinal inflammation.

Some or all of the following diagnostic tests may be ordered:

- Colonoscopy is a procedure that examines both the big and small intestines.
- Endoscopic ultrasonography (EUS) is used to look for inflammation and ulcers in the digestive tract.
- The inside of the rectum and anus are examined using flexible sigmoidoscopy.

- A CT scan or an MRI is used to look for symptoms of infection or an abscess.
- Upper endoscopy is a procedure that involves looking into the digestive tract from the mouth to the beginning of the small intestine.
- Capsule endoscopy involves swallowing miniature camera equipment miniature camera equipment. As it goes through the digestive tract, the webcam records photographs.

1.5.1.4 TREATMENT

Treatment for IBD varies depending on the kind and symptoms. Medications can help you manage inflammation so that you don't have any symptoms (remission). The following medications are used in the treatment of IBD:

- Aminosalicylates (anti-inflammatory drugs such as sulfasalazine, mesalamine, or balsalazide) reduce gastrointestinal (GI) irritation.
- Infections and abscesses are treated with antibiotics.
- Biologics work by interrupting the immune system's inflammation-causing messages.
- Prednisone and other corticosteroids keep the immunity in balance and manage flare-ups.
- Immunomodulators are drugs that help to calm down a hyperactive immune system.

These over-the-counter IBD medicines may also help you:

- Medication to treat diarrhea.
- Nonsteroidal anti-inflammatory medicines (NSAIDs) are medications that are used to treat inflammation (NSAIDs).
- Vitamins and supplements, such as probiotics, are beneficial (https://my.clevelandclinic.org/health/diseases/15587-inflammatory-bowel-disease-overview).

1.6 FATAL DISEASES

1.6.1 CANCER

Cancer is a term used to describe a group of disorders characterized by uncontrolled cell growth and spread. It is possible to die if the growth of cancerous cells in this stage, known as metastasis, is not controlled. Cancer is caused by a number of external and internal sources, such as cigarettes, chemicals, radioactivity, and infectious organisms. Cancer has a wide range of causes that are complex and only half understood. Dietary variables, some illnesses, insufficient physical activity, overweight, and environmental contaminants are all known to raise the risk of cancer. These factors may interact to produce or promote malignancy in the human body, making cancer the main cause of mortality (Pinquart & Sorensen, 2001).

1.6.1.1 TYPES OF CANCER

- **On the Basis of Tissue Affected:**
  - Carcinomas are cancers that develop on the inside and outside of the body, such as the lungs, breasts, and colon.
  - Sarcomas are tumors that grow in bone, cartilage, adipose, tissue, muscle, as well as other supporting tissues.
  - Lymphomas are malignancies that start in the lymphatic system and immune response tissues and spread throughout the body.
  - Leukemias are malignancies that start in the stem cells and spread throughout the body.
  - Cancers of the thyroid, pituitary gland, adrenal gland, as well as other glandular organs are known as adenomas.
- **On the Basis of Organ Affected:**
  - Colorectal cancer (CRC) affects the colon;
  - Lung cancer affects the lungs;
  - Cancer of the liver;
  - Cancer of the stomach;
  - Bladder cancer affects the bladder;
  - Esophageal cancer affects the esophagus;
  - Non-Hodgkin lymphoma (NHL) affects the immune system;
  - Lip and oral cavity cancers affect the lips and mouth;
  - Cancer of the nasopharynx.

1.6.1.2 SYMPTOMS

Cancer symptoms differ depending on the species and location of the disease. Coughing, difficulty breathing, and chest pain are all symptoms of LC. Diarrhea, constipation, and bleeding are common symptoms of colon cancer. Some cancers don’t show any signs or symptoms at all. Symptoms of some malignancies, such as colon cancer, do not appear until the disease has progressed to an advanced stage.

Most malignancies can cause the following symptoms:

- Chills;
- Fatigue;
- Fever;
- Sweats at night;
- In the body, there is a thickening or bump;
- Coughing or hoarseness that persists;
- Bleeding or discharge that isn't explained;
- Any ache that won't go away;
- Unusual stomach discomfort or difficulty;
- Swallowing.

1.6.1.3 CAUSES

Cancer develops from the body's normal cells. When the system needs more cells, they multiply, and when the system does not, they die. Cancer appears to develop when the body's cell development is uncontrolled, and cells divide very rapidly. It can also happen when cells lose their ability to perish.

There are numerous types of cancer. Cancer can occur in nearly every organ or tissue, including the lungs, colon, breast, skin, bones, and nerve tissue. Cancer can be caused by a variety of factors, including:

- Other compounds, such as benzene;
- Excessive alcohol consumption;
- Toxins in the environment, such as toxic mushrooms;
- Excessive exposure to the sun;
- Genetic issues;
- Obesity;
- Viruses.

1.6.1.4 MDIAGNOSIS

The diagnostic tests for cancer are – imaging tests; laboratory tests; and other testing information.

1. **Imaging Tests:**

   i. **Mammograms:** An information sheet that explains the difference between screening and diagnostic mammograms, as well as mammography screening guidelines. Discusses the advantages and disadvantages of mammography screening.

ii. **Computed Tomography (CT):** A fact sheet that explains the CT scan procedure and technology, as well as its diagnostic and treatment applications.

2. **Laboratory Tests:**

i. **The Role of Screening and Diagnostic Laboratory Testing:** A fact sheet describing the role of screening and diagnostic laboratory tests. A brief overview of the elements that influence the results is included.

ii. **Cervical Cancer Screening, including the Pap Test and HPV Testing:** A fact sheet describing cervical cancer screening, which covers the Pap test and HPV testing. Testing and treatment guidelines are included in the fact sheet.

iii. **PSA (Prostate-Specific Antigen) Test:** Facts and figures that explain the PSA prostate cancer (PC) testing and discuss its advantages and limits (Pinquart & Sorensen, 2001).

1.6.1.5 TREATMENT

1. **Gene Therapy:** Several projects involving cancer gene therapy are currently being developed. According to the Chinese Medical Journal (2002), one of these efforts was a team of scientists from Shanghai Second Medical University that effectively introduced a gene into human tumor cells through a retrovirus. Antigens on the surface of tumor cells usually distinguish them from normal cells. Noncancerous cells that are similar to cancerous cells also carry these antibodies, albeit at a much lower level.

2. **Surgery:** In some circumstances, surgery can be used to diagnose, control, or even prevent cancer. The majority of cancer patients will require surgery. It has the best chance of curing cancer, especially if it hasn't progressed to other areas of the body.

3. **Chemotherapy (Chemo):** It is the treatment of cancer with medicines or treatments. Chemotherapy is a terrifying prospect for many individuals. However, understanding what chemo is, how it works, and what to expect might help to alleviate your anxieties. It can also help you feel more in charge of your cancer treatment.

4. **Radiation Therapy:** It is a type of cancer treatment that uses high-energy electrons or waves to kill or harm cancer cells. It is among the most commonly used cancer treatments, either alone or in combination with other treatments (Mathur et al., 2015).

1.6.2 HIV/AIDS

The human immunodeficiency virus (HIV) affects immune cells known as CD4 cells. T cells are white blood cells that travel throughout the system, detecting infections as well as defects and abnormalities in other cells. HIV seeks out and infects CD4 cells, which the infection subsequently exploits to multiply. As a result, the cells are killed, compromising the body's natural defenses against infections and diseases. This increases the chance of secondary infections and malignancies, as well as the severity of such infections and cancers.

The acronym AIDS refers to acquired immunodeficiency syndrome. HIV has progressed to this stage. A CD4 count of less than 200 cells/mm^3 is used by doctors to diagnose AIDS. They may also suspect AIDS if a person has a history of opportunistic infections, cancers consistent with AIDS, or both.

1.6.2.1 MSYMPTOMS

Some patients get flu-like signs a couple of months after contact with the virus. This is referred to as the acute stage. In a week to a month, these symptoms usually disappear. You can have HIV for a long time without becoming unwell. The chronic stage is also known as clinical latency. The most serious stage of HIV infection is AIDS. The HIV virus has damaged the immune system, making it less capable of fighting infections at this time. Opportunistic infections are infections that a healthy immune system can usually fight off. You must have less than 200 CD4 cells/cubic millimeter of blood to be identified with AIDS, or you must have experienced opportunistic infections.

1.6.2.2 CAUSES

A virus is responsible for the transmission of HIV. Sexual contact, illicit injectable drug use or reusing needles, contact with the infected blood, transmission from mother to child, childbirth, or lactation are all ways it might spread. HIV kills CD4 T cells, which are WBCs that assist your system fight disease. Your immune system grows weaker as your CD4 T cells decrease.

1.6.2.3 *DIAGNOSIS*

HIV antibodies can be detected in a person's blood by healthcare professionals. Before verifying a positive result, they will analyze the blood. Testing kits for use at home are also available. Current HIV testing technologies can identify HIV in as little as two weeks. Trusted Source. People who have known potential risks should be tested more frequently. A quick test is available for anyone who believes they will be at threat of obtaining HIV. If the number is negative, the test supplier will normally prescribe a follow-up test in a few weeks.

The following are the several types of HIV tests:

- NATs, or nucleic acid amplification tests, can identify HIV as soon as 10 days after exposure.
- HIV can be detected in a sample of blood as soon as 18 days after exposure using an antigen/antibody blood test.
- Antibody tests are the most common quick testing and self-tests, and they can identify HIV antibodies as soon as 21 days after exposure.

1.6.2.4 *TREATMENT*

Although there is no curative therapy, there are therapies that can slow the disease's progression. Antiretrovirals, or antiretroviral drugs, can help to minimize the risk of HIV transmission. They can also improve a person's quality of life and lengthen their life expectancy. Many HIV patients enjoy long and healthy lives because of their therapy. These drugs are getting more effective, and the majority of people handle them well. It is possible that a person will only need to take one pill each day:

1. **Emergency HIV Pills: PEP:** Anyone who may have been exposed to the virus within the last 72 hours trusted Source should get medical advice on post-exposure prophylaxis (PEP).
2. **Antiretroviral Drugs:** These, which combat the disease and slow the transmission of the infection, are used to treat HIV. People usually take a cocktail of drugs known as highly active antiretroviral

therapy (HAART) or combination antiretroviral therapy (CART). HAART or cART are two terms that might be used to describe the method.

3. **Protease Inhibitors:** HIV requires the enzyme protease to replicate. Protease inhibitors bind to the active site and stop it from working, stopping HIV from replicating (Kasper et al., 2015; https://www.mayoclinic.org/diseases-conditions/hiv-aids/symptoms-causes/syc-20373524).

1.7 CONCLUSION

Chronic diseases are now the leading cause of death in practically every country, and the danger to people's lives, health, and countries' economic development is expanding rapidly.

In all WHO areas except Africa, chronic conditions are the major cause of death and burden of disease. Men and women have equal death and disease burden rates, which rise with age. Low and middle-income nations have a greater chronic disease mortality rate than high-income countries. People under the age of 70 account for 45% of chronic disease fatalities and 86% of chronic disease burden. There is sufficient information to cope with this menace and save thousands of lives. Many countries have demonstrated that effective and cost-effective solutions, as well as the expertise to implement them, work. The global objective of preventing illness can be met if existing treatments are combined as part of a complete, integrated approach. The issue is how governments, businesses, and civilized society can collaborate to put such ideas into practice.

ACKNOWLEDGMENT

We thank the MET's Institute of Pharmacy, Bhujbal Knowledge City, Adgaon, Nashik, India. We extend our sincere appreciation to Principal Dr. Sanjay J. Kshirsagar Sir and our Guide Dr. Neelam L. Dashputre Ma'am for their unwavering support and provision of all necessary facilities to successfully accomplish this endeavor. Additionally, we would like to extend our heartfelt thanks to all those who have assisted us, both known and unknown, throughout the entire process, from the initial drafting to the final publication of this chapter.

KEYWORDS

- **chronic diseases**
- **diabetic retinopathy**
- **diagnosis**
- **digestive system**
- **fatal diseases**
- **inflammatory bowel diseases**
- **risk factors**

REFERENCES

Ahire, E. D., Surana, K. R., Patil, C. D., Shah, H. S., Sonwane, G. B., & Talele, S. G., (2020). Role of omega-3 fatty acids in different neurodegenerative disorders. In: *Applied Pharmaceutical Science and Microbiology* (pp. 173–194). Apple Academic Press.

Alexander, D. W., Berson, F. G., & Epstein, D. L., (1988). A clinical trial of timolol and epinephrine in the treatment of primary open-angle glaucoma. *Ophthalmology, 95*(2), 247–251.

Allen, A., Carville, S., & Mckenna, F., (2018). Diagnosis and management of rheumatoid arthritis in adults: Summary of updated nice guidance. *BMJ, 362.*

Azermai, M., Petrovic, M., Elseviers, M. M., Bourgeois, J., Van, B. L. M., & Vander, S. R. H., (2012). Systematic appraisal of dementia guidelines for the management of behavioral and psychological symptoms. *Ageing Research Reviews, 11*(1), 78–86.

Bains, J., Birks, J., & Dening, T., (2002). Antidepressants for treating depression in dementia. *Cochrane Database of Systematic Reviews*, (4).

Ballard, C., Neill, D., O'brien, J., Mckeith, I. G., Ince, P., & Perry, R., (2000). Anxiety, depression, and psychosis in vascular dementia: Prevalence and associations. *Journal of Affective Disorders, 59*(2), 97–106.

Banerjee, S., Hellier, J., Dewey, M., Romeo, R., Ballard, C., Baldwin, R., Bentham, P., et al., (2011). Sertraline or mirtazapine for depression in dementia (HTA-SADD): A randomized, multicenter, double-blind, placebo-controlled trial. *The Lancet, 378*(9789), 403–411.

Belbasis, L., Dosis, V., & Evangelou, E., (2018). Elucidating the environmental risk factors for rheumatic diseases: An umbrella review of meta-analyses. *International Journal of Rheumatic Diseases, 21*(8), 1514–1524.

Berson, F. G., Cohen, H. B., Foerster, R. J., Lass, J. H., Novack, G. D., & Duzman, E., (1985). Levobunolol compared with timolol for the long-term control of elevated intraocular pressure. *Archives of Ophthalmology, 103*(3), 379–382.

Brodaty, H., & Arasaratnam, C., (2012). Meta-analysis of nonpharmacological interventions for neuropsychiatric symptoms of dementia. *American Journal of Psychiatry, 169*(9), 946–953.

Bunn, F., Goodman, C., Sworn, K., Rait, G., Brayne, C., Robinson, L., Mcneilly, E., & Iliffe, S., (2012). Psychosocial factors that shape patient and carer experiences of dementia diagnosis and treatment: A systematic review of qualitative studies. *Plos Medicine, 9*(10), e1001331.

Burgener, S. C., Yang, Y., Gilbert, R., & Marsh-Yant, S., (2008). The effects of a multimodal intervention on outcomes of persons with early-stage dementia. *American Journal of Alzheimer's Disease & Other Dementias®, 23*(4), 382–394.

Burns, A., Guthrie, E., Marino-Francis, F., Busby, C., Morris, J., Russell, E., Margison, F., et al., (2005). Brief psychotherapy in Alzheimer's disease: Randomized controlled trial. *The British Journal of Psychiatry, 187*(2), 143–147.

Deeks, J. J., (2001). Statistical methods for examining heterogeneity and combining results from several studies in meta-analysis. *Systematic Reviews in Health Care*, 285–312.

Dennis, R. J., Solarte, I., & Rodrigo, G., (2010). Asthma in adults. *BMJ Clinical Evidence, 2010.*

Devanand, D. P., Pelton, G. H., Marston, K., Camacho, Y., Roose, S. P., Stern, Y., & Sackeim, H. A., (2003). Sertraline treatment of elderly patients with depression and cognitive impairment. *International Journal of Geriatric Psychiatry, 18*(2), 123–130.

Drawz, P., & Rahman, M., (2015). Chronic kidney disease. *Annals of Internal Medicine, 162*(11), itc1–itc16.

Enache, D., Winblad, B., & Aarsland, D., (2011). Depression in dementia: Epidemiology, mechanisms, and treatment. *Current Opinion in Psychiatry, 24*(6), 461–472.

Epstein, D. L., Krug, Jr. J. H., Hertzmark, E., Remis, L. L., & Edelstein, D. J., (1989). A long-term clinical trial of timolol therapy versus no treatment in the management of glaucoma suspects. *Ophthalmology, 96*(10), 1460–1467.

Eriksdotter, M., Vedin, I., Falahati, F., Freund-Levi, Y., Hjorth, E., Faxen-Irving, G., Wahlund, L. O., et al., (2015). Plasma fatty acid profiles in relation to cognition and gender in Alzheimer's disease patients during oral omega-3 fatty acid supplementation: The OmegAD study. *Journal of Alzheimer's Disease, 48*(3), 805–812.

Faxén-Irving, G., Freund-Levi, Y., Eriksdotter-Jönhagen, M., Basun, H., Hjorth, E., Palmblad, J., Vedin, I., et al., (2013). Effects on transthyretin in plasma and cerebrospinal fluid by DHA-rich n–3 fatty acid supplementation in patients with Alzheimer's disease: The omegAD study. *Journal of Alzheimer's Disease, 36*(1), 1–6.

Ferretti, L., Mccurry, S. M., Logsdon, R., Gibbons, L., & Teri, L., (2001). Anxiety and Alzheimer's disease. *Journal of Geriatric Psychiatry and Neurology, 14*(1), 52–58.

Freund-Levi, Y., Basun, H., Cederholm, T., Faxén-Irving, G., Garlind, A., Grut, M., Vedin, I., et al., (2008). Omega-3 supplementation in mild to moderate Alzheimer's disease: Effects on neuropsychiatric symptoms. *International Journal of Geriatric Psychiatry: A Journal of the Psychiatry of Late Life and Allied Sciences, 23*(2), 161–169.

Geyer, O., Lazar, M., Novack, G. D., Lue, J. C., & Duzman, E., (1986). Levobunolol compared with timolol for the control of elevated intraocular pressure. *Annals of Ophthalmology, 18*(10), 289, 290.

Gibbons, L. E., Teri, L., Logsdon, R., Mccurry, S. M., Kukull, W., Bowen, J., Mccormick, W., & Larson, E., (2002). Anxiety symptoms as predictors of nursing home placement in patients with Alzheimer's disease. *Journal of Clinical Geropsychology, 8*(4), 335–342.

Gitlin, L. N., Hauck, W. W., Dennis, M. P., & Winter, L., (2005). Maintenance of effects of the home environmental skill-building program for family caregivers and individuals with Alzheimer's disease and related disorders. *The Journals of Gerontology Series A: Biological Sciences and Medical Sciences, 60*(3), 368–374.

Glanville, J. M., Lefebvre, C., Miles, J. N., & Camosso-Stefinovic, J., (2006). How to identify randomized controlled trials in Medline: Ten years on. *Journal of the Medical Library Association, 94*(2), 130.

González-Salvador, M. T., Arango, C., Lyketsos, C. G., & Barba, A. C., (1999). The stress and psychological morbidity of the Alzheimer patient caregiver. *International Journal of Geriatric Psychiatry, 14*(9), 701–710.

Higgins, J. P., (2011). *Cochrane Handbook for Systematic Reviews of Interventions*. Version 5.1. The Cochrane collaboration. http://www.cochrane-handbook.org (accessed on 6 December 2023).

Kane, R. L., Butler, M., Fink, H. A., Brasure, M., Davila, H., Desai, P., Jutkowitz, E., et al., (2017). *Interventions to Prevent Age-Related Cognitive Decline, Mild Cognitive Impairment, and Clinical Alzheimer's -type Dementia.*

Lindqvist, E., Saxne, T., Geborek, P., & Eberhardt, K., (2002). Ten-year outcome in a cohort of patients with early rheumatoid arthritis: Health status, disease process, and damage. *Annals of the Rheumatic Diseases, 61*(12), 1055–1059.

Livingston, G., Johnston, K., Katona, C., Paton, J., Lyketsos, C. G., & Old age task force of the world federation of biological psychiatry, (2005). Systematic review of psychological approaches to the management of neuropsychiatric symptoms of dementia. *American Journal of Psychiatry, 162*(11), 1996–2021.

Lopez, O. L., Becker, J. T., & Sweet, R. A., (2005). Non-cognitive symptoms in mild cognitive impairment subjects. *Neurocase, 11*(1), 65–71.

Loscalzo, J., Fauci, A., Kasper, D., Hauser, S., Longo, D., Jameson, J., & Harrison, T., (n.d.). *Harrison's Principles of Internal Medicine.*

Lu, P. H., Edland, S. D., Teng, E., Tingus, K., Petersen, R. C., & Cummings, J. L., (2009). Donepezil delays progression to ad in mci subjects with depressive symptoms. *Neurology, 72*(24), 2115–2121.

Lyketsos, C. G., Delcampo, L., Steinberg, M., Miles, Q., Steele, C. D., Munro, C., Baker, A. S., et al., (2003). Treating depression in Alzheimer's disease: Efficacy and safety of sertraline therapy, and the benefits of depression reduction: The DIADS. *Archives of General Psychiatry, 60*(7), 737–746.

Lyketsos, C. G., Steele, C., Baker, L., Galik, E., Kopunek, S., Steinberg, M., & Warren, A., (1997). Major and minor depression in Alzheimer's disease: Prevalence and impact. *Journal of Neuropsychiatry and Clinical Neurosciences, 9*(4), 556–561.

Mathur, G., Nain, S., & Sharma, P. K., (2015). Cancer: An overview. *Acad. J. Cancer Res, 8*(1).

Monastero, R., Mangialasche, F., Camarda, C., Ercolani, S., & Camarda, R., (2009). A systematic review of neuropsychiatric symptoms in mild cognitive impairment. *Journal of Alzheimer's Disease, 18*(1), 11–30.

Mottram, P. G., Wilson, K., Scally, L., & Vassilas, C., (2000). Psychotherapeutic treatments for older depressed people. *Cochrane Database of Systematic Reviews*, (4).

Noecker, R. J., Awadallah, N. S., & Kahook, M. Y., (2007). Travoprost 0.004%/timolol 0.5% fixed combination. *Drugs of Today, 43*(2), 77–83.

Öhman, A., Josephsson, S., & Nygaard, L., (2008). Awareness through interaction in everyday occupations: Experiences of people with Alzheimer's disease. *Scandinavian Journal of Occupational Therapy, 15*(1), 43–51.

Olazarán, J., Reisberg, B., Clare, L., Cruz, I., peña-Casanova, J., Del, S. T., Woods, B., et al., (2010). Eficacia de las terapias no farmacológicas en la enfermedad de Alzheimer: Una revisión sistemática. *Dement Geriatr Cogn Disord, 30*(2), 161–178.

Orgeta, V., Qazi, A., Spector, A., & Orrell, M., (2015). Psychological treatments for depression and anxiety in dementia and mild cognitive impairment: Systematic review and meta-analysis. *The British Journal of Psychiatry, 207*(4), 293–298.

Palmer, K., Berger, A. K., Monastero, R., Winblad, B., Bäckman, L., & Fratiglioni, L., (2007). Predictors of progression from mild cognitive impairment to Alzheimer's disease. *Neurology, 68*(19), 1596–1602.

Petersen, R. C., (2003). *Mild Cognitive Impairment: Aging to Alzheimer's Disease*. Oxford University Press.

Petersen, R. C., Smith, G. E., Waring, S. C., Ivnik, R. J., Tangalos, E. G., & Kokmen, E., (1999). Mild cognitive impairment: Clinical characterization and outcome. *Archives of Neurology, 56*(3), 303–308.

Petracca, G. M., Chemerinski, E., & Starkstein, S. E., (2001). A double-blind, placebo-controlled study of fluoxetine in depressed patients with Alzheimer's disease. *International Psychogeriatrics, 13*(2), 233–240.

Pinquart, M., & Sorensen, S., (2001). How effective are psychotherapeutic and other psychosocial interventions with older adults? In: *Database of Abstracts of Reviews of Effects (dare): Quality-Assessed Reviews [Internet]*. Center for reviews and dissemination (UK).

Quinn, J. F., Raman, R., Thomas, R. G., Yurko-Mauro, K., Nelson, E., Dyck, C., et al., (2010). Docosahexaenoic acid supplementation and cognitive decline in Alzheimer disease: A randomized trial. *JAMA, 304*(17), 1903–1911. doi: 10.1001/jama.2010.1510.

Ravalico, G., Salvetat, L., Toffoli, G., Pastori, G., Croce, M., & Parodi, M. B., (1994). Ocular hypertension: A follow-up study in treated and untreated patients. *New Trends in Ophthalmology, 9*(2), 97–101.

Reifler, B. V., Teri, L., Raskind, M., Veith, R., Barnes, R., White, E., & Mclean, P., (1989). Double-blind trial of imipramine in Alzheimer's disease patients with and without depression. *Am. J. Psychiatry, 146*(1), 45–49.

Rolland, Y., Pillard, F., Klapouszczak, A., Reynish, E., Thomas, D., Andrieu, S., Rivière, D., & Vellas, B., (2007). Exercise program for nursing home residents with Alzheimer's disease: A 1-year randomized, controlled trial. *Journal of the American Geriatrics Society, 55*(2), 158–165.

Rozzini, L., Chilovi, B. V., Peli, M., Conti, M., Rozzini, R., Trabucchi, M., & Padovani, A., (2009). Anxiety symptoms in mild cognitive impairment. *International Journal of Geriatric Psychiatry: A Journal of the Psychiatry of Late Life and Allied Sciences, 24*(3), 300–305.

Schultz, S. K., Hoth, A., & Buckwalter, K., (2004). Anxiety and impaired social function in the elderly. *Annals of Clinical Psychiatry, 16*(1), 47–51.

Shinto, L., Quinn, J., Montine, T., Dodge, H. H., Woodward, W., Baldauf-Wagner, S., Waichunas, D., et al., (2014). A randomized placebo-controlled pilot trial of omega-3 fatty acids and alpha lipoic acid in Alzheimer's disease. *Journal of Alzheimer's Disease, 38*(1), 111–120.

Smolen, J. S., Aletaha, D., Barton, A., Burmester, G. R., Emery, P., Firestein, G. S., Kavanaugh, A., et al., (2018). Rheumatoid arthritis. *Nature Reviews Disease Primers, 4*(1), 1–23.

Solfrizzi, V., D'introno, A., Colacicco, A. M., Capurso, C., Del Parigi, A., Caselli, R. J., Scapicchio, P. L., et al., (2007). Incident occurrence of depressive symptoms among patients with mild cognitive impairment–the Italian longitudinal study on aging. *Dementia and Geriatric Cognitive Disorders, 24*(1), 55–64.

Spector, A., Orrell, M., Lattimer, M., Hoe, J., King, M., Harwood, K., Qazi, A., & Charlesworth, G., (2012). Cognitive behavioral therapy (CBT) for anxiety in people with dementia: Study protocol for a randomized controlled trial. *Trials, 13*(1), 1–7.

Stanley, M. A., Calleo, J., Bush, A. L., Wilson, N., Snow, A. L., Kraus-Schuman, C., Paukert, A. L., et al., (2013). The peaceful mind program: A pilot test of a cognitive–behavioral therapy–based intervention for anxious patients with dementia. *The American Journal of Geriatric Psychiatry, 21*(7), 696–708.

Stern, Y., Tang, M. X., Albert, M. S., Brandt, J., Jacobs, D. M., Bell, K., Marder, K., et al., (1997). Predicting time to nursing home care and death in individuals with Alzheimer's disease. *JAMA, 277*(10), 806–812.

Tappen, R. M., & Williams, C. L., (2009). Therapeutic conversation to improve mood in nursing home residents with Alzheimer's disease. *Research in Gerontological Nursing, 2*(4), 267–275.

Visser, P. J., Scheltens, P., & Verhey, F. R. J., (2005). Do mci criteria in drug trials accurately identify subjects with predementia Alzheimer's disease? *Journal of Neurology, Neurosurgery & Psychiatry, 76*(10), 1348–1354.

Waldorff, F. B., Buss, D. V., Eckermann, A., Rasmussen, M. L. H., Keiding, N., Rishøj, S., Siersma, V., et al., (2012). Efficacy of psychosocial intervention in patients with mild Alzheimer's disease: The multicenter, rater blinded, randomized Danish Alzheimer intervention study (daisy). *BMJ, 345*.

Weiner, M. F., Doody, R. S., Sairam, R., Foster, B., & Liao, T. Y., (2002). Prevalence and incidence of major depressive disorder in Alzheimer's disease: Findings from two databases. *Dementia and Geriatric Cognitive Disorders, 13*(1), 8–12.

Wishart, P. K., & Batterbury, M., (1992). Ocular hypertension: Correlation of anterior chamber angle width and risk of progression to glaucoma. *Eye, 6*(pt 3), 248–256.

Woods, B., O'philbin, L., Farrell, E. M., Spector, A. E., & Orrell, M., (2018). Reminiscence therapy for dementia. *Cochrane Database of Systematic Reviews*, (3).

CHAPTER 2

Overview on Novel Drug Delivery Systems for the Betterment of Chronic Diseases

KAMLESH D. MALI, NIKITA S. KUMAWAT, and VIVEK S. KUMAWAT

R. C. Patel Institute of Pharmaceutical Education and Research, Shirpur, Dhule, Maharashtra, Indial

ABSTRACT

A large number of persons are affected by a chronic illness. It is unlikely that regular drug delivery alone will be able to cure some drugs due to their poor solubility, short half-life, and limited bioavailability. The challenge can be resolved through the development of innovative drug delivery systems (DDSs). For the purpose of enhancing therapeutic indices while simultaneously reducing undesirable effects, new DDSs and formulations are realistic and promising possibilities. Unquestionably, a new technical idea may be transformed into a unique technology for use in a pharmaceutical delivery system, but this requires lengthy testing, revision, and development before it can be put into use. Because it makes use of a variety of nanocarriers, carrier-based innovative drug delivery, which includes oral, intranasal, ocular, transdermal, vaginal, and other drug delivery routes, is more effective than traditional dosage forms at targeting specific organs of the body than traditional dosage forms. When it comes to the treatment of various ailments, carrier-based medication delivery has been proven to be a successful drug delivery technique that also improves patient compliance.

2.1 INTRODUCTION

Many new chemical entities are now being synthesized; however, their solubility and absorption are proving difficult. Various solutions to this issue have

Novel Drug Delivery Systems in the Management of Chronic Diseases.
Raj K. Keservani, Umesh D. Laddha, Eknath D. Ahire, and Sanjay J. Kshirsagar (Eds.)

been presented, but each one has its own set of problems (Gowda, Moin, & Godugu, 2016). As a consequence, a carrier-based nanoparticle drug delivery system (DDS) is being developed, which employs a range of techniques and nanocarriers to improve efficiency, drug loading capacity, and release profile length. Only a few of the benefits of the new drug delivery approach over traditional/conventional dosage forms include excellent biocompatibility, sustained-controlled drug release, and greater bioavailability. Nano-carriers also offer a range of administration methods, including oral, intravenous, pulmonary, and topical drug delivery. If drugs are lowered to a nanosize range, they may provide various advantages, such as improved blood circulation, increased drug affinity, and therapeutic effectiveness. Polymeric nanoparticles (PNPs), dendrimers (Gautam et al., 2015), magnetic nanoparticles, liposomes (Keservani et al., 2010; Keservani & Gautam, 2020), metallic nanoparticles, and other pharmaceutical nanosystems may alter the drug delivery pathway to obtain maximal therapeutic effect. Carrier-based drug administration is a different approach to increasing medication bioavailability while keeping drug molecules intact. Encapsulated (biodegradable polymer-coated) drugs also have a better release profile than standard dose forms. Carrier-based delivery is used to deliver therapeutic medications or diagnostic equipment to specified locations in a controlled way. Due to its site-specific and target-oriented distribution, it also provides a variety of benefits in terms of detecting chronic ailments and diseases (Patra et al., 2018). Different stages of the novel drug discovery system for drug release (Ahire et al., 2021a) (NDDS) are mentioned in Figure 2.1.

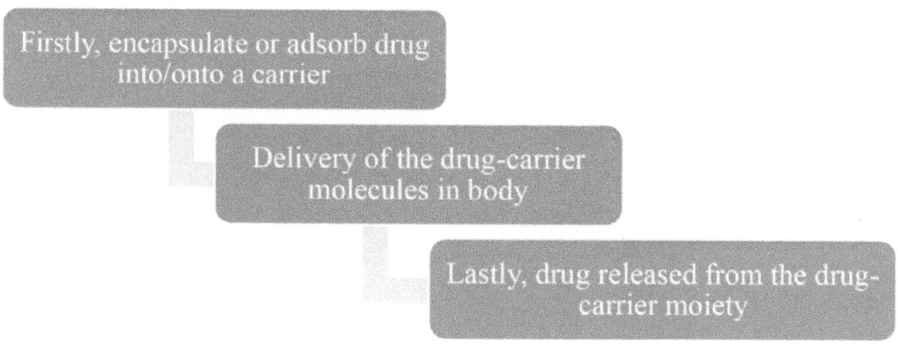

FIGURE 2.1 Different stages of NDDS.

Targeting is the ability to guide a drug-loaded system to the desired location. Two basic techniques for addressing the intended areas for drug release can be distinguished: There are two types of targeting: passive and active. Passive targeting is when chemotherapeutic drugs accumulate preferentially in solid tumors due to increased vascular permeability of tumor tissues compared to healthy tissue. Surface modification of drug carriers – ligand complexes that are selectively recognized by receptors – is one strategy that could allow active targeting. Because the interaction of ligand-receptor can be extremely selective, more specific delivery to the target area may be possible.

The development of excellent formulations necessitates controlled delivery and biodegradation. All possible release mechanisms include desorption of surface-bound/adsorbed pharmaceuticals, diffusion through the carrier matrix, diffusion (in the case of nanocapsules) through the carrier wall, carrier matrix erosion, and a combination erosion/diffusion process. Because how a drug is delivered has such a large impact on how it is chosen, the mode of delivery can mean the difference between success and failure. In sustained (or continuous) drug release, polymers that release a medication at a predetermined rate due to diffusion out of its polymer or polymer degradation throughout time are used in sustained (or continuous) drug release (Devi, Jain, & Valli, 2010).

2.1.1 MERITS OF NOVEL DRUG DELIVERY SYSTEM (NDDS)

It contributes to the improvement of the solubility of a poorly soluble medication. The reduction in size results in an increase in surface area. Bitter taste can be masked by coating the drug with a suitable polymer and applying it to the tongue. Patient compliance is improved as a result of a reduction in dosing frequency. It lessens the occurrence of side effects while increasing the bioavailability. It helps to keep the drug safe from enzymatic degradation. The nano-size of the particle reduces the dose and toxicity of the drug. NDDS delivers the drug to the targeted site, increasing the effectiveness and safety of the treatment.

2.2 PROBLEM IN CONVENTIONAL DRUG DELIVERY

The problem of conventional DDS was overcome by a novel drug delivery system (NDDS). Provide safe efficacy and stable drug delivery through it (Figure 2.2).

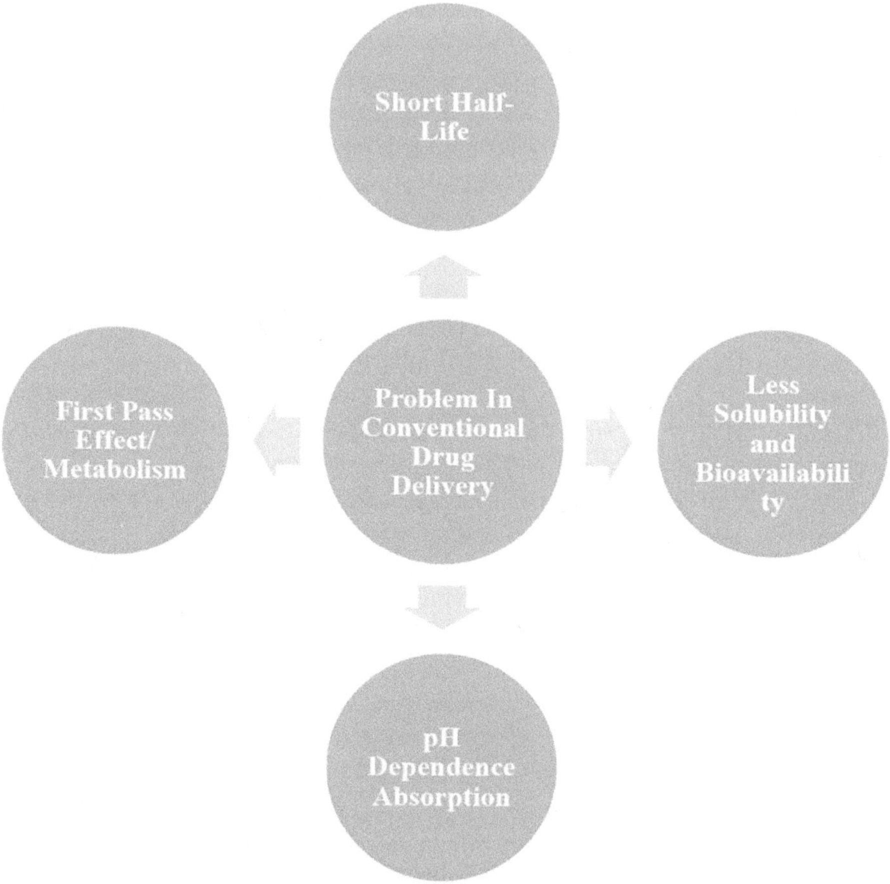

FIGURE 2.2 Problem of conventional drug delivery system.

2.2.1 *SHORT HALF-LIFE*

Some drugs have a short half-life, so their action is only for a short time. To maintain desirable drug concentrations, drugs must be administered on a regular basis. In that case, polymer coating/encapsulating the drug with a biodegradable polymer is critical for the controlled release of the drug over a long period of time. Using novel delivery development/encapsulation will reduce dosing frequency and improve patient compliance. Obeidat WM and Price JC investigated piroxicam and theophylline microspheres for controlling drug delivery and minimizing or eradicating local side effects. Drug microspheres were entrapped with the pH-sensitive Eudragit S 100 polymer at a theoretical loading of 25% using the emulsion-solvent evaporation process.

The size distribution of microsphere batches ranged from 125 to 500 microns, with geometric means close to 300 microns. The microsphere shape was identified using optical light microscopy. The USP Dissolution Apparatus II was used to perform *in vitro* dissolution tests on the microspheres at 37 +/–0.5°C at 100 rpm following the protocol for enteric-coated goods at two sequential varied pH mediums (1.2 and 6.5). The drug release of both preparations was sustained at pH 6.5 with theophylline microspheres, indicating a more prolonged release (Obeidat & Price, 2006).

2.2.2 LESS SOLUBILITY AND BIOAVAILABILITY

The bioavailability of a drug is primarily determined by its solubility and rate of dissolution. Because some drugs have a slow solubility/dissolution rate (e.g., famotidine, valsartan, and simvastatin), they enter the systemic circulation slowly and with a lower concentration. If a drug has the property of dissolving slowly OR poorly soluble, i.e., BCS class II, Class IV, its conventional/traditional form has lower bioavailability and increases the time of course of treatment. Some suitable techniques for solubility enhancement, drug encapsulation into mucoadhesive polymers such as solid dispersion (SD), nano-sizing, and micronization are available in novel delivery (Shirsath & Goswami, 2019). A novel nanoparticulate drug form will always be beneficial for increasing the bioavailability of the same drug. Curcumin is insoluble, unstable at different pH levels, and susceptible to photodegradation. We prosperously formulated Curcumin entrapped lipidic nanoparticle having 15 mg curcumin per ml of the SLN dispersion, with a 1.4 × 106 increase in curcumin solubility in an aqueous system compared to its intrinsic solubility of 11 ng/ml and high drug loading (Gupta et al., 2020; Ahire et al., 2018).

2.2.3 LESS SOLUBILITY AND BIOAVAILABILITY

For absorption and solubilization, some drugs require a specific pH (Lornaxicam is less soluble in lower pH than higher pH). When a drug formulation is administered in its traditional form, it degrades and loses its effectiveness. Biocompatible nanocarriers are used to encapsulate the medicine into nanoparticulate form, which protects the drug and improves absorption at the desired location. The use of nanoparticles (NPs) as chemotherapeutic DDSs is frequently impeded by reasons such as poor circulation, instability, and ineffective targeting. As a result, we develop biocompatible and biodegradable pH-responsive hybrid NPs that address these concerns. To achieve better target recognition, the NPs are composed of polylactic-co-glycolic acid (PLGA) contain drug encircled by a crosslinked bovine serum albumin (BSA) shell. The shell is functionalized with the acidity-triggered rational membrane (ATRAM) peptide to allow the chemotherapeutic agent to be released into cancer cells in a pH-responsive manner (acidic tumor microenvironment). Following uptake, the NPs are degraded by the cancer cells' unique intracellular conditions, allowing the chemotherapeutic agent to be released. Based on our findings, the pH-responsive hybrid NPs ATRAM-BSA-PLGA are a promising targeted cancer DDS (Palanikumar et al., 2020; Ahire et al., 2020).

2.2.4 FIRST PASS EFFECT/METABOLISM

Sanjay & Krutika (2009) developed and optimized alginate mucoadhesive carvedilol microspheres for nasal administration to reduce first-pass metabolism and improve therapeutic efficacy to treat hypertension and angina pectoris. The microspheres were created using a water-in-oil (W/O) emulsification method. Regression analysis was used to determine the best formulation conditions. The particle size ranged between 26.36 and 54.32 m, making it ideal for intranasal absorption. SEM was used to determine the shape and surface properties of the microspheres, such as spherical and smooth surfaces. The percent encapsulation efficiency was found to range between 36.62% and 56.18%. The adhesion number ranged from 69.25 to 85.28 when *in vitro* mucoadhesion was assessed using sheep nasal mucosa. Differential scanning calorimetry analysis and X-ray diffraction analysis results revealed carvedilol dispersion in the microspheres at the molecular level. Carvedilol was found to be safe in *in vitro* tests in a pH 6.2 phosphate buffer.

2.2.5 DRUG DELIVERY TECHNIQUE/ADMINISTRATION ROUTE

Various unique drug delivery practices were created by the pharmaceutical industry to improve selectivity, potency, and affordability (Figure 2.3 and Table 2.1).

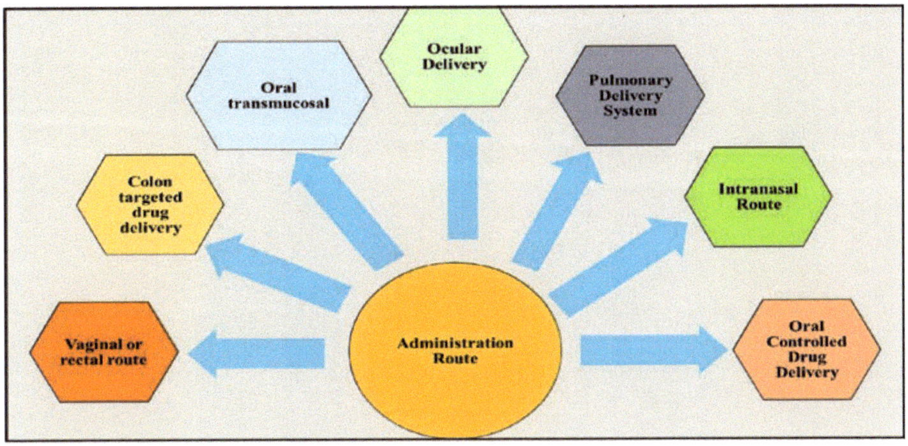

FIGURE 2.3 Administration route for NDDS.

2.3 ROLE OF NOVEL DRUG DELIVERY SYSTEM (NDDS) IN CHRONIC DISEASE

The World Health Organization (WHO) defines chronic disease as one that lasts a long time, progresses slowly, and does not spread from person to person (Reynolds et al., 2018). The WHO classifies chronic diseases into four categories: cardiovascular diseases (like heart attacks and strokes), cancer, diabetes, and chronic respiratory diseases (like asthma and chronic obstructive pulmonary disease (COPD)) (Grover & Joshi, 2015) (Figure 2.4).

2.3.1 ULCERATIVE COLITIS (UC)

Ulcerative colitis (UC) is a chronic inflammatory condition of the intestine. UC is a type of inflammatory bowel disease (IBD) characterized by high levels of IL-13 and continuous mucosal inflammation that spreads from the rectum to the cecum. Polymeric microparticles (MPs) have been shown to be a promising strategy for the delivery of targeted drugs in cancer treatment. MPs were used as a controlled medication delivery carrier to treat UC. The primary goal of the targeted drug strategy is to boost therapeutic effectiveness while reducing side effects by selectively delivering active drugs to the highest concentrations in inflamed intestinal tissues. The DDS with controlled release is one of the most effective methods for overcoming

TABLE 2.1 Different Novel Drug Delivery Systems Used in Chronic Diseases

Route of Administration	Novel Drug Delivery	Formulation	Outcome	References
Oral-controlled drug delivery	Nanoparticle, Microspheres Microencapsles	Ketoprofen PCL microspheres	Prolonged action for 12 hrs. and also shows a high %EE of about 96.81%	Wagh & Naik (2015)
Intranasal route	Microspheres Nanoemulsions Nanoparticles	Almotriptan mucoadhesive mi-crospheres	In spite of having a short half-life about 3–4 hrs. Show long-lasting action. Also, show good *In vitro* muco-adhesion in the range of 79.45 ± 1.69%–95.48 ± 1.27%.	Abbas & Marihal (2014)
Pulmonary delivery system	Liposomes Micelles Nanoparticles Dendrimers	Terbutaline sulfate (TS) microspheres	Microspheres with a 33% drug loading exhibited sustained release of 32.7% over 180 min in phosphate buffer.	Cook, Pannu, & Kellaway (2005)
Ocular delivery	Nanogel Nanoemulsion SLNs	Acyclovir-loaded chitosan nanoparticles	Sustained release action over a period of 24 hours.	Rajendran et al. (2010)
Oral transmucosal	Liposomes Microsphere	Atenolol containing microsphere	Sustained release profile from the microspheres; the concentrations at 24 h were 0.75 ± 0.1 μg/ml. Moreover, the absolute bioavailability of microsphere formulations was higher.	Patel, Liu, & Brown (2011)
Colon-targeted drug delivery	–	Diltiazem HCl and indomethacin tablet.	–	Novel Colon Targeted Drug Delivery System Using Natural Polymers (n.d.)
Transdermal delivery	Noisomes Liposomes Microemulsions Transfersomes Protransfersomes	Ketoprofen transfersome	–	Sachan et al. (n.d.); Dubey et al. (2006)
Vaginal or rectal route	Mucoadhesive tablet, thermo-sensitive Gel	Clotrimazole thermosensitive gel	–	Yun Chang et al. (2002)

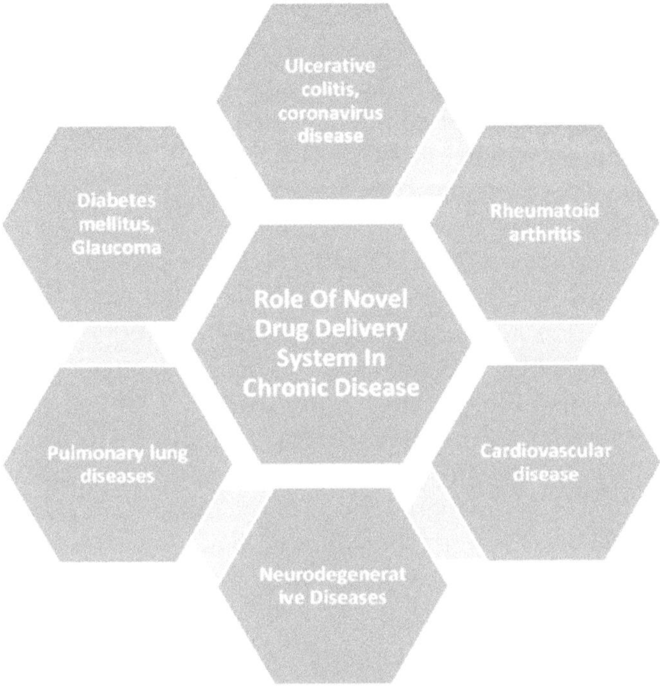

FIGURE 2.4 Role of novel drug delivery system in chronic disease.

the majority of the issues associated with other administration systems. Regulated release drug delivery encapsulates the medication and releases it at a controlled manner over an extended period of time using polymer-based discs, MPs, nanoparticles, and pellets (Nidhi et al., 2016).

2.3.2 RHEUMATOID ARTHRITIS (RA)

Rheumatoid arthritis (RA) is a common autoimmune disease characterized by severe joint inflammation that causes severe disability and premature death. Different drug classes, such as corticosteroids, nonsteroidal anti-inflammatory drugs (NSAIDs), disease-modifying anti-rheumatic drugs (DMARDs), and biologics, are discussed, with examples of the most commonly used drugs in each class. Microspheres, nanoparticles, dendrimers, liposomes, and other NDDSs are promising tools because they have overcome the shortcomings of traditional drug delivery methods (Thakur et al., 2018). Many researchers use PLGA nanoparticles to increase the circulation duration of encapsulated drugs and to control the rate of release. When administered intravenously to arthritic rats and mice, PLGA Betamethasone systems were found to be far more effective at reducing inflammation than free glucocorticoids. Nagei et al. created a nanogel ointment containing ketoprofen solid nanoparticles with a mean particle size of 83 nm using the bead mill method, which had an anti-inflammatory effect in AIA rats (Higaki et al., 2005). Vandana and colleagues developed a Nimesulide aloe vera gel. *In vitro* and *in vivo* studies were conducted to assess the anti-inflammatory potential of the formulations developed. Nimesulide emulgel penetration was 54% at 30 minutes, compared to 44% for commercialized Nimesulide gel, indicating better drug release without skin irritation. It also has a higher drug loading capacity of 86% compared to 70% in commercial formulations, making it suitable for having a strong anti-inflammatory effect (Vandana, 2014). When micelles made from phospholipid preparations (phosphogliv) are loaded with methotrexate (MTX) and administered to arthritic rats, they show an increased ability to reduce inflammation. When arthritic mice were given Camptothecin micelles instead of free Camptothecin, they showed less inflammation (Montazeri Aliabadi, Brocks, & Lavasanifar, 2005; Mohanty et al., 2019).

2.3.3 CARDIOVASCULAR DISEASES

Cardiovascular disease affects the cardiovascular system, the vascular systems of the brain and kidneys, as well as the peripheral arteries (CVD). The main goals of nanoparticle-based formulations in the treatment of cardiovascular disorders are targeted distribution and increased bioavailability for vascular restenosis. Liposomes made of phosphatidylcholine (PC) and cholesterol were loaded with the nanoparticle medication sirolimus and coated with chitosan as a nanoparticle medicine for restenosis. The liposomal sirolimus produced has been demonstrated to significantly reduce vascular restenosis. Carvedilol, a-blocker, is another medication that is commonly used to treat myocardial infarction (MI), hypertension, congestive heart failure (CHF), and post-MI left ventricular dysfunction. Its bioavailability is limited by its low water solubility and significant pre-systemic metabolism. When compared to the free medication, the niosome-based nanoparticle formulation encapsulating carvedilol achieved 1.7–2.3-fold higher plasma concentrations, resulting in greater bioavailability and therapeutic impact.

2.3.4 NEURODEGENERATIVE DISEASES (NDS)

Neurodegenerative diseases (NDs) are defined by a progressive loss of neuronal function that results in neuronal death. Patients with NDs, such as

Alzheimer's disease (AD), Parkinson's disease (PD), and multiple sclerosis (MS), experience symptoms related to movement, memory, and dementia as a result of the slow loss of neurons. Because of the BBB's high selectivity, a little amount of therapeutic medications can reach the brain. As a result, higher doses are required, which has a negative systemic impact. The key goals of nanoparticle-based treatment methods in NDs are targeted delivery and sustained local release of therapeutic drugs into the afflicted area of the brain after crossing the BBB. Several studies have used vascular or PNPs to encapsulate various medications (apomorphine, ropinirole, mitoapocynin, bromocriptine) to promote to promote prolonged drug release and reduce the negative effects of standard PD therapy. Anti-inflammatory methods such as PNPs or PEGylated liposomes are being developed to reduce neuronal cell death in PD.

2.3.5 PULMONARY LUNG DISEASES

Pulmonary lung illnesses (IPF) include asthma, COPD, cystic fibrosis (CF), pulmonary tuberculosis (TB), and idiopathic pulmonary fibrosis (Yetisgin et al., 2020). The authors describe a novel method for producing sustained release (SR) particles for use in pulmonary medication administration. Using a spray-drying method, high purity nanoparticles of a hydrophilic, ionized medication are encapsulated within lipophilic microspheres. Terbutaline sulfate (TS), the model drug, was found to be directly proportional to drug loading and phospholipid concentration in the microspheres. Around 33% drug loaded Microspheres demonstrated SR of 32.7% over 180 minutes in phosphate buffer. The release of simulated lung fluids was not significantly different. There was no significant burst release, indicating that the nanoparticles were adequately coated during the spray-drying process. Confocal microscopy confirmed that nanoparticles free surface of the microsphere. With a fine particle fraction (FPF) of 46.5F 1.8%, and a mass median aerodynamic diameter (MMAD) of 3.93F 0.12 Am, the SR microspheres were created as a dry powder (carrier-free) for inhalation (Cook, Pannu, & Kellaway, 2005). Gels containing nanocarriers the administration of nanocarrier-loaded *in situ* gels with stimuli-responsive behavior has piqued the interest of many researchers. The in *situ* gel for ophthalmic is composed of environmentally responsive polymers, which change structurally in response to minute changes in environmental conditions such as ionic strength, pH, and temperature. Furthermore, incorporating nanoparticles into gel reduces the burst release issues associated with nanoparticles, allowing for longer drug release. Temperature-sensitive polymers like Pluronic (PF-127 and PF-68), poloxamer 407 and poloxamer 188, Carbopol 934P, sodium alginate, and hydroxypropyl methylcellulose (HPMC) K4M have been shown in studies to provide prolonged drug release, increased ocular bioavailability, and extended drug release for up to 8 hours. Morsi et al. investigated acetazolamide delivery using nanoemulsion (NE)-based ion-sensitive *in situ* gels (Gorantla et al., 2020).

2.3.6 TYPE 1 DIABETES MELLITUS (DM)

Diabetes mellitus (DM) is a chronic hyperglycemic condition characterized by improper carbohydrate, lipid, and protein metabolism that can be caused by a number of factors. Niosomes are non-ionic surfactant vesicles with a lamellar structure formed by the self-assembly of surfactant molecules. Niosomal vesicular systems are being developed to improve the oral bioavailability of medications with low water solubility, such as Glibenclamide, as well as to reduce the frequency of dosing and dose-dependent toxicity of pharmaceuticals with shorter half-lives, such as Metformin. The use of niosomes to encapsulate active ingredients may help to reduce antidiabetic drug toxicity while also improving absorption, stability, and drug action (Rai et al., 2016).

2.3.7 GLAUCOMA

Glaucoma is regarded as one of the most serious health issues in the world. It's the leading cause of preventable blindness because of its asymptomatic nature in the early stages and patients' non-adherence. The results of trials that confirmed the efficacy of the aforementioned DDSs were also discussed briefly (Rahić et al., 2021). Acetazolamide-filled cationic NEs are more effective in the treatment of glaucoma because they reduce aqueous humor production and thus lower IOP more effectively. The size of the droplets ranged from 240 to 443. Anionic NEs had a negative zeta potential of 36.9 mV and cationic NEs had a positive zeta potential of 41.4 mV (Shunmugaperumal, 2011). According to an in vitro study (Chen, Gupta, & Cheung, 2010), colonels nanosuspension reduced IOP by 31%, with pharmacological effects lasting up to 12 hours, which was significantly longer than standard formulations. Diclofenac nanosuspensions with a 105 nm diameter and an 8-mV zeta potential improved drug retention and penetration in ocular tissues. The initial burst release was followed by continuous drug release (47% at 30 minutes) (73% at 8 h). *In vivo* research in rabbits revealed that the AUC was 3.06-0.57 gh/mL, with a maximal concentration (Cmax) of 0.78–0.11 g/mL and a tmax of 2 h (Shi et al., 2015). Positively charged timolol-loaded liposome formulations with a size of 136.00±18.00 nm and a zeta potential of 2.75±2.74 mV were integrated into eye drops in rabbits' eyes and reduced IOP to 11.96±0.74 mmHg 1 h and 13.61±0.95 mmHg 2 h after treatment (Yu et al., 2015). Timolol-loaded CS-coated niosomes with a size of 2±3 m had 1.7 and 2.34 times greater Cmax and AUC in aqueous humor, respectively, than timolol solution (Kaur et al., 2010; Aggarwal & Kaur, 2005). Gupta et al. developed a timolol maleate-loaded CS/HPMC-based polymer matrix to improve ocular retention. Because of CS's viscosity and bioadhesion, the proposed formulation was non- to mildly irritating, well-tolerated, and demonstrated better transcorneal drug permeation, longer corneal retention, and sustained drug release for 12 hours (Gupta, 2009). Nanoparticles In a variety of ways, antiglaucoma medications have been incorporated into nanoparticles. Hyaluronic acid-modified CS nanoparticles could be a potential medication delivery mechanism for glaucoma (Wadhwa et al., 2010). Li et al. created betaxolol-loaded CS/montmorillonite (MT) nanoparticles with a size of 460±0.6 nm and a zeta potential of 290.18 mV. The IOP decrease of nanoparticles (4.44 mmHg) was greater than that of the solution due to the sustained drug release and extended retention on the preocular surface (5.04 mmHg). Nanoparticles could significantly reduce IOP in rabbits and have a therapeutic effect (Li et al., 2018). Warsi et al. used polyvinyl alcohol (PVA) and tocopheryl PEG 1000 succinate (TPGS) as emulsifiers to make 129±12.3 nm dorzolamide-loaded poly(D,L-lactide-co-glycolide) (PLGA) nanoparticles. After 4 hours, nanoparticles with TPGS showed a maximum IOP drop of 29.12%, with a 22.8% IOP drop maintained for the next 20 hours. In an *in vivo* study with rabbits, nanoparticles containing PVA resulted in a maximum IOP drop of 22.8% after 4 hours, with a 16% IOP reduction maintained for 16 hours. Both formulations were ophthalmologically safe and did not irritate the eyes. They may be developed into once-daily eye drops that are safe for long-term use and can improve patient adherence (Warsi et al., 2014).

2.3.8 COVID-19

Given the potential for using polymeric-based nanoparticles to treat a variety of human ailments, the benefits of using PNPs to combat SARS-CoV-2 infection and disease are numerous (Ahire et al., 2020a). PLA and PLGA have received FDA approval due of their numerous benefits, and as a result, they have been widely studied. The most commonly used approaches in the fabrication of PLA or PLGA-based drug delivery nano-sized systems are nanoprecipitation and emulsification. Pfizer/BioNTech and Moderna, for example, have developed cationic lipid nanoparticle-based mRNA-loaded vaccines for COVID-19. Because mRNA is non-infectious, has a short and immunogenic half-life, and can be safely and efficiently transported *in vivo*, it is preferable to DNA or whole virus vaccines (Ahire & Kshirsagar, 2021). By 2020, nanomedicine and COVID-19 vaccines will be available (Ahire et al., 2020b; Talele et al., 2022).

2.4 CONCLUSION

The NDDS involves different drug delivery routes, such as oral, intranasal, ocular, pulmonary, transdermal, vaginal, etc., for target in organ for specific therapeutic activity. These is achieved by means of carriers, which are used to improve bioavailability, improved solubility, avoid first-pass metabolism, and minimized adverse effects. These overall results achieve better Pharmacological activity of the particular drugs, which is a crucial parameter, especially in terms of Chronic diseases. From the various studies, it is proven that NDDS works more effectively in chronic diseases than conventional drug delivery.

KEYWORDS

- **cardiovascular disease**
- **chronic disease**
- **liposomes**
- **microspheres**
- **nanocarrier**
- **neurodegenerative disease**
- **niosomes**
- **novel drug delivery system**
- **ocular disease**
- **rheumatoid arthritis**

REFERENCES

Abbas, Z., & Marihal, S., (2014). Gellan gum-based mucoadhesive microspheres of almotriptan for nasal administration: Formulation optimization using factorial design, characterization, and *in vitro* evaluation. *Journal of Pharmacy and Bioallied Sciences*, *6*(4), 267–277. doi: 10.4103/0975-7406.142959.

Aggarwal, D., & Kaur, I. P., (2005). Improved pharmacodynamics of timolol maleate from a mucoadhesive niosomal ophthalmic drug delivery system. *International Journal of Pharmaceutics*, *290*(1, 2), 155–159. doi: 10.1016/j.ijpharm.2004.10.026.

Ahire, E. D., & Kshirsagar, S. J., (2021). Immune responses induced by different vaccine platforms against coronavirus disease-19. *Journal: Exploration of Immunology*, 243–257.

Ahire, E. D., Sonawane, V. N., & Surana, K. R., (2020). Role of drug repurposing in current treatment strategies against COVID-19; systemic review. *Pharm. Reson.*, 24–29.

Ahire, E. D., Sonawane, V. N., Surana, K. R., & Talele, G. S., (2021). Drug discovery, drug-likeness screening, and bioavailability: Development of drug-likeness rule for natural products. In: *Applied Pharmaceutical Practice and Nutraceuticals* (pp. 191–208). Apple Academic Press.

Ahire, E. D., Sonawane, V. N., Surana, K. R., Jadhav, K. R., Sonawane, D. D., & Shah, A. A., (2020). Convalescent plasma therapy: A promising approach in the treatment of COVID-19. *Int. J. Pharm. Sci. Res.*, *11*, 4078–4086.

Ahire, E., Thakkar, S., Borade, Y., & Misra, M., (2020). Nanocrystal based orally disintegrating tablets as a tool to improve dissolution rate of Vortioxetine. *Bulletin of Faculty of Pharmacy, Cairo University*, *58*(1, 2), 11–20.

Ahire, E., Thakkar, S., Darshanwad, M., & Misra, M., (2018). Parenteral nanosuspensions: A brief review from solubility enhancement to more novel and specific applications. *Acta Pharmaceutica Sinica B*, *8*(5), 733–755.

Chen, M. C. W., Gupta, M., & Cheung, K. C., (2010). Alginate-based microfluidic system for tumor spheroid formation and anticancer agent screening. *Biomedical Microdevices*, *12*(4), 647–654. doi: 10.1007/s10544-010-9417-2.

Cook, R. O., Pannu, R. K., & Kellaway, I. W., (2005). Novel sustained release microspheres for pulmonary drug delivery. *Journal of Controlled Release*, *104*(1), 79–90. doi: 10.1016/j.jconrel.2005.01.003.

Devi, V., Jain, N., & Valli, K., (2010). Importance of novel drug delivery systems in herbal medicines. *Pharmacognosy Reviews*, 27–31. doi: 10.4103/0973-7847.65322.

Dubey, V., et al., (2006). Transdermal delivery of a pineal hormone: Melatonin via elastic liposomes. *Biomaterials*, *27*(18), 3491–3496. doi: 10.1016/j.biomaterials.2006.01.060.

Gautam, S. P., Keservani, R. K., Gautam, T., Gupta, A. K., & Sharma, A. K., (2015). An alternative approach for acetylation of amine terminated polyamidoamine (PAMAM) dendrimer. *ARS Pharmaceutica, 56*(3), 155–159.

Gorantla, S., et al., (2020). Nanocarriers for ocular drug delivery: Current status, and translational opportunity. *RSC Advances* (pp. 27835–27855). Royal Society of Chemistry. doi: 10.1039/d0ra04971a.

Gowda, D. V., Moin, A., & Godugu, K., (2016). *Nano Structured Lipid Carrier-Based Drug Delivery System Gastro RDDS-Review View Project Pectin-HPMC Mesalamine Tablets View Project, Article in Journal of Chemical Research*. Available at: https://www.researchgate.net/publication/301520149 (accessed on 6 December 2023).

Grover, A., & Joshi, A., (2015). An overview of chronic disease models: A systematic literature review. *Global Journal of Health Science*, 210–227. doi: 10.5539/gjhs.v7n2p210.

Gupta, H., Aqil, M., Khar, R. K., Ali, A., Bhatnagar, A., Mittal, G., & Jain, S., (2009). Development and characterization of 99mTc-timolol maleate for evaluating efficacy of in situ ocular drug delivery system. *AAPS Pharm Sci Tech, 10,* 540–546.

Gupta, T., et al., (2020). Enhancing bioavailability and stability of curcumin using solid lipid nanoparticles (CLEN): A covenant for its effectiveness. *Frontiers in Bioengineering and Biotechnology*, 8. doi: 10.3389/fbioe.2020.00879.

Higaki, M., et al., (2005). Treatment of experimental arthritis with poly(D, L-lactic/glycolic acid) nanoparticles encapsulating betamethasone sodium phosphate. *Annals of the Rheumatic Diseases*, *64*(8), 1132–1136. doi: 10.1136/ard.2004.030759.

Kaur, I. P., et al., (2010). Improved ocular absorption kinetics of timolol maleate loaded into a bioadhesive niosomal delivery system. *Graefe's Archive for Clinical and Experimental Ophthalmology*, *248*(10), 1467–1472. doi: 10.1007/s00417-010-1383-0.

Keservani, R. K., & Gautam, S. P., (2020). Formulation and evaluation of baclofen liposome vesicles using lecithin. *ARS Pharmaceutica, 61*(3), 175–180.

Keservani, R. K., Sharma, A. K., & Ramteke, S., (2010). Novel vesicular approach for topical delivery of baclofen via niosomes. *Lat. Am. J. Pharm.*, *29*, 1364–1370.

Li, J., et al., (2018). Montmorillonite/chitosan nanoparticles as a novel controlled-release topical ophthalmic delivery system for the treatment of glaucoma. *International Journal of Nanomedicine*, *13*, 3975–3987. doi: 10.2147/IJN.S162306.

Mohanty, S., et al., (2019). Novel drug delivery systems for rheumatoid arthritis: An approach to better patient compliance. *Biomedical and Pharmacology Journal*, *12*(1), 157–170. doi: 10.13005/bpj/1624.

Montazeri, A. H., Brocks, D. R., & Lavasanifar, A., (2005). Polymeric micelles for the solubilization and delivery of cyclosporine A: Pharmacokinetics and biodistribution. *Biomaterials*, *26*(35), 7251–7259. doi: 10.1016/j.biomaterials.2005.05.042.

Nidhi, et al., (2016). Microparticles as controlled drug delivery carrier for the treatment of ulcerative colitis: A brief review. *Saudi Pharmaceutical Journal*, pp. 458–472 Elsevier B.V. doi: 10.1016/j.jsps.2014.10.001.

Obeidat, W. M., & Price, J. C., (2006). Preparation and evaluation of Eudragit S 100 microspheres as pH-sensitive release preparations for piroxicam and theophylline using the emulsion-solvent evaporation method. *Journal of Microencapsulation, 23*(2), 195–202. doi: 10.1080/02652040500435337.

Palanikumar, L., et al., (2020). pH-responsive high stability polymeric nanoparticles for targeted delivery of anticancer therapeutics. *Communications Biology*, *3*(1). doi: 10.1038/s42003-020-0817-4.

Patel, V. F., Liu, F., & Brown, M. B., (2011). Advances in oral transmucosal drug delivery. *Journal of Controlled Release*, 106–116. doi: 10.1016/j.jconrel.2011.01.027.

Patra, J. K., et al., (2018). Nano based drug delivery systems: Recent developments and future prospects 10 technology 1007 nanotechnology 03 chemical sciences 0306 physical chemistry (incl. structural) 03 chemical sciences 0303 macromolecular and materials chemistry 11 medical and health sciences 1115 pharmacology and pharmaceutical sciences 09 engineering 0903 biomedical engineering Prof Ueli Aebi, Prof Peter Gehr. *Journal of Nanobiotechnology*. BioMed Central Ltd. doi: 10.1186/s12951-018-0392-8.

Rahić, O., et al., (2021). Novel drug delivery systems fighting glaucoma: Formulation obstacles and solutions. *Pharmaceutics* (pp. 1–58). MDPI AG. doi: 10.3390/pharmaceutics13010028.

Rai, V. K., et al., (2016). Novel drug delivery system: An immense hope for diabetics. *Drug Delivery* (pp. 2371–2390). Taylor and Francis Ltd. doi: 10.3109/10717544.2014.991001.

Rajendran, N., et al., (2010). Acyclovir-loaded chitosan nanoparticles for ocular delivery. *Asian Journal of Pharmaceutics*, *4*(4), 220–226. doi: 10.4103/0973-8398.76749.

Ravi, V., & Kumar, T. P., (2008). Novel colon targeted drug delivery system using natural polymers. *Indian journal of pharmaceutical sciences*, *70*(1), p.111.

Reynolds, R., et al., (2018). A systematic review of chronic disease management interventions in primary care. *BMC Family Practice*. BioMed Central Ltd. doi: 10.1186/s12875-017-0692-3.

Sachan, R., et al., (n.d.). *Drug Carrier Transfersomes: A Novel Tool for Transdermal Drug Delivery System, 2*(2), 309–316. Available at: www.ijrdpl.com (accessed on 6 December 2023).

Shi, S., et al., (2015). Chitosan grafted methoxy poly(ethylene glycol)-poly(ε-caprolactone) nanosuspension for ocular delivery of hydrophobic diclofenac. *Scientific Reports*, *5*. doi: 10.1038/srep11337.

Shirsath, N. R., & Goswami, A. K., (2019). Nanocarriers based novel drug delivery as effective drug delivery: A review. *Current Nanomaterials*, *4*(2), 71–83. doi: 10.2174/2405461504666190527101436.

Shunmugaperumal, T., & Balakrishnan, A. K., (2011). Influence of acetazolamide loading on the (*in vitro*) performances of non-phospholipid-based cationic nanosized emulsion in comparison with phospholipid-based anionic and neutral-charged nanosized emulsion. *Informa Healthcare*, 1003–1015.

Talele, S. G., Ahire, E. D., Surana, K. R., Sonawane, V. N., & Talele, G. S., (2022). Corona virus disease (COVID-19): A past and present prospective. *Asian Journal of Pharmaceutical Research, 12*(1), 45–53.

Thakur, S., et al., (2018). Novel drug delivery systems for NSAIDs in management of rheumatoid arthritis: An overview. *Biomedicine and Pharmacotherapy* (pp. 1011–1023). Elsevier Masson SAS. doi: 10.1016/j.biopha.2018.07.027.

Vandana, K. R., Yalavarthi, R. P., Sundaresan, C. R., Sriramaneni, N. R., & Vadlamudi, C. H., (2014). In-vitro assessment and pharmacodynamics of nimesulide incorporated Aloe vera transemulgel. *Current Drug Discovery Technologies*, *11*(2), 162–167.

Wadhwa, S., et al., (2010). Hyaluronic acid modified chitosan nanoparticles for effective management of glaucoma: Development, characterization, and evaluation. *Journal of Drug Targeting*, *18*(4), 292–302. doi: 10.3109/10611860903450023.

Wagh, P., & Naik, J., (2015). Formulation and characterization of ketoprofen embedded polycaprolactone microspheres using solvent evaporation method. *ADMET and DMPK*, *3*(2), 141–153. doi: 10.5599/admet.3.2.167.

Warsi, M. H., et al., (2014). Dorzolamide-loaded PLGA/vitamin E TPGS nanoparticles for glaucoma therapy: Pharmacoscintigraphy study and evaluation of extended ocular hypotensive effect in rabbits. *Colloids and Surfaces B: Biointerfaces*, *122*, 423–431. doi: 10.1016/J.COLSURFB.2014.07.004.

Yetisgin, A. A., et al., (2020). Therapeutic nanoparticles and their targeted delivery applications. *Molecules*. MDPI AG. doi: 10.3390/molecules25092193.

Yu, S., et al., (2015). Liposome incorporated ion sensitive in situ gels for opthalmic delivery of timolol maleate. *International Journal of Pharmaceutics*, *480*(1, 2), 128–136. doi: 10.1016/j.ijpharm.2015.01.032.

Yun, C. J., et al., (2002). Prolonged antifungal effects of clotrimazole-containing mucoadhesive thermosensitive gels on vaginitis. *Journal of Controlled Release, 82*(1), 39–50. doi: 10.1016/S0168-3659(02)00086-X.

CHAPTER 3

Insight into Suitable Polymers and Materials for the Preparation of Novel Drug Delivery Systems

AMIT VITTHAL KAKAD,[1] LAYBA SAJJADALI SAYYAD,[1] POOJA SHANKAR NIPHADE,[1] BHADANE ROSHANI GIRIDHAR,[1] ANUJA SHIVAJI NALKAR,[2] UMESH DILIP LADDHA,[1] and SANJAY J. KSHIRSAGAR[1]

[1]*Pharmaceutics Department, MET's Institute of Pharmacy, Bhujbal Knowledge City, Affiliated to Savitribai Phule Pune University, Nashik, Maharashtra, India*

[2]*Bachelor of Pharmacy, MET's Institute of Pharmacy, Bhujbal Knowledge City, Affiliated to Savitribai Phule Pune University, Nashik, Maharashtra, Indial*

ABSTRACT

A polymer is a material with a large molecular weight that consists of millions of repeating connected units, each of which is a comparatively light and simple molecule. Various polymers are used in the drug delivery process. The sources and applications of natural and synthetic biopolymers in particulate in novel drug delivery systems (NDDSs), such as microparticles (MPs) and nanoparticles, are discussed in this chapter. The present state of research into novel drug delivery technologies is discussed. For decades, drug transport and pharmacokinetics have been significant aspects of therapeutic and biomedical practice. In recent years, the globe has seen rapid progress in the creation of various medication delivery protocols, particularly when it comes to the usage of biopolymers that degrade. In the field of medication delivery,

Novel Drug Delivery Systems in the Management of Chronic Diseases.
Raj K. Keservani, Umesh D. Laddha, Eknath D. Ahire, and Sanjay J. Kshirsagar (Eds.)

polymers are becoming increasingly significant. Polymers are used in a variety of pharmaceutical applications, from tablet binders to viscosity and flow control agents in liquids, suspensions, and emulsions. In this chapter, we go over the essential polymers employed in innovative drug delivery, as well as their pharmaceutical applications, stability, chemistry, and applications. The versatility of polymers in novel dosage forms has significant advantages over the barriers that emerge in traditional dosage forms.

3.1 INTRODUCTION

Both pharmacologically active chemicals and excipients are added to pharmaceutical dosage forms to aid in the formulation and production of the final dosage form for delivery to the patients. Indeed, the final dosage form's features (i.e., the majority of the time, the excipients determine the bioavailability and stability of the drug chosen, their concentration, and how they interact with the active molecule and each other). Over the last few years, significant progress has been achieved in the creation of new medication delivery systems (NDDS). Polymeric niosomes, nanoparticles, nanoemulsions (NEs), nanocapsules, phytosomes, liposomes, microspheres, transferosomes, and ethosomes are only a few of the innovative herbal and synthetic formulations that have been reported (Ajazuddin & Saraf, 2010). The term "Novel drug delivery system" (NDDS) refers to the development of new pharmacological forms with improved properties such as reduced particle size, greater penetration characteristics, and site target selection. When compared to their effect in traditional dosage forms, NDDSs can be employed to improve the effectiveness of biotherapeutic drugs (Faheem & Abdelkader, 2019). The goal of innovative drug delivery is to keep the drug in the appropriate effective concentration range for a long time while also delivering it in a targeted manner. One method of obtaining NDDS necessitates the use of a carrier, or an additional inert component known as a matrix or polymer (Talele et al., 2021).

The word "polymer" comes from the Greek word "poly" meaning many and "meros" which means parts. So, polymer refers to a compound which composed of many components or parts. More specifically, polymers are defined as molecules consisting of repeated units of smaller molecules. Polymers such as cellulose, DNA, rubber, and cotton have been in nature since the dawn of time. Natural polymers have been processed since the dawn of time, when early humans woven and dyed wool, silk, and carbohydrates from cotton and flax. Early South American civilizations used natural rubber

(*Hevea brasiliensis*) for waterproofing (Olatunji, 2015). Polymer is used as an excipient in a variety of sectors, including veterinary meals, wood and paper, fabrics and clothing, cosmetics, and pharmaceuticals. Furthermore, more emphasis should be directed to carrier material research in order to improve more acceptable carriers that can reduce drug toxicity, improve activity, and improve the overall efficacy of the agents (Ajazuddin & Saraf, 2010). Polymers are categorized into two categories based on their origin: synthetic and natural. Biopolymers are natural polymers with the prefix "bio" indicating that their source is a living organism, and they are commonly biodegradable. Biopolymers can be chemically synthesized starting from organic monomers, extracted from plants or produced by certain microorganisms (Oprea & Voicu, 2020).

Natural polymers can be split into six categories based on their source: Proteins, polynucleotides, polysaccharides, polyisoprenes, polyphenols, and polyesters. Natural polymers are the most promising category of biomaterials, their number of applications and significant use in the biomedical sector being increasingly higher due to their diversity when it comes to composition and properties (Kulshrestha & Mahapatro, 2008). Green polymers, made from renewable resources, are an eco-friendly more efficient alternative for synthetic materials since they can help minimize the quantity of non-biodegradable waste generated by the medical industry on a daily basis (Song et al., 2018).

Natural polymers are a preferable option in TDD as they are readily available, inexpensive, potentially biodegradable and biocompatible and can undergo various chemical and surface modifications to fit the requirement of the TDD system. A TDD system uses a combination of polymers and an embedded medication that is given in a controllable and sustainable manner into or through the skin (Tojo, 2005). These compounds are used in a variety of pharmaceutical applications, including extended and SR coated dosage forms, osmotic drug delivery systems (DDSs), extended, and controlled release matrices, mucoadhesive, and bio-adhesive, compression tablets as a compressibility enhancer, liquid preparations as a thickening agents and stabilizers, tablets, and granules as a binder, semisolid preparations as gelling agents, and many others. These polymeric materials have also been employed in transdermal patches as a filler, free-flowing agents, taste masker, and pressure sensitive adhesives (Shokri & Adibki, 2013). Biopolymers like a polylactic acid (PLA) and chitosan are currently gaining popularity in areas such as 3D printing and tissue engineering (Olatunji, 2015).

Polymers have been an important component in the formulation and innovation of NDDSs because of their ability to release active ingredients for a longer period of time without fluctuations in plasma levels. Immediate or controlled delivery of the drug from preparation can be achieved by proper selection of excipients. Polymer nature, its characteristics, applications, advantages, disadvantages are always taking consideration while selection of polymer. Several importance of polymers in NDDS and their application are mentioned as follows (Misra & Shahiwala, 2020).

Polymers have been an important component in DDSs because of their ability to release active ingredients for a longer period of time without fluctuations in plasma levels. For ophthalmic DDSs, many polymeric systems have been developed that use mucoadhesion, viscosity modification *in situ* gelling to increase retention period and decrease administration frequency. Biodegradable and nonbiodegradable polymeric implants for sustainable drug delivery for chronic vitreoretinal diseases have been developed. Blood-brain-barrier (BBB) provides limits for the transportation of therapeutic agents to the brain. Efforts to target the brain directly via the olfactory and trigeminal nerve pathways, bypassing the BBB, which is favorable in many CNS conditions. That's happen only by considering the use of polymers. Comprising use of adhesive polymer, rate-controlling polymers, a backing membrane, and a release liner make up transdermal delivery system. These all components are type of polymeric system that's why the choice of an appropriate polymer component is important for the pharmacological effectiveness of transdermal DDSs. To suit the dosage-specific objectives, a variety of natural and synthetic polymers are available, ranging from providing a barrier function in the form of a basic film coating, conferring mucoadhesive qualities, or incorporating the API into matrix for delayed or extended or SR effects. The several types of gastric DDSs evolved, to tackle these challenges including swelling, floating, bioadhesive, and combinational delivery systems, all of which are made with distinct polymers and formulation procedures. Polymers derived from natural and synthetic sources can be utilized to provide targeted and controlled medication administration in the colon and play important to the efficacy of these gastric dosage forms. Bioadhesive polymers have now been utilized widely in buccal medication delivery systems to promote greater durability of the dosage form in the targeted site (Misra & Shahiwala, 2020).

3.2 EUDRAGIT

- **Synonyms:** Methacrylic acid-methyl acrylate copolymer.
- **IUPAC Name:** Methyl prop-2-enoate; 2-methylprop-2-enoic acid.

- **Molecular Formula:** $C_8H_{12}O_4$.
- **Molecular Weight:** 172.18 (https://pubchem.ncbi.nlm.nih.gov/, n.d.)

Eudragit polymers are adaptable polyacrylate polymers with different degrees of solubility, making them ideal for long-term formulation. It was made by polymerizing acrylic and methacrylic acids, or their esters, whose physicochemical properties are controlled by the R functional group. USPNF, BP, Ph. Eur, and The Handbook of Pharmaceutical Excipients have all included Eudragit. Eudragit polymers are adaptable polyacrylate polymers with different degrees of solubility, making them ideal for long-term formulation (Sukhbir Singh, 2015). It was manufactured in Darmstadt, Weiterstadtand, and Evonik Industries was the first to market it in the 1950s. Rohm Gmbh and Co. KG, Darmstadt from Germany, owns the trademark.

3.2.1 EUDRAGIT POLYMER CLASSIFICATION

Eudragit polymers are categorized according to their intended purpose or the sort of formulation they generate. Eudragit polymers are among them:

- Sustained release formulations allow for time-controlled medication release;
- Enteric formulations treat gastro-resistance and gastrointestinal (GI) disorders;
- Protective formulations provide moisture protection and odor/taste masking.

Poly(meth)acrylates are soluble in digestive juices due to the production of salt. Eudragit L, S, FS, and E polymers are examples. These polymers with alkaline or acidic groups allow the active substance to be released in a pH-dependent manner. From basic taste masking to gastric resistance to controlled drug release in all parts of the gut, there are several applications.

Poly(meth)acrylates are insoluble in water but dissolve in digestive juices. pH-independent swelling of Eudragit RL grade and RS grade polymers having alkaline groups, as well as Eudragit NE grade polymers having neutral groups, allows for controlled temporal release of the active ingredient. Drug release pattern that will be delayed and continuous (Thakral et al., 2013) (Table 3.1).

TABLE 3.1 Different Trade Types of Eudragit and its Characteristics

Trade Name	Tg (°C)	Form	Physical Property	IUPAC
Eudragit RL 100	63 ± 5	Granule	Colorless, clear to cloudy granules.	Poly(ethyl acrylate-co-methyl
Eudragit RL PO	63 ± 5	White powder	White powder having mild amine-like odor.	Methacrylate-co-trimethyl Ammonia
Eudragit RL 30 D	55	30% aq. dispersion	Milky-white liquid	ethyl methacrylate chloride
Eudragit RL 12.5	–	12.5% organic solution	Faint yellow liquid with low viscosity.	1:2:0.2
Eudragit RS 100	65	Powder	White powder	Poly(ethyl
Eudragit RS PO	65	Powder	White powder and amine-like odor.	acrylate-co-methyl meth acrylate-co-
Eudragit RS 30 D	55	30% aq. dispersion	Milky-white liquid having low viscosity with a faint odor.	trimethyl ammonia ethyl methacrylate chloride)
Eudragit RS 12.5	–	12.5% organic solution	Yellow liquid of low viscosity.	1:2:0.1
Eudragit L 100	>130	Powder	White powder having faint odor.	Poly(methacrylic acid-co-methyl
Eudragit L 12.5	>130	12.5% organic solution.	Colorless to slightly cloudy liquid, odor of isopropyl alcohol.	methacrylate) 1:1
Eudragit S 100	>130	Powder	Powder having faint odor.	Poly(methacrylic
Eudragit S 12.5	>130	12.5% organic solution	Colorless to slightly cloudy liquid, odor of isopropyl alcohol.	acid-co-methyl methacrylate) 1:2
Eudragit L 30D-55	96	30% aq. dispersion	Milky-white liquid, less viscosity, faint odor.	Poly(methacrylic acid-co-ethyl
Eudragit L 100-55	96	Powder	White powder and faint odor	acrylate) 1:1
Eudragit E 100	–	Granules	Colorless to yellow-tinged granules and characteristic amine-like odor.	Poly(butyl meth-acrylate-co-(2-di-methylaminoethyl) methacry-
Eudragit E 12.5	–	12.5% organic solution	Light yellow liquid, less viscosity, appearance is clear to slightly cloudy, and characteristic odor.	late-co-methylmeth-acrylate) 1:2:1
Eudragit E PO	–	Powder	White powder having mild amine-like characteristic odor.	

3.2.2 PHARMACEUTICAL PROPERTIES OF EUDRAGIT POLYMERS

The brand name Eudragit is well-known in the industry for poly(meth) acrylates. These polymers allow the active ingredient in a solid dose form to work while passing through the human body. The ability to combine multiple polymers allows for the creation of a desired drug release profile by releasing the medication at the appropriate place, at the correct time, and over a specified period of time. Protection from external effects (moisture) and taste/odor masking to enhance patient compliance are also significant features. The product line offers complete flexibility for customized medication release patterns, with the best performance for enteric, protective, and sustained-release features (https://healthcare.evonik.com/, n.d.).

3.2.3 PHARMACEUTICAL APPLICATIONS

Eudragit has properties that make it suited for ophthalmic applications, such as no toxicity, a positive charge, and a regulated release profile. Expansion of the GI system by swelling or unfolding to a large size, which inhibits emptying of the dose form through the pyloric sphincter, slowed motility of the GI tract by concurrent administration of medications or pharmaceutical excipients these all procedures can be accomplished using various grades of eudragit. By coating with eudragit polymer, sustained intestinal delivery of medications was established, bypassing the stomach and releasing the loaded drug for long periods of time into the intestine. Eudragit L and Eudragit S are types of enteric acrylic resins that are commercially accessible. Both forming a film which provide resistant to stomach fluid. At pH 6 and 7, Eudragit L grade and S grade shows solubility in intestinal fluid. Availability of Eudragit L grade in solid, aqueous dispersion, or an organic solution (Isopropanol). Eudragit S is only available as a solid and an organic solution (Sukhbir Singh, 2015).

Irritable bowel syndrome was treated with Tegaserod maleate, while Eudragit L 100 with S100 by mixture of 1:1, 1:2, and 1:3 ratio. Sodium para-amino salicylate Pellets were coated with Eudragit L 30 D-55 using a fluidized bed processor and tested *in vitro* for two hours using 0.1 N HCl before being switched to phosphate buffer pH 6.8. The most acceptable effects against stomach attack were obtained with a 60% w/w Eudragit L30 D 55 coating solution (https://www.pharmaexcipients.com/, n.d.).

The mechanical characteristics of casted Eudragit E-100 films was investigated by the combined effect of two cohesive promoters, succinic or citric acid, and triacetin as a plasticizer. The created films were transparent, elastomeric, self-adhesive, and pale yellow in color. The Eudragit E100 polymer produced wrinkle-free transparent films with good skin adherence. Because of the erosion of the hydrophilic Eudragit E100 polymer, release kinetics from the transdermal therapeutic system were examined, and 100% release was seen within 20 minutes 40 (Sukhbir Singh, 2015).

3.3 CYCLODEXTRIN

The synonyms of cyclodextrin are Cavitron, cyclic oligosaccharide, cycloamylose, cycloglucan, encapsin, and Schardinger dextrin.

3.3.1 DESCRIPTION

Cyclodextrins are oligosaccharides containing at least six sugar molecules that are cyclic in nature. A (14)-glucoside connects D-glucopyranose unit's bonds. There are three natural cyclodextrins, a, b, and g, differ in their molecular structure. Cyclodextrins are white, odorless, fine powders. A-cyclodextrin and g-cyclodextrin are listed as a-cyclodextrin and g-cyclodextrin, respectively, in the PhEur 2005. Impurities in b-cyclodextrin that could cause problems. β-Cyclodextrin and other cyclodextrins are stable in solid condition state if kept away from high humidity. Cyclodextrins should be kept cool and dry in a firmly sealed container (Raymond, 2006).

3.3.2 PHARMACEUTICAL FORMULATION OR TECHNOLOGY APPLICATIONS

In contrast, B-cyclodextrin is nephrotoxic and should be avoided in parenteral formulations. When used orally, B-cyclodextrin is considered harmless and is commonly found in tablet and capsule formulations. When administered orally or parenterally, B-cyclodextrin derivatives are generally benign, and the derivatives 2-hydroxypropyl-b-cyclodextrin and

3-hydroxypropyl-b-cyclodextrin are becoming increasingly relevant in pharmaceutical formulations.

B-cyclodextrin can be employed in both wet-granulation and direct-compression methods in oral tablet formulations. Depending on the manufacturer, b-cyclodextrin has different physical qualities. When directly compressed, however, b-cyclodextrin has poor flow characteristics; hence lubricant is needed, such as 0.1% w/w magnesium stearate.

Cyclodextrins have been utilized in parenteral formulations to make stable preparations of pharmaceuticals that would be made with a nonaqueous solvent. It forms water-soluble compounds with lipophilic medicines like corticosteroids in eye drop compositions. They've been shown to help improve permeation of drug into the eye., improve aqueous stability, and lessen local irritation by increasing the drug's water solubility (Raymond, 2006; https://www.pharmaexcipients.com/, n.d.) (Table 3.2).

TABLE 3.2 Types of Cyclodextrin and its Characteristics

Type	Empirical Formula	Molecular Weight	Melting Points
α-Cyclodextrin	$C_{36}H_{60}O_{30}$	972	250–260C
β-Cyclodextrin	$C_{42}H_{70}O_{35}$	1,135	255–265°C
γ-Cyclodextrin	$C_{48}H_{80}O_{40}$	1,297	240–245°C

3.3.3 SOLUBILITY

- α-cyclodextrin is soluble in 1 part water in 7 parts waters at 208°C and 1 part water in 3 parts waters at 508°C.
- β-cyclodextrin is soluble in propylene glycol (1:200), water (1:50), and methylene chloride (1:20) at 208°C and 508°C, but nearly insoluble in acetone, ethanol (95%), and methylene chloride.
- γ-cyclodextrin is soluble in 1 part water in 4.4 parts water at 208C and 1 part water in 2 parts waters at 458°C (https://www.pharmaexcipients.com/, n.d.).

3.4 LYCOAT NG 73

Lycoat NG 73® is a granular polymer of hydroxypropyl starch, created specifically for Oro-dispersible films (ODFs). Lycoat NG 73 dissolves quickly in cold water and does not create lumps. Its ability to create films will be enhanced by simple heating; it will rise the viscosity and form homogeneous solution while cooking, preventing lumps and agglomerates from developing. Without the inclusion of a film-forming agent, it can be used as the only film-forming polymer in the preparation of ODF with outstanding functioning (El-Setouhy & El-Malak, 2010).

LYCOAT® (Hydroxypropyl pea starch) gives formation of thin film and coating properties in various oral dosage form of pharmaceutical and nutraceutical preparations. Our product line includes a variety of grades based on the solubility in hot and cold water and molecular weight. It's made for aqueous procedures and is ready to use right away. In wet granulation operations, these excipients also utilized as a binder. In all types of tablets, hard capsules, soft capsules, and orally dissolving films, LYCOAT® can be majorly used as film forming agent and coating agent (https://www.pharma-excipients.com/, n.d.).

Characteristics and uses:

- Excellent mechanical properties;
- Available in different grades with different viscosity;
- Specifically designed for aqueous film coatings of immediate release;
- No foam formation, high gloss and smoothness with no logo bridging and a neutral taste;
- Used for film formation;
- Most commonly used in tablet coating;
- Applicable for other oral dosage for pharmaceutical and/or nutraceuticals like swallowable tablet, hard capsules, soft capsules, orally dissolving films.

3.5 COMPRITOL

- **Synonyms:** E471, glycerol behenate, glyceryl monobehenate, Compritol 888 ATO, 2,3-dihydroxypropyl docosanoate, and docosanoic acid.

3.5.1 DESCRIPTION

Glyceryl behenate is available in fine white powder or a hard waxy mass with a slight odor. Having 65–78°C melting point. Soluble in chloroform and dichloromethane when heated, nearly insoluble in ethanol (95%), hexane, mineral oil, and water when cold. The most common applications are coating agents, tablet binders, and lubricants for tablets and capsules. Used as a lubricant in capsules and tablets larger than 10. Lipophilic matrix or coating for tablets and capsules with extended release 1.0–3.0. In cosmetics it is used as a viscosity-increasing component in the range of 1.0–5.0 (Raymond, 2006).

3.5.2 PHARMACEUTICAL FORMULATION OR TECHNOLOGY APPLICATIONS

Cosmetics, foods, and oral medicinal formulations all include glyceryl behenate. It is mostly employed in cosmetics as a viscosity-increasing ingredient in emulsions. Glyceryl behenate is largely used in pharmaceutical formulations as a lubricant for capsules and tablets as well as a lipidic coating excipient. It has been investigated as a means of encapsulating retinoids and other medications. Rather than a lubricant, it's also been studied as a hot-melt coating compound sprayed onto a powder, and as a matrix-forming additive for the controlled release of water-soluble pharmaceuticals (Mona, 2014; Ahire et al., 2020).

3.5.3 SAFETY AND STORAGE

Glyceryl behenate is largely non-irritant and benign substance that is utilized in cosmetics, meals, and oral medicinal formulations. Shows 5 g/kg LD50 on mouse through oral route. Glyceryl behenate store in a tightly sealed container at temperatures below 35°C (Raymond, 2006).

3.6 POLYETHYLENE GLYCOL (PEG)

- Synonyms: Carbowax, Lipoxol, Carbowax Sentry, Pluriol E, poly-oxyethylene glycol, and LutrolE
- Chemical Name: a-Hydro-o-hydroxypoly(oxy-1,2-ethanediyl) (Table 3.3).

TABLE 3.3 Various Polymer Grade and its Molecular Weight

Polymer Grade	Average Molecular Weight
PEG 200	190–210
PEG 300	285–315
PEG 400	380–420
PEG 540 (blend)	500–600
PEG 600	570–613
PEG 900	855–900
PEG 1000	950–1,050
PEG 1450	1,300–1,600
PEG 1540	1,300–1,600
PEG 2000	1,800–2,200
PEG 3000	2,700–3,300
PEG 3350	3,000–3,700
PEG 4000	3,000–4,800
PEG 4600	4,400–4,800
PEG 8000	7,000–9,000

3.6.1 DESCRIPTION

Polyethylene glycol (PEG) generally used as Suppository base, Lubricant in tablet and capsule, Plasticizer, Ointment base. According to the USPNF, PEG is an addition polymer of ethylene oxide and water. At room temperature, PEG 200–600 grades are liquids, while grades 1,000 and higher are solids. Liquid grades of PEG 200–600 are viscous liquids that are clear, colorless, or slightly yellow in color. They are faint but distinct odor as well as a bitter and burning flavor. At room temperature, PEG 600 can form a solid. The solid grade (PEG > 1000) comes in a variety like pastes, light sweet scent, waxy flakes, and is off-white in color. PEG 6000 and higher grades of PEG are available as milled powders and they flow freely (https://www.pharma-excipients.com/, n.d.).

3.6.2 SOLUBILITY

All grades of PEG are water soluble and miscible with other PEGs in all quantities. Gels can form in aqueous solutions of greater molecular weight classes. Acetone, benzene, alcohols, glycols, and glycerin are all soluble in liquid PEGs. Acetone, dichloromethane, and ethanol are all soluble in solid PEGs (95%). Density is 11–1.14 g/cm^3 for liquid PEGs at 258°C; 1.15–1.21 g/cm^3 for solid PEGs at 258°C.

3.6.3 CONDITIONS OF STABILITY AND STORAGE

Although grades with a molecular weight less than 2,000 are hygroscopic, PEGs are chemically stable in air and in solution. PEGs are resistant to microbial development and do not go rancid. Autoclaving, filtration, or gamma irradiation can all be used to sterilize PEGs and aqueous PEG solutions (Raymond, 2006).

3.6.4 PHARMACEUTICAL FORMULATION OR TECHNOLOGY APPLICATIONS

PEGs are commonly employed in pharmaceutical formulations, including parenteral, topical, ophthalmic, oral, and rectal treatments. It's been employed in biodegradable polymeric matrices in controlled-release systems in the lab. PEGs are non-irritating, stable, hydrophilic compounds. Although water-soluble and easily washed from the skin, PEGs do not readily penetrate the skin. PEG Solid grades are commonly used in ointments with added liquid PEG grade to modify the viscosity of the base (https://www.pharmaexcipients.com/, n.d.).

PEGs can improve the efficiency as a binder in tablet and provide granules more plasticity. However, when employed alone, they have a limited binding effect and can extend disintegration if used at quantities more than 5% w/w. A mixture of powdered ingredients with 10–15% w/w PEG 6000 is heated to 70–75°C. When utilized for thermoplastic granulation. If the mixture is agitated during cooling, it turns paste-like and produces granules. When prolonged disintegration is necessary, this technique is suitable for the manufacture of formulations like lozenges. PEGs can also be used to increase the water solubility or dissolution of insufficiently soluble compounds by producing solid dispersions (SDs) with the suitable PEG. PEGs are also used as steroid solvents in animal research. Dry heat sterilization at 150°C for 1 hour may cause oxidation, darkening, and formation of acidic breakdown products in solid grades. In an ideal world, sterilization would take place in an inert environment. The use of an appropriate antioxidant can also help to prevent PEG oxidation (Raymond, 2006).

3.7 POLYVINYL PYRROLIDINE (PVP)

The water-soluble polymer polyvinylpyrrolidone (PVP), often known as povidone or polyvidone is produced from N-vinylpyrrolidone monomer. Walter Reppe was the first to synthesis PVP, and a patent of acetylene chemistry derivatives was filed in 1939. PVP was first employed as substitute for blood plasma, then in medical, pharmacy, cosmetics, and industrial applications.

3.7.1 PROPERTIES

Polar solvent and water as well dissolve the PVP. It's soluble in a different alcohol, including ethanol, methanol, and unusual solvents like the eutectic mixture solvent created by choline chloride and urea (Relin). This powder is light flaky and hygroscopic in nature. When dry, it absorbs near about 40% of its total weight in atmospheric water. It has great wetting capabilities in solution and rapidly create the films. That's why it is suitable for used coating or as a coating additive. PVP and its oxidized hydrolysate were discovered to have luminous characteristics in a 2014 study (Kurakula & Rao, 2020).

3.7.2 USES

After the 1950s, PVP was utilized as an expander of plasma volume for trauma patients. Because of its tendency to cause histamine release and interfere with blood grouping, it is not recommended as a volume expander. It is used as a binder in tablet dosage form, and when taken orally, it just flows through the body. Autopsies have demonstrated, however, that cross povidone (PVPP) causes pulmonary vascular damage by injecting medications meant for oral consumption. When PVP is combined with iodine, it creates the povidone-iodine complex, which has disinfectant qualities. This compound can be found in good variety, including solutions, ointments, pessaries, liquid soaps, and surgical scrubs. Pyodine and Betadine are two of the many trade names for this substance (https://www.pharmaexcipients.com/, n.d.).

Pleurodesis is a pleural fusion caused by recurrent pleural effusions. Povidone iodine is as safe and effective as talc for this purpose, and it may be favored due to its ease of use and inexpensive cost. Some contact lenses and their packaging solutions include PVP. Action of lubricant or wetting substance integrated into the lens, reducing friction. Bausch and Lomb's Ultra contact lenses with MoistureSeal Technology and Air Optix contact lenses are two examples of this utilization (Kurakula & Rao, 2020).

3.7.3 TECHNICAL USES

PVP is also utilized in a variety of technological applications:

- Used as adhesive in glue sticks and hot-melt adhesives.
- Suitable for the production of water purification and dialysis membranes, Applicable for agricultural areas like seed treatment, crop protection.
- To improve photoresist resolution for cathode ray tubes.
- Used as a thickening agent in teeth whitening gels.
- Applicable in chemical and mechanical planarization processes and as a specific additive used for ceramics, batteries, inks, inkjet paper, and fiberglass.

3.7.4 OTHER APPLICATIONS

Because of its polarity, PVP bonds to polar molecules particularly well. As a result, its way found that into photo-quality inkjet papers and transparencies, as well as inks for inkjet printers.

PVP is also suitable for personal and cosmetic goods care like shampoos and toothpastes, as well as paints and moist adhesives like envelopes and old postage stamps. It's also been utilized in steel-quenching solutions and contact lens solutions. PVP was used in the early formulations of hair sprays and gels, and it is still used in some today. PVP is a stabilizer and has the E number E1201 as a food additive. E1202 is PVPP (crospovidone). It's also utilized in the wine business as a preservative (Franco & de Marco, 2020).

3.7.5 SAFETY

This chemical is USFDA-approved for a variety of purposes, and it is widely regarded as safe (GRAS). However, allergic reactions to PVP/povidone have been observed, particularly when the drug is used subcutaneously (under the skin) and when it comes into contact with autologous serum (internal blood fluids) and mucosal membranes. Allergic responses that have been documented include: After using PVP-Iodine to treat impetigo, a boy developed an allergic reaction. He was discovered to be allergic to the solution's PVP component (Kurakula & Rao, 2020).

3.8 POLYCAPROLACTONE (PCL)

3.8.1 DESCRIPTION

PCL is made through ring opening polymerization of -caprolactone with stannous octoate as a catalyst. The ring-opening polymerization (ROP) can be done with a variety of catalysts (Labet & Thielemans, 2009). Polycaprolactone (PCL) is a hexanoate repeat unit-based aliphatic and synthesized biodegradable polyester made from crude oil (Labet & Thielemans, 2009; Woodruff & Hutmacher, 2010). PCL has been authorized by the FDA authority for the human use as a drug carrier, suture, or adhesion barrier, among other things (Li & LaBarbera, 2017). Formulations with various functionalized characteristics for medication administration and tissue engineering. However, PCL's inherent characteristics make it inappropriate for many medicinal formulations and it lacks functionalization (Dash & Konkimalla, 2012). Following the recent launch of a PCL-based microparticles (MPs) dermal filler belongs to the collagen stimulator class (Ellanse), PCL is being employed in the fast-increasing area of human esthetics (Moers-Carpi & Sherwood, 2013).

PCL has been utilized in the administration of drugs for a variety of ailments, but has also stretched to proteins, peptides, vaccines, and other bioactive compounds (Espinoza et al., 2020). It is non-toxic and tissue friendly, making it ideal for resorbable sutures, regenerative treatment scaffolds, and drug administration. PCL takes the longest to breakdown (2–3 years) and is destroyed by microbes by breakdown of its aliphatic ester bond under physiological circumstances. Of all the polyesters, PCL is the slowest to decompose (Dwivedi et al., 2020). PCL has a high permeability to plenty of medicines and great biocompatibility, making it ideal for regulated drug administration (Mohamed & Yusoh, 2015). PCL beads used to encapsulate a range of medicines for slow release and targeted medication delivery (Bhavsar & Amiji, 2008). It's utilized in root canal filling and as a constituent of "night guards" (dental splints) in dentistry (as the Resilon composite). It functions similarly to gutta-percha, has comparable handling qualities, and may be softened with heat or dissolving with solvents like chloroform for re-treatment. The main difference between PCL-based filling materials for root canal (Resilon and Real Seal) and gutta-percha is that PCL-based filling materials are biodegradable (Hiraishi et al., 2007).

3.8.2 NATURE AND CHEMISTRY

IUPAC name of PCL (1,7)-Polyoxepan-2-one, IUPAC name is poly(hexano-6-lactone), where PCL is also known as 2-Oxepanoner, 6-Caprolactone polymer and CAS no is 24980-41-4. Chemical formula of PCL is $(C_6H_{10}O_2)_n$ and density is 1.145 g/cm^3 where melting point is 60°C (140°F). PCL is super soluble in chloroform, dichloromethane, carbon tetrachloride, benzene, toluene, cyclohexanone, and 2-nitropropane at room temperature; slightly soluble in acetone, 2-butanone, ethyl acetate, dimethyl formamide, and acetonitrile at room temperature; and insoluble in alcohols, petroleum ether, diethyl ether, and water (Labet & Thielemans, 2009). Water, oil, solvent, and chlorine are all resistant to it (Woodruff & Hutmacher, 2010).

PCL is semi-crystalline polymer, hydrophobic in nature and having glass transition temperature (Tg) of 60°C and a melting point of 59–64°C, as required by its crystalline nature, which allows for simple formability at low temperatures. PCL samples can have a molecular weight ranging from 3,000 to 90,000 g/mol, and they can be categorized based on this value (Mohamed & Yusoh, 2015). Crystallinity reduces as molecular weight increases (Woodruff & Hutmacher, 2010).

This polymer is both flexible and robust. Several natural polymers (chitosan, starch, hydroxyl appetite) as well as synthetic polymers (polyurethane (PU), PEG, oxazolines, PVA, polyethylene oxide (PEO), polylactic-co-glycolic acid (PLGA), and polylactic acid) have been found compatible with PCL, which can change its thermal, rheological, and biophysical properties (Mandal & Shunmugam, 2020). In physiological settings (such as the human body), PCL is damaged by ester bonds hydrolysis and as a result, it has attracted a lot of interest for usage as an implanted biomaterial. It is particularly intriguing development of long-term implanted devices, as its degradation is even slow than those of polylactide (Hajiali et al., 2018).

3.9 POLYLACTIC ACID (PLA)

Biobased, biodegradable, and bioabsorbable aliphatic polyester can be made from 100% reusable bio-resources like sweet potato, rice, and wheat through fermentation and polymerization (Fattahi et al., 2019). PLA, a renewable-source thermoplastic that is compostable and biodegradable (Singhvi & Gokhale, 2013). The FDA has approved PLA for use in biomaterials such as wound dressings, sutures, bone plates, and medication delivery devices (Toncheva et al., 2014). PLA is a biopolymer that can be semicrystalline or completely amorphous in nature. PLA is made from lactic acid, which is obtained by fermenting agricultural items like as maize. PLA is synthesized by ROP of the cyclic lactide or direct condensation of lactic acid (Bajpai et al., 2014). PLA has some evident drawbacks, such as its slow degradation rate, which makes it unsuitable for varieties of applications; and the absence of cell recognition sites, which are critical for tissue compatibility, on the surface of PLA used in tissue engineering (Cheng et al., 2009).

It is most common and interesting biodegradable polymers is poly(lactic acid). To generate high crystalline poly(lactic acid), lactic acid is required which should be optically pure and leads to the polymer's excellent strength, chemical, and heat resistance qualities (Singhvi & Gokhale, 2013). The pace at which PLA degrades is determined by its crystallinity. The breakdown of the graft polymer reduces as the lactide concentration rises (Ha & Gardella, 2005). It is one of drawback that properties of polymer can change after processing, because of thermal and mechanical stresses. PLA degrades at temperatures above 200°C, although the rate and extent of degradation are dependent on factors such as temperature, time, catalyst quantity and low molecular weight impurities (Omathanu Pillai & Ramesh, 2001).

3.9.1 NATURE AND CHEMISTRY

PLA is a semi-crystalline or amorphous polymer and having glass transition temperature (Tg) and melting temperature (Tm) of approximately 55°C and 180°C, respectively (Singhvi & Gokhale, 2013). The CAS no of PLA is 26100-51-6, density is 1.210–1.430 $g \cdot cm^{-3}$; melting point is 150 to 160°C.

PLA is soluble in chloroform, acetonitrile, dioxane, methylene chloride, dichloroacetic acid and 1,1,2-trichloroethane at room temperature, but only slightly soluble in toluene, ethyl benzene, acetone, and tetrahydrofuran at room temperature.

PLA is insoluble in alcohols, water, and linear hydrocarbons and another way PLA in crystalline nature cannot dissolved by tetrahydrofuran, acetone, and ethyl acetate (Misra & Shahiwala, 2020). Different structural characteristics, such as molecular weights and composition, could influence

PLA's thermal properties (Omathanu Pillai & Ramesh, 2001). PLA and its copolymers in nanoparticle form employed in the encapsulating process of numerous drugs like antipsychotics. PLA stents indicated a potential future in treatment management of ureteral combat injuries (Ha & Gardella, 2005).

PLA can be mixed with certain biodegradable or non-biodegradable polymers like polypropylene, polyethylene, polystyrene, polycarbonates, polyethylene terephthalate, and chitosan, to make composite materials, or it can be combined with natural fibers, carbon nanotubes, ceramic nanoparticles and cellulose to make composite materials. PLA-hydroxyapatite (HA) suitable for bone healing devices, in which HA microparticle or nanoparticles disseminated in the polymer matrix, are a good example (Kaseem, n.d.).

3.10 PLGA

Poly(lactic-co-glycolic acid), also known as PLGA, PLG, or poly(lactic-co-glycolic acid), is a biodegradable and biocompatible copolymer that is utilized in FDA-approved medicinal devices. The cyclic dimers of lactic acid and glycolic acid (1,4-dioxane-2,5-diones) are co-polymerized to form PLGA via ring-opening co-polymerization. Polymers can be made as random or block copolymers which gives them additional polymer characteristics and 2-ethylhexanoate, alkoxides, and aluminum isopropoxide are common catalysts utilized in the production of this polymer. During polymerization, ester linkages bond form in consecutive monomeric units (of glycolic or lactic acid) in PLGA, resulting in a linear, aliphatic polyester (Astete & Sabliov, 2006).

Different types of PLGA generated depending on the lactide to glycolide ratio used for polymerization; they are generally characterized by the monomers molar ratio utilized (e.g., the copolymer PLGA 75:25 is made up of 75% indicates lactic acid and 25% of glycolic acid.). Depending on the block structure and molar ratio, PLGAs can range from entirely amorphous to fully crystalline (Samadi et al., 2013). Among the numerous biodegradable polymers, PLGA is the most popular due to its clinical experience, favorable degradation characteristics, and potential for long-term drug administration like sustained drug delivery.

Such polymers are FDA-approved polymers that are strong physically and extensively biocompatible have been widely explored as drug, protein, and numerous other macromolecules like RNA, DNA, and peptides delivery vehicles (Makadia & Siegel, 2011; Quinten Ruhé, 2006).

PLGA has been deemed a viable polymer for the administration of numerous active medications because of its superior features, including biocompatibility, biodegradability, and mechanical strength (Mohammadi-Samani & Taghipour, 2015). The polyester PLGA is a copolymer of PLA and polyglycolic acid (PGA) (Makadia & Siegel, 2011). The releasing and degradation rates of integrated medicinal molecules are affected by changes in PLGA characteristics during polymer biodegradation. The initial molecular weight, device size, lactide-to-glycolide, ratio, water exposure (surface shape), and storage conditions have all been found to have an impact on PLGA physical properties (Houchin & Topp, 2009). PLGA copolymers are glassy in form and have a reasonably stiff chain structure since there physiological temperature above than of 37°C. Tg of PLGAs is also reported to decrease when the lactide percentage in composition of copolymer and molecular weight of the copolymer decrease (Makadia & Siegel, 2011).

PLGA is made up of molecules that are $(C_3H_6O_3)_n.(C_2H_4O_3)_n$. It has a density of 1.53 g/mL at 25°C. PLGA can be shaped into almost any shape or size, and it can hold molecules of any size. It may be soluble in a various of common solvents (Nair & Laurencin, 2007). PLGA copolymers are amorphous in nature and having glass transition temperature is 40 to 60°C (Astete & Sabliov, 2006). Dichloromethane, tetrahydrofuran, ethyl acetate, chloroform, hexafluoro isopropanol, acetone, and benzyl alcohol are among the solvents that they are soluble in. They degrade at a quicker rate than their homo polymers, PLA or PGA, on their own. Even though PLGA is hydrophobic, it softens in a damp environment (Kapoor et al., 2015). For the controlled delivery of medicines or proteins, PLGA-based drug delivery vehicles such as microspheres, microcapsules, nanospheres, and nanofibers have been created. The medication or protein has been evidence that to have variable degrees of contacts with the base polymer, resulting in quick or protracted release patterns, depending on the type of the PLGA employed (Nair & Laurencin, 2007).

The presence of water, PLGA degrades through ester bonds hydrolysis. When compared to predominantly lactide materials, the time required for PLGA degradation is proportional to the monomer ratio used in production; the greater the rate of percent of glycolide units used in production, the quicker the time needed for degradation. The copolymer with a 50:50 monomer ratio is exceptions to the general norm, as it degrades more quickly (about two months). Furthermore, polymers that are close with esters (rather than free carboxylic acid (CO_2H)) have longer half-lives of breakdown (Samadi et al., 2013).

Many medical products, such as grafts, threads, implants, prosthetic devices, surgical sealing films, micro- and nanoparticles, have benefited from this flexibility in degradation (Pavot et al., 2014). In the body, PLGA is hydrolyzed to yield the main monomers: lactic acid and glycolic acid. These two monomers are by-products of several metabolic enzymatic pathways in the body under normal physiological conditions. The tricarboxylic acid cycle metabolizes lactic acid, which is then excreted as CO_2 and water. Glycolic acid is processed similarly and eliminated by the kidneys as well (Xu et al., 2017).

There is no systemic cytotoxicity associated with employing PLGA for biomaterial uses since the two monomers may be metabolized by the body. The acidic breakdown of PLGA has shown lower the local pH low enough to generate an autocatalytic environment. It has been demonstrated that the pH within a microsphere may reach pH 1.5 (Karen, 1999).

3.11 CARBOMERS (CARBOPOL® POLYMERS)

Carbomer, also known as Carbopol, is a high molecular weight polymer that is cross-linked with acrylic acid and either allyl sucrose ether or allyl pentaerythritol ether. Carbomer research has matured and is now widely used, primarily in cosmetics and pharmaceutical research and production. Carbomer analysis began in the 1950s, and Goodrich was the first to develop and produce it in the United States. Carbomer was developed and patented in 1957. After that various number of carbomer-based extended-release tablet formulations have been patented.

Carbopol polymers are now widely used as ingredients in almost every type of pharmaceutical dosage system, from controlled release tablets to oral suspensions to other Novel Delivery Systems, as well as a variety of topical agents. Despite being very mild acids (much weaker than acetic acid), these polymers react directly with alkali to form salts. The pH of Carbopol polymer aqueous dispersions ranges from 2.8 to 3.2, depending on polymer concentration (Khamanga & Walker, 2011). The greater the concentration, the greater the carboxyl concentration and, as a result, the lower the pH. Carbomer quickly absorbs water, becomes hydrated, and swells. Because of its hydrophilic nature, cross-linked structure, and essentially poor solubility in water, Carbopol is a most capable component for controlled drug release systems. Carbopol's are made from primary particles with average diameters ranging from 0.2 to 6.0 microns. The flocculated agglomerates can't be degraded into the final particles once they're made. Each particle can be thought of as a network structure made up of polymer chains that are linked together by cross-linking.

Carbopol polymers are created through a cross-linking process. Various grades of Carbopol are available due to its cross-linking characteristics and production conditions. Carbopol 934 P is polymerized in benzene after being cross-linked with allyl sucrose. Carbopol 974 P, 971 P, and 71G are polymerized in ethyl acetate after being cross-linked with allyl pentaerythritol. Polycarbophil is a cross-linked polymer that is polymerized in the solvent benzene. Carbopol 971 P and Carbopol 974 P are made in the same way under the same conditions however, Carbopol 971 P has a slightly lower degree of cross-linking agent over Carbopol 974 P. Carbopol 71 G is the granular version of the Carbopol grade (Gupta et al., 2021).

3.11.1 REGULATORY STATUS OF CARBOPOL POLYMERS

Carbopol polymers, Noveon polycarbophils, and Pemoline polymeric emulsifiers have all been thoroughly reviewed and tested for toxicity. Both Noveon polycarbophils and calcium polycarbophils are GRAS (Generally Recognized as Safe) materials. Noveon, Inc. has additionally decided that Carbopol 934, 934P, 971P, and 974P are GRAS whilst utilized in nutrition tablets (assumes a use stage of our merchandise starting from 8% to 30% of pill mass, with a normal use of 15% to 20%; the common weight of the nutrition pill is 500 mg primarily based totally on records from the Physician's Desk Reference, 48th edn., 1994). The Cosmetic Ingredient Review Expert Panel summarized the toxicity of Carbopol polymers in their assessment of the safety of Carbopol polymers for cosmetic ingredients. This assessment, in addition to next toxicology testing, discovered a low toxicity and infection potential (Berney, 1979; Prabhakar, 2010).

3.11.2 USES

Carbomers are acrylate polymers used for a variety of applications. They are used as emulsifiers, suspending agents, binding agents and gel bases in pharmaceutical processes and they also used to make fake tears. Carbomer 974P is a main non-aqueous component (5% polymer, 94% water) of Buffer Gel, a proprietary formulation that also contains dibasic sodium phosphate, dibasic potassium phosphate, sorbic acid, magnesium sulfate, disodium EDTA and monobasic sodium phosphate. It functions as a buffer, microbicide, and spermicide. It's made out of an aqueous gel with a pH of 3.9 that acidifies double as much semen to a pH of 5 while keeping the vaginal acidity intact. It is used vaginal delivery as a spermicide also prevents sexually transmitted diseases like HIV infection and others (Aronson, 2010).

Carbomers are high-molecular-weight acrylic acid homo- and copolymers that have been cross-linked with such a polyalkvenyl polyether. They are weakly acidic in nature, and their thickening ability must be neutralized with an appropriate base. Aqueous, anhydrous, and hydroalcoholic systems (Vitthal & Charles, 2016).

To avoid precipitation, carbomer, as an anionic adhesive, should not be used in conjunction with divalent basic drugs. The bioadhesive drug delivery system (BDDS) acts on a variety of cavity epidermal cell mucosa or skin surfaces, including the GI tract, oral cavity, vagina, nasal cavity, epidermis, and so on. Membranes, tablets, and gels are the available dosage forms. Agent, stick, and so on. Gel is a new type of bioadhesive drug release system that has grown in popularity in recent years. It has a good dispersion, strong adhesion, good stability, a long-lasting drug effect, is easy to apply, and can prevent the first pass metabolism and the site of administration.

3.11.3 RHEOLOGICAL PROPERTIES

The rheological houses of Carbopol polymers range relying on particle size, molecular weight among crosslinks, and the fraction of general devices that arise as terminal, loose chain ends. Table 3.4 suggests the viscosity variety of diverse Carbopol polymers.

TABLE 3.4 Viscosity Range of Different Carbopol Polymers

No.	Polymer	Viscosity
1	Carbopol 934 NF	30,500–39,400
9	Carbopol 934 P NF	29,400–39,400
12	Carbopol 71 G NF	4,000–11,000

3.11.3.1 CARBOPOL POLYMERS APPLICATIONS

Controlled release in tablets, bio-adhesion in oral cavity, ophthalmic, intestinal, nasal, vaginal, and rectal programs are all made viable via way of means of the simple water swellable Carbopol polymers. Thickening at extraordinarily low concentrations to provide a wide type of obvious viscosity and glide situations in topical, lotions, creams, and gels, oral suspensions, and transdermal gel reservoirs. Several properties of Carbopol make it potentially valuable as a pharmaceutical excipient in numerous applications such as:

1. Gel-forming capacity,
2. Viscoelastic nature in a final formulation,
3. Thermal stability.

In controlled release and solid dosage formulations Carbopol is used for the last four decades. In recent years, the number of manufacturers commercializing controlled-release tablets based on carbomers has increased significantly. Carbopol polymer-primarily based totally pill formulations have proven zero-order and near-zero-order launch kinetics. At low concentrations, those polymers are effective (much less than 10%). Nonetheless, they showcase extraordinarily fast and green swelling in each simulated gastric fluid (SGF) and simulated intestinal fluid (SIF). Carbopol polymers yield tablet with excessive hardness and less friability. These polymers may be efficaciously formulated right into a huge variety of pill forms, inclusive of conventional swallowable tablet, chewable tablet, buccal tablet, sublingual tablet, bubbling tablet, and suppositories, in addition to supplying controlled-launch properties. Carbomers dissolve quicker at decrease doses than different excipients. Because of those factors, Carbopol polymers are greater extensively used inside the components of dosage forms.

Because Carbopol polymers swell quickly in water and soak up a big quantity of it, curler compaction is getting used to make a brand-new shape of Carbopol polymer 71G NF with a purpose to keep away from use of flammable solvents. Carbopol polymer 71G NF is a flexible managed-launch additive that may be utilized in pill dosage shape in direct compression. As a result, carbomers function as powerful managed launch marketers for matrix pills whilst additionally enhancing drug bioavailability. As the best granular polymer (Carbopol 71G NF) handy for direct compression formulation, they act as green binders in each dry and moist granulation processes (Alexander, 2010).

Oral Suspension Applications: Carbopol polymers were great utilized in oral suspensions for decades to thicken, alternate waft properties, droop insoluble ingredients, and offer bio-adhesion. These polymers are sizable due to the fact they take away the settling hassle even at low concentrations. When Carbopol polymers are hydrated and neutralized, they swell and shape colloidal dispersion. They are tremendously green at low degrees and offer long-time period suspension balance throughout an extensive pH range. They also can be used to growth viscosity and masks the sour flavor of sure drugs (Prabhakar, 2010).

3.12 POLY(METHYL METHACRYLATE) (PMMA)

Because of its biocompatibility, poly(methyl methacrylate) (PMMA) is one of the maximums broadly researched biomedical materials, and latest guides have proven a growing hobby in its packages as a drug carrier. Polymerization strategies or pre-shaped polymer-primarily based totally strategies may be used to create PMMA-primarily based totally particulate carriers (PMMAP).

These debris have the capacity for use as an adjuvant for vaccines in addition to a service of many capsules which includes antibiotics and antioxidants through diverse routes of administration. The launch of medication from PMMAP is generally biphasic, with incomplete drug launch. Recent techniques for enhancing launch profiles have targeted on growing polymer hydrophilicity with the aid of using synthesizing functionalized PMMA microspheres or formulating PMMA composites with hydrophilic polymers. This review looks at the current state of nano/micro-PMMA-based particulate carriers' preparation techniques, biomedical applications, drug release kinetics, and toxicity.

Since the works of Prof. Speiser and collaborators, the use of PMMA as a biocompatible polymer for particle preparation has been considered a watershed moment in the history of microencapsulation (Kreuter, 1976). PMMA was first used as a particulate carrier material in formulation of nanoparticles for vaccination purposes (Kreuter, 1978) and also used on beads to fill holes caused by surgical removal of chronic osteomyelitis.

Poly(methyl 2-methylpropenoate) is (IUPAC name) of PMMA, is a synthetic homopolymer of methylmethacrylate monomer that is non-biodegradable (MMA). It is a hard, rigid, however brittle fabric with a pitcher transition temperature of 105°C. PMMA is surprisingly water insoluble however will become barely extra water soluble when exposed to water, as evidenced by a reduction in contact angle and pronounced contact angle hysteresis. Partially substituted hydrocarbons such as trichloroethylene are good organic solvents for PMMA.

3.12.1 BIOMEDICAL APPLICATIONS OF PMMA PARTICULATE CARRIERS

Several research have shown that for vaccines PMMAP used as adjuvants and can transport a variety of drugs, including antioxidants, anti-inflammatory agents, antihistamines, antihypertensives, antidiabetics, and antibiotics. Vaccine preparation One of the first uses for PMMAP was as a polymeric particulate adjuvant in vaccines. PMMAP was used to adjuvant split influenza vaccines with varying antigen content. The virus or virus subunits could be partially encapsulated by polymerization, and the antibody responses obtained with this preparatory work were much higher and the safety against infection was much better than with the conventional adjuvants permitted for human vaccination, such as aluminum hydroxide or aluminum phosphate. The current resurgence of interest in PMMAP-based vaccines arose in the 1990s due to their potential use as a human immunodeficiency virus adjuvant (HIV). PMMA nanoparticles are an increasing study subject in the look for secure surfactant-unfastened vaccines because of their correct protection profile, biocompatibility, low cost, ease of manufacturing and administration, loss of immunogenicity, and applicability to special vaccine antigens. New vaccine formulations primarily based totally on PMMA-based functional nano/microspheres and the HIV-1 Tat nutrients have been developed in this scenario. The findings indicate that these functional PMMA nanoparticles have a high potential for novel Tat nutrient vaccines against HIV-1 and represent a hopeful delivery system for vaccination with other nutrient vaccines, especially when a native protein formation is required. However, PMMA's non-biodegradability may continue to impede the produced vaccine formulations (Bettencourt & Almeida, 2012; Ahire et al., 2018).

3.12.2 TOXICITY OF PMMA PARTICULATE CARRIERS

PMMA is used in biomedical applications as a biocompatible material that does not typically elicit an immune response from the host. This is demonstrated by the fact that for above 60 years; this material mostly preferred in humans as bone cement and device material. As a result, particles of this material with the size of nanoparticles are frequently found in the bodies of patients. So far, the majority of studies on PMMAP toxicity have been related to their use as vaccine adjuvants. A year after intramuscular (i.m.) injection of a PMMA nanoparticle-containing influenza vaccine, no histological abnormalities were found at the injection site in guinea pigs – said that core shell nanosphere- and microsphere-PMMA-primarily based totally vaccine formulations have been now no longer cytotoxic *in vitro* and have been nicely tolerated in mice following i.m. or intranasal administration. Some current courses have already addressed the issue of the toxicity of drug-loaded PMMAP. PMMA-CS microspheres containing ampicillin were found to be Heme compatible and non-cytotoxic to mouse fibroblast cells. *In vivo* toxicity testing on albino rats revealed no significant changes in biochemical or pathological parameters. The acute toxicity results obtained after administering the nanoparticles orally to healthy Wistar male albino rats were satisfactory. Although it Is now broadly prevalent that PMMA is a non-poisonous polymer as well as a safe drug delivery material with hemocompatibility and cytocompatibility, there is some concern about polymer non-biodegradability, and some undesirable biological effects such as suppression of phagocytic and antibacterial interest of human polymorphonuclear leukocytes and inflammatory responses have been described. Filler injections for cosmetic surgery have been linked to granuloma formation. As a result, the question of whether these products should be injected for cosmetic purposes remains unresolved. Overall, there is a scarcity of information about PMMA particulate carrier toxicity. As a result, it is recommended that toxicological evaluation become an important part of the design of such particles in the future (Bettencourt & Almeida, 2012).

3.13 POLY(ACRYLIC ACID)

Poly(acrylic acid) (PAA; also known as Carbomer) is a polymer with the formula $(CH_2–CHCO_2H)_n$. It is an acrylic acid derivative $(CH_2=CHCO_2H)$. Aside from homopolymers, plenty of copolymers and crosslinked polymers, in addition to in part deprotonated derivatives, are recognized and commercially valuable. PAA is an anionic polymer in water at impartial pH, because of this that that a lot of its facet chains lose their protons and collect a poor charge. PAAs which have been in part or absolutely deprotonated are polyelectrolytes which could soak up and preserve water and swell to generally their unique volume. Many packages depend on those properties, which might be acid-base and water-attracting.

3.13.1 STRUCTURE, PRODUCTION, AND DERIVATIVES

Structure production and derivatives A polyolefin is polyacrylic acid. Polyethylene with CO_2H substituents on alternating carbons is what it is. Alternating carbon atoms withinside the spine is stereogenic due to those groups (colloquially: chiral). As a result, even though this issue is not often discussed, acrylic acid exists in atactic, syndiotactic, and isotactic forms. Polymerization begins with radicals and is assumed to be stereo random. Crosslinking can be implemented in different of ways (Kaiser et al., 2009). PAA is created through the process of free radical polymerization. AIBN and Potassium persulfate are examples of initiators. PAA used in dispersants, and the molecular mass has a significant effect on the rheological characteristics and dispersion capacity, and the applications. In 2008, approximately 1,600,000,000 kg were produced. Polyacrylic acid is a vulnerable anionic polyelectrolyte whose diploma of ionization is laid low with the pH of the solution. At low pH, PAA can shape hydrogen-bonded interpolymer complexes with loads of non-ionic polymers (which include PEO, poly-N-vinyl pyrrolidone, polyacrylamide, and a few cellulose ethers (Khutoryanskiy, n.d.). PAA also can shape polycomplexes in aqueous answers with contrary costs polymers including surfactants, chitosan, and drug molecules (e.g., streptomycin). PAAs are offered as white, fluffy powders whilst dry. The undoubtedly charged sodium ions are certain to the polyacrylate withinside the dry powder shape, however they are able to dissociate in aqueous answers. Because of the presence of many steel cations, the polymer can soak up a huge quantity of water (Nurkeeva, n.d.).

3.13.2 APPLICATIONS

PAA's primary application is as a superabsorbent. PAA is used in detergents and dispersants to the tune of 25% (Khutoryanskiy, n.d.). Disposable diapers contain polyacrylic acid and its derivatives. Acrylic acid is likewise a key factor of super adsorbent polymers (SAPs), which can be cross-connected polyacrylates able to soak up and preserving extra than 100 instances their very own weight in liquid. SAPs have been approved for use in wrapping with indirect food contact by the US Food and Drug Administration (FDA). Detergents frequently contain copolymers of acrylic acid, which aid in dirt sequestration. Cross-linked polyacrylic acid is also used in the manufacture of household products such as floor cleaners. PAA has the potential to deactivate the antiseptic chlorhexidine gluconate (Kaiser et al., 2009). The neutralized polyacrylic acid gels are suitable biocompatible matrices for medical applications such as gels for skin care products. PAA films can be deposited on orthopedic implants to protect them from corrosion. Crosslinked hydrogels (HGs) of PAA and gelatine have also been used as medical glue. Other applications involve paints and cosmetics. They stabilize suspended stable in liquids, save you emulsions from separating, and manipulate the consistency in glide of cosmetics. Carbomer codes (910, 934, 940, 941, and 934P) are an illustration of molecular weight and the precise additives of the polymer. For many programs PAAs are utilized in shape of alkali metal or ammonium salts, e.g., sodium polyacrylate. Emerging applications HGs derived from PAA had been appreciably researched to be used as bandages and wound restoration aids (Mogoşanu & Grumezescu, 2014).

3.14 POLOXAMERS (PLURONICS®)

Poloxamers are triblock copolymers sold under the brand names Pluronic® (BASF), Lutrol® (BASF), Kolliphor® (BASF), Synperonic® (Croda), and Antarox® (Rhodia). They are prepared by two hydrophilic PEO blocks that are linked to a hydrophobic PPO block and are arranged in the chemical structure PEOxPPOyPEOx, where x and y represent the number of units of PEO and PPO blocks, respectively. In this sense, poloxamers are synthesized by sequentially adding monomer PO and EO with alkaline catalyst, like sodium or potassium hydroxide, yielding various poloxamers with varying numbers of hydrophilic EO and hydrophobic PO units, which are also distinguished by their distinct hydrophilic-lipophilic balance (HLB) value. Poloxamer 188, that's bought as poloxamer P188 however additionally as Pluronic® F68 and Kolliphor® P188, is an instance of an industrial product this is utilized in mobileular culture (mammalian, insect, and plant) due to the fact it may manage shear forces and confer cryoprotection in suspension cultures (Vitthal & Charles, 2016).

Pluronics® are recognized for his or her amphiphilic properties, however they may be additionally recognized for his or her temperature sensitivity, being capable of produce thermoreversible gels, a number of that have already been accepted with the aid of using the FDA and the Environmental Protection Agency (EPA) for a lot of applications, such as meals additives, drug transport providers in cosmetics, pharmaceutical ingredients, and tissue engineering. These copolymers have a sol-gel transition temperature round 37°C (physiological temperature) and a gel-sol transition temperature round 50°C (Almeida et al., 2018).

The polymers have surfactant residences because of their amphiphilic shape it is able to be used to enhance the water solubilization of hydrophobic, oily materials or to preserve growing the dispersibility of materials with special hydrophobicity's (Vitthal & Charles, 2016).

3.14.1 POLOXAMER CHEMISTRY

Poloxamer is a commercially to be had ABA tri-block copolymer composed of a hydrophilic block PEO (A) and a hydrophobic block PPO (B). BASF evolved nomenclature for Pluronic grades, which are to be had in 3 types [liquid (L), paste (P), and flake (F)] via way of means of bringing up the sort earlier than the digits, for example, L61 and P181 denote liquid and paste types, respectively. The first one or digits extended via way of means of 300 approximate the PO molecular weight, and the final one explains the one-10th EO weight percent withinside the copolymer. Poloxamer reveals a lower in zeta capacity with growing molecular weight from Pluronic F68 to F127. Furthermore, there may be a trendy fashion in the direction of more discount with growing concentration (specifically for F127). The creation of a sterically stabilizing adsorbed polymer layer is demonstrated by a decrease in zeta potential. When the pH is reduced, the polyelectrolyte concentration rises, causing the zeta potential to fall (Mishra, n.d.).

Different varieties of poloxamer have various applications, ranging from surfactants to drug carriers, based on its segmental weight. Because PEO has a high solubility in water and PPO has less solubility in such media, their block copolymer functions as an amphiphilic structure with surface-active properties. The synthesis of poloxamer has been performed thru the sequential anionic ring beginning polymerization of ethylene oxide and propylene oxide withinside the presence of potassium and sodium hydroxide (as activators). PPO segments are shaped first, observed via way of means of PEO polymerization to the chain. Choi et al. additionally created image crosslinking Pluronic the usage of diacrylate functions, which verified a surprising quantity transition throughout a temperature variety of 4–40°C To goal the tumor, Kim et al. used chitosan-conjugated Pluronic F68 and F127 that were image-crosslinked. The addition of chitosan to the nanoparticles did not change its parameter like temperature and dimensions responsiveness, but it did significantly improve *in-vitro* cellular internalization. Because of the amphiphilic structure of poloxamer, various molecules can be conjugated to it for encapsulation and delivery. The amphiphilic nature of the poloxamers, and other side, can be combined with other intriguing properties of emerging biomaterials to create multifunctional materials with more versatile applications. For example, mussel-inspired chemistry can improve the poloxamers' water solubility and wet adhesion. In this regard, the mussel-inspired moiety 3,4-dihydroxyphenyl-l-alanine (DOPA), which contains a catechol group, utilized as modify poloxamers. Controlling polymerization traditionally is difficult, and tailoring polymer structure is impossible. Atom transfers radical polymerization, one of the living polymerization methods that uses metal complexes, produces architected structures with controlled topology and dispersity (Matyjaszewski, 2012).

Pluronic become used withinside the ARTP response as a macromonomer/macroinitiator to provide the architected polymer thermosensitivity and micellar form. ATRP of isopropyl acrylamide become used to create a twin thermosensitive polytrioxane triblock copolymer, which become initiated from 2-bromoisobutyryl end-capped Pluronic 17R4. It became counseled that this type of clever shape may be used as a DDS and a biosensor. Pluronic as a macromonomer may be used to synthesize numerous sorts of block copolymers. Because of the variety of the blocks, a huge variety of morphology variety become determined at special pH and temperature. Temperature and pH versions may be used to music the polymer's hydrophobicity. The hydrophobicity of terminal blocks become elevated via way of means of growing the temperature or being in an unprotonated nation at an excessive pH value. This kind of behavior is extraordinarily beneficial in interface tissue engineering. Cationic polymers have formerly been used as a nonviral gene shipping system. The important aggregation attention of block polymer reduced as temperature elevated, ensuing in dehydration of polymer blocks and elevated hydrophobicity. The LCST temperature of block polymer become better than Pluronic and elevated as PEGMA attention elevated. The clouding factor become stimulated via way of means of inorganic components which includes KBr and K_2SO_4, the ions of which affected the LCST primarily based totally on their function withinside the Hofmeister series (Zarrintaj et al., 2020).

3.14.2 STABILIZATION PROPERTIES

Poloxamer 407 promotes the steadiness of protected capsules in unique proteins (35Y40). Poloxamer 407 additionally reduces peptide unfolding inside the presence of low CMC and a loss of electrolytic binding. The third-dimensional shape of protein molecules is higher preserved whilst Poloxamer 407 is present (10,35Y38). The addition of Poloxamer 407 advanced the micro- or nanoparticles. Poloxamer 407 delivered to protein-loaded poly(epsilon-caprolactone) MPs elevated their hydrophilicity and averted them from aggregating. Similarly, urease integrated into poly(lactide-co-glycolide) microspheres changed into stabilized and its bioactivity changed into retained. There have additionally been a few exciting findings concerning interactions among Poloxamer 407 and liposomes. Poloxamer 407 is both co-solubilized with lipid at some point of liposome training or delivered after the liposomes had been formed. The inclusion of Poloxamer 407 in a liposomal shipping gadget sterically stabilized liposomes, extending their half-life. The steric aggregation and fusion of thawed egg phosphatidylcholine (PC) multilamellar vesicles had been averted with the aid of using poloxamer 407. Using quasi-elastic mild scattering, poloxamer 407 considerably reduces the dimensions of multilamellar vesicles. Poloxamer 407-liposome interactions had been studied the usage of differential scanning calorimetry and photon correlation spectroscopy. Poloxamer 407 is concept to be able to penetrate liquid crystalline bilayers. Liposome stabilization is at once associated with the incorporation or adsorption of Poloxamer 407, a phenomenon that happens above the CMT. Poloxamer 407 molecules, on the alternative hand, stay wonderful beneath CMT (non-related or universe). Poloxamer 407 has additionally been used to enhance the stableness of W/O/W a couple of emulsions (Dumortier et al., 2006).

3.15 CELLULOSE DERIVATIVE (CD)

Cellulose is the maximum ample natural compound at the planet, and it's far commonly produced with the aid of using plants. It is the maximum structurally critical issue of natural cells and tissues. Cellulose is an obviously taking place lengthy chain polymer that in a roundabout way contributes to the human meals cycle.

This polymer has an extensive variety of applications, consisting of veterinary foods, wooden and paper, fibers and clothing, beauty, and pharmaceutical industries as an excipient. Cellulose has a number of semi-artificial derivatives which can be broadly used withinside the pharmaceutical and beauty industries. Cellulose ethers and cellulose esters are the 2 major varieties of cellulose derivatives (CDs), every with its personal set of physicochemical and mechanical properties. These polymers are broadly used inside the improvement of pharmaceutical formulations and pharmaceutical products. These compounds are utilized in numerous pharmaceutical applications, consisting of prolonged and osmotic drug transport systems, not on time launch lined dosage bureaucracy, prolonged and managed launch matrices, bio-adhesives and mucoadhesive, compression drugs as compressibility enhancers, liquid dosage bureaucracy as thickening marketers and stabilizers, granules, and drugs as binders, semisolid arrangements as gelling marketers, and lots of others. These polymeric substances have additionally been utilized in transdermal patches as fillers, flavor maskers, free-flowing agents, and strain touchy adhesives. Nowadays, cellulose-primarily based totally polymers and cellulose are very famous in pharmaceutical industries and have become more and more critical on this area due to pharmaceutical researchers' efforts to create new derivatives and locate new packages for existing compounds (Shokri & Adibki, 2013).

Cellulose is one of the maximums not unusual place polymers utilized in ophthalmic formulations. Methylcellulose (MC) turned into first used as a viscosity enhancer in ocular formulations withinside the 1940s. Since then, cellulosic polymers had been substantially researched for ocular management in each animal and humans. Because natural cellulose is water-insoluble, cellulosic derivatives are extensively utilized in ocular formulations. Cellulosic derivatives have precious viscosity-growing residences that may be used to enhance bioavailability in polymer-primarily based totally ophthalmic formulations. These macromolecules have a clean capacity as a drug service inside the eye. Furthermore, the swelling residences, chemical residences, and structural morphology of those derivatives have a sizable effect on the discharge mechanism of the medication loaded in those systems. Cellulose is an herbal polymer that may be a predominant factor of plant life and is an herbal and sustainable polymer withinside the world (Gupta et al., 2021).

Carboxymethyl cellulose (CMC) is a biocompatible, biodegradable, nontoxic, and water-soluble cellulose ether by-product that is used for numerous purposes. CMC is a white to cream-colored, odorless, tasteless powder, and sodium CMC is normally called CMC. It is a water-soluble anionic, linear anhydrous-glucose polysaccharide this is normally appeared as safe (GRAS). At any temperature, CMC is soluble in water. CMC hydrates quick because of its exceptionally hygroscopic nature. When CMC powder is delivered into water, fast hydration might also additionally motive agglomeration and lump formation. Lump formation may be averted with the aid of using the usage of excessive agitation at the same time as including the powder to the water, or with the aid of using preblending the CMC powder with different dry elements which include sugar earlier than including it to the water. As the attention of the answer increases, so does its viscosity.

Hydroxypropyl methylcellulose (HPMC) is a cellulose ether wherein hydroxyl agencies were substituted with one or greater of the cellulose ring's 3 hydroxyl agencies. HPMC is a hydrophilic (water-soluble), biodegradable, and biocompatible polymer with several packages which includes drug delivery, dyes, and paints, cosmetics, adhesives, coatings, agriculture, and textiles. HPMC is likewise soluble in polar natural solvents, permitting it for use in each aqueous and nonaqueous environments. It has great solubility residences, being soluble in each warm and bloodless natural solvents.

3.16 GELATIN

Gelatin is an herbal polymer derived from the hydrolytic of protein from collagen, and its particular amino acid shape confers numerous scientific benefits (Kumosa et al., 2018). Gelatin is typically observed withinside the shape of tablets, granules, or powders, and it can occasionally be soluble in water earlier than use (Yang et al., 2016). Gelatin is being studied appreciably with the aid of using scientists as a matrix for 3-dimensional mobileular tradition and as a thing of tissue-engineering scaffolds (Ha & Gardella, 2005). In positive nutritionally balanced foods, gelatin has an excessive protein content material and may be used to update fats and carbohydrates (Lv et al., 2019). Collagen is the maximum clearly happening protein in people and animals as a supply of protein-matrix of gelatin. Collagen may be observed everywhere withinside the frame, however it's far maximum considerable withinside the skin, bones, tendons, and ligaments (Bou-Gharios et al., 2019).

3.16.1 CHEMISTRY

Gelatin is created with the aid of using in part hydrolyzing collagen. The maximum considerable protein inside the frame is collagen. It is a vital structural thing of lots of our tissues, which includes tendons and bones. This might also additionally seem contradictory at first: how can collagen offer shape to tough tissues which include bone while gelatin is, with the aid of using definition, a gel with the aid of using analyzing the secondary shape of each protein, we will provide an explanation for this phenomenon. Gelatin and collagen are polypeptide chains held collectively with the aid of using hydrogen bonds among adjoining amino acids. The secondary shape of the protein is the structural conformation (shape) that effects from those interactions. Collagen is made from 3 polypeptide chains that tightly intertwine to shape an exceptionally strong triple-helix conformation. These triple-helical molecules are then organized into collagen fibril bundles. This excessive diploma of company explains why collagen has advanced mechanical residences and is well-applicable to supplying shape and aid to a number of the frame's (https://chembam.com/resources-for-students/the-chemistry-of/gelatin/, n.d.).

3.16.2 USAGE

Because of its gelling houses, gelatin is a vital element of the meal's enterprise in present day cuisine. The meals enterprise, for example, makes use of gelatin for texture, foaming, and clarity, in addition to stabilize the meals shape (Calvarro et al., 2016). Gelatin has additionally been utilized in canned meat merchandise which include sausages, frankfurters, loaves, and hams to maintain misplaced juices and to offer an awesome warmth switch medium at some point of cooking, while safe to eat gelatin is utilized in confectionery merchandise which include fruit salads, ice cream, foam, and cottage cheese (Abdalbasit, 2013). Gelatin may be used as a safe to eat movie and coating cloth because of its brilliant movie-forming functionality and dietary value (Ramos et al., 2016). Gelatin also can be determined in cosmetics and fitness merchandise as a gelling agent, which include tub-tub salts, shampoos, sunscreens, frame lotions, hair spray, and facial cream (Anchana, 2016). In the sector of skincare, collagen, and gelatin-primarily based totally purposeful and dietary ingredients had been extensively used. Gelatin is used withinside the clinical enterprise for hydrogel, nanomicrosphere containers, nanofibers, pharmaceutical additives, and mobileular transplantation carriers. Gelatin is likewise used withinside the encapsulation of diverse drug merchandise thru its micro/nanoparticles. Gelatin is used as a matrix inside the clinical and pharmaceutical industry for intravenous infusions, injection drug transport microspheres, and implants. In addition, Gelatin taken orally can enhance bone and joint fitness (Abdalbasit, 2013).

3.16.3 CHARACTERISTICS

Gelatin consists of 18 one-of-a-kind sorts of complicated amino acids, with glycine, proline, and hydroxyproline accounting for about 57% of the total, at the same time as the closing 43% includes different first-rate amino acid households which include glutamic acid, alanine, arginine, and aspartic acid. Gelatin is made from 25.2% oxygen, 6.8% hydrogen, 50.5% carbon, and 17% nitrogen, with a combination of unmarried and double spread-out chains of hydrophilic person g/mol. Heat increments for gelatin will dissolve the shape into colloids at some point of processing, however at temperatures beneath 35°C to 40°C, its kingdom becomes gelatinous (Abdalbasit, 2013).

3.16.4 ADVANTAGES

- Collagen derivative;
- Cheaper;
- Biocompatible;
- Biodegradable;
- Antithrombogenic.

3.16.5 DISADVANTAGES

- Necessitates cross-linking;
- Less reliable;
- Subpar mechanical houses.

3.16.6 STABILITY

Isothermal (annealing at 15, 20, and 25°C for 15 to a 120 min) and nonisothermal (cooling at 1°C min^{1}) aqueous gelatin gels of 10, 20, and 40 wt.% have been organized. The activation power various extensively at some point of the process, in line with isoconversional kinetic evaluation of DSC facts on gel melting (gel-sol transition) of all sorts of gels. The activation power barrier to melting of isothermally organized gels changed into determined to boom with growing annealing temperature, which changed into the predominant impact discovered. The activation power barrier to melting of non-isothermally organized gels changed into determined to be round a 120–140 kJ mol^{1} and improved with concentration. The use of temperature modulated DSC gel melting established nearby reversibility (Dranca & Vyazovkin, 2009).

3.17 POLYVINYL ALCOHOL (PVA)

Polyvinyl alcohol (PVA) is a water-soluble polymer that has determined enormous use in a number of industries because of its brilliant mechanical and chemical houses. The consequences of PVA with various stages of hydrolysis and polymerization at the rheological houses of cement mortar are investigated in this chapter. The effects display that the changed Bingham version can describe the rheological houses of PVA-changed cement mortar. When the diploma of polymerization (DP) and PVA content material are low, the yield strain of changed cement mortar is decrease than that of unmodified mortar. With the boom of PVA content material and polymerization diploma, the yield strain and plastic viscosity of changed cement mortar boom sharply, which can be large than the ones of the unmodified cement mortar. The impact of PVA hydrolysis diploma on yield strain and plastic viscosity of changed cement mortar, on the alternative hand, isn't always clear (Liu et al., 2020).

3.17.1 CHEMISTRY

The houses of polyvinyl acetate rely upon the quantity or diploma of its hydrolysis, mainly whether or not it's miles complete or partial which in flip dictates its categorization into groups, namely – (i) in part hydrolyzed; and (ii) absolutely hydrolyzed. The molecular weights of PVA merchandise can variety from 20,000 to 4,00,000, relying at the period of the preliminary vinyl acetate polymer, the diploma of hydrolysis used to cast off the acetate corporations, and whether or not the hydrolysis takes place below alkaline or acidic conditions. The PVA shape. Hydrolysis stages variety from what's taken into consideration an ordinary cost of 80% to greater than 99%. Crosslinking of the linear polymers outcomes inside the formation of PVA HGs with tunable homes in almost completely hydrolyzed forms, which then bring about polymer (gel)-fluid (sol) species. Polymer contents impact the bodily kingdom of the ensuing material: low polymer content material outcomes in smooth substances due to the fact fluid flows freely thru the matrix, while excessive polymer content material outcomes in sizeable stiffening and strengthening of the material's matrix. Because of specially on movie-forming, chemical stability, and hydrophobicity, research at the diffusive permeability of solutes in PVA gel membranes and their software for separation had been delineated (Gaaz et al., 2015).

3.17.2 USAGE

PVA, that's derived often from polyvinyl acetate through hydrolysis, is effortlessly degradable through organic organisms and is a water-soluble polymer with a crystalline shape (Razzak & Darwis, 2001). PVA is a synthetic polymer that changed into broadly used within the first 1/2 of the

20th century. It has been used inside the business, commercial, medical, and meals sectors to provide an extensive variety of cease merchandise, inclusive of lacquers, resins, surgical threads, and meals packaging substances that come into touch with meals (DeMerlis & Schoneker, 2003). PVA is a biodegradable alternative for herbal polymers which might be utilized in paper coating and fabric sizing (Liu et al., 2007). Because of its likeminded shape and hydrophilic homes, this polymer is broadly utilized by mixing with different polymer compounds, along with biopolymers and different polymers with hydrophilic homes; it's far utilized in loads of business packages to enhance the mechanical homes of films (Limpan et al., 2012). There are a few man-made polymers to be had which might be crafted from non-renewable and non-biodegradable sources, along with PVA (Maria et al., 2008). Biodegradable polymers are derived from petroleum-primarily based totally artificial substances that decompose evidently below aerobic (composting) or anaerobic (landfill) conditions (Qiu & Netravali, 2013; Tang et al., 2013) PVA is an extensively used thermoplastic this is riskless and innocent to dwelling tissues. This polymer is being studied considerably because of its use in cross-connected nanofillers and merchandise. PVA is a biodegradable polymer, and the presence of hydroxyl corporations at the carbon atoms will increase its degradability thru hydrolysis (Qiu & Netravali, 2013).

3.17.3 CHARACTERISTICS

PVA, additionally called PVOH, PVA, or PVAL, is a water-soluble artificial polymer. It is beneficial for movie formation, emulsification, and adhesion. It has no odor and is non-toxic, in addition to being proof against grease, oils, and solvents. It is ductile, however robust and flexible, and it serves as a remarkable oxygen and aroma barrier.

3.17.4 ADVANTAGES

- Easy to use and versatile [paper, plastic, metal foil, leather, wood];
- Low cost and accessibility;
- There is no aging issue;
- Extreme joint pliability;
- Effective insulator;
- Long-term dependability;
- Glass adhesion;
- Oil resistance.

3.17.5 DISADVANTAGES

- Low modulus;
- Production moisture sensitivity;
- Inadequate oxygen permeability.

3.18 CHITOSAN

Chitin is the second one maximum considerable polysaccharide (after cellulose) on Earth, and it became first defined in 1811 through Henri Braconnot. Nature finds it as ordered microfibrils inside the exoskeletons of mollusks and crustaceans, as well as fungi and insect cuticles (Kaur & Dhillon, 2014). Because of its natural abundance, available more than 1,000 tons per year, with approximately 70% coming from marine species (Islam et al., 2017).

Chitin is a poly(-(1-4)-N-acetyl-d-glucosamine) with (14)-linkages, and it is thought to have a cellulose-like structure, with the hydroxyl group in the C_2 position replaced by an acetamido group. Chitin has three polymorphs, denoted as, α, β and γ depending on the orientation of the polysaccharide chains. This polymorph is antiparallel in structure, with each chain strongly interacting with the next via hydrogen bonds, resulting in high thermochemical stability and insolubility (Fereidoon, 2005; Khoushab & Yamabhai, 2010). As a result, chitin is a highly insoluble polymer that is poorly biodegradable. As a result, only the action of chitinase enzymes, which are abundant in nature, can degrade chitin.

3.18.1 CHEMISTRY

Chitin is N-deacetylated in the following ways when it is turned into chitosan: With the help of a controlled alkali solution and high temperatures, it should be possible to keep the level of acetylation and molecular weight by controlling: Keep an eye on the alkali, temperature, and time. The higher the level of acetylation of chitosan, the better it is at stopping the growth of germs.

3.18.2 IN THE MAKING OF CHITOSAN

The quantity of chitosan produced at some point of the producing technique can boom because of the addition of enzymes: lipases–phospholipases Chitosan also can be determined withinside partitions of fungi because it occurs naturally through enzymatic deacetylation.

Enzymatic digestion is used to isolate the fungal chitosan extraction from mycelia (Jiménez-Gómez & Cecilia, 2020).

The usage of chitosan are:

- Some chitosan applications in biomedical and pharmaceutical materials;
- Preparation of artificial skin;
- Surgical sutures;
- Contact lenses;
- Blood dialysis membranes;
- Artificial blood vessels;
- Antitumor;
- Blood anticoagulant;
- Anti-Gastritis;
- Hemostatic;
- Hypocholesterolemia agent;
- Antithrombogenic agent;
- Drug and gene-delivery systems;
- Dental therapy.

The characteristics of chitosan are:

- Deprotonated amino institution can act as sturdy nucleophile.
- Enable to shape intermolecular hydrogen bonds: excessive viscosity.
- Existence of reactive companies for chemical activation and cross-linking.
- Insoluble in water and natural solvents, however soluble in dilute aqueous acid solutions.
- Its paperwork salts with natural and inorganic acids.
- Complexing and chelating houses.
- Cationic biopolymer with excessive price density (one fantastic price according to glucosamine residue).
- Flocculating agent (interacts with negatively charged molecules).
- Entrapment and adsorption houses (filtration and separation).
- Film-forming ability (adhesive substances for isolation of biomolecules).

3.18.3 ADVANTAGES

- Non-poisonous and biodegradable;
- Biodegradable;
- Hydrocolloid (biopolymer);

- While maximum hydrocolloids are impartial or negatively charged, chitosan has a fantastic price.

3.18.4 DISADVANTAGES

1. Drug launch control is limited because the drug dissolves quickly in the stomach and the pH levels are better.
2. Problems with using drugs made from chitosan.
3. Stop the absorption of different fat-soluble substances.

3.18.5 STABILITY

Chitosan, an herbal multifunctional polymer, is appeared as a beneficial compound in scientific and pharmaceutical generation because of its specific and flexible organic houses. Significant studies attempt has these days been made with a purpose to expand secure and green chitosan merchandise. However, the hassle of bad balance of chitosan-primarily based totally structures limits its sensible applicability; thus, setting up enough shelf-existence for chitosan formulations has emerge as a large challenge. Controlling environmental elements, manipulating processing conditions (e.g., temperature), introducing a right stabilizing compound, growing chitosan blends with any other polymer, or editing the chitosan shape the usage of chemical or ionic marketers can all enhance balance. The effect of internal, environmental, and processing elements at the long-time period balance of chitosan merchandise is mentioned on this review. Despite being one of the maximum plentiful biopolymers at the planet, chitin's makes use of, and packages are confined because of its low solubility. Chitosan is shaped whilst chitin is deacetylated. This biopolymer, which consists of randomly distributed (1-4)-related D-units, has higher physicochemical houses due to the fact it may be dissolved below acidic conditions, it may undertake numerous conformations or structures, and it may be functionalized with a huge variety of purposeful companies to modulate its superficial composition to a particular application. Because of its biodegradability, bio-adhesiveness, and bioactivity, chitosan is taken into consideration a notably biocompatible biopolymer with a huge variety of packages. As a result, chitosan is a promising biopolymer for quite a few biomedical programs (skin, bone, tissue engineering, synthetic kidneys, nerves, livers, wound healing). This biopolymer is likewise used to entice natural compounds and dyes, in addition to selectively separate binary mixtures. Furthermore, chitosan may be used as a catalyst or as a beginning molecule to provide high-cost products. Taking those assumptions into account, this overview makes a specialty of the shape and change of chitosan, in addition to its makes use of and programs (Szymańska & Winnicka, 2015).

3.19 PECTIN

3.19.1 GENERAL INFORMATION

Apples and citrus end result are used to extract pectin. Once the right concentrations of acid and sugar are reached, soluble pectin can shape a gel. This is beneficial for thickening syrups used to make jams and jellies.

3.19.2 CHEMISTRY

Pectin is a gelatin-like carbohydrate. In cooking, pectin behaves like a gel and is occasionally known as a "fragile solid." Pectin is semi-soluble in liquids; this means that it may soak up a few liquids. This is especially critical whilst cooking end result and greens as it lets in them to soften. Pectin is a polysaccharide that has programs in meals, pharmaceuticals, and quite a few different industries. Its importance withinside the meals enterprise stems from its cap potential to shape gel withinside the presence of Ca^{2+} ions or a solute at low pH. Despite the reality that the precise mechanism of gel formation is unknown, sizable development has been made on this direction. Depending at the pectin, gel formation includes coordinate bonding with Ca^{2+} ions, hydrogen bonding, and hydrophobic interactions. Gelation in low methoxyl pectin is as a result of ionic linkage thru calcium bridges among carboxyl corporations from specific chains which are in near touch with every different. Crosslinking of pectin molecules in high methoxyl pectin includes a mixture of hydrogen bonds and hydrophobic interactions among the molecules. A range of things impact pectin gelation (Thakur et al., 1997), together with pH, the presence of different solutes, molecular size, diploma of methoxylation, the wide variety and association of facet chains, and fee density at the molecule.

3.19.3 USAGE

Pectin is used as a fat and/or sugar replacer in jams, jellies, frozen foods, and, more recently, low-calorie foods. It is used in the pharmaceutical industry to decrease blood levels of cholesterol and treat GI disorders. Other uses for pectin include edible films, paper substitutes, foams, and plasticizers, etc. Pectins are vulnerable to heat degradation during processing, in addition to pectolytic degradation, and the nature of the ions and salts present in the system affects the degradation. Due to the negative gelling behavior of pectin from different sources, apple pomace and orange peel are the two main sources of commercial pectin despite being present in the cellular walls of most plants.

3.19.4 CHARACTERISTICS

Pectin's dissolve in natural water. Monovalent cation salts of pectinic and pectic acids are generally soluble in water; divalent and trivalent cation salts are both weakly and absolutely insoluble. When dry powdered pectin is blended with water, it hydrates speedy and paperwork clumps. These clumps are made of semidry packets of pectin enclosed in an incredibly hydrated outer coating envelope. Clump formation may be averted with the aid of using dry blending pectin powder with a water-soluble service fabric or with the aid of using the usage of pectin with stepped forward dispersibility because of unique production treatment. Dilute pectin answers are Newtonian, however at low concentrations, they showcase non-Newtonian, pseudoplastic behavior. Viscosity, solubility, and gelation are all associated in a few ways. Factors that growth gel strength, for example, can even growth the tendency to gel, lower solubility, and growth viscosity, and vice versa. Pectin's' residences are decided with the aid of using their shape, that's a linear polyanion. As a result, monovalent cation salts of pectin in answer are incredibly ionized, and the distribution of ionic fees alongside the molecule has a tendency to maintain it in a prolonged shape because of coulombic repulsion (Pornsak n.d.).

3.19.5 ADVANTAGES

Pectin supplementation might also additionally have some of fitness benefits:

- Reduces blood sugar and fats levels;
- Reduces the chance of colon cancer;
- Supports a healthful weight;
- Aids with gastrointestinal problems (Sundarraj & Raj, 2012).

3.19.6 DISADVANTAGES

Pectin has few facet effects; however, as it impacts digestion, it could purpose fueloline or bloating in a few people. Furthermore, you need to keep away from it in case you are allergic to the meals from which it became derived. The majority of business merchandise and dietary supplements comprise apples or citrus peels (Sundarraj & Raj, 2012).

3.20 SODIUM ALGINATE

3.20.1 GENERAL INFORMATION

One of the maximum famous individuals of the hydrogel institution is sodium alginate. The hydrogel is a water-swollen, cross-related polymeric community shaped with the aid of using an easy response among one or greater monomers (Ahmed, 2015). HGs take in water due to hydrophilic purposeful businesses connected to the polymeric backbone, and that they face up to dissolution due to cross-hyperlinks among community chains. Sodium alginate is an obviously happening anionic polymer derived from brown seaweed this is composed of mannuronic (M) and guluronic (G) acids organized in numerous combinations (fig), consisting of blocks wealthy in both M or G devices or blocks of alternating G and M devices. The guluronic acids from close by chains shape ionic crosslinks withinside the presence of divalent Ca^{2+} cations, ensuing in alginate hydrogel. The physicochemical houses of the hydrogel are described with the aid of using the ratio of M and G devices (Ouwerx et al., 1998).

3.20.2 CHEMISTRY

Sodium alginate ($NaC_6H_7O_6$) is a linear polysaccharide alginic acid spinoff composed of 1, four-d-mannuronic (M) and l-guluronic (G) acids. Sodium alginate is a mobileular wall element of marine brown algae that includes 30 to 60% alginic acid. The conversion of alginic acid to sodium alginate will increase its water solubility, which aids in extraction. Bacterial alginates are produced with the aid of using handiest bacterial genera, pseudomonas, and azotobacter, and are used for environmental safety in addition to the synthesis of biofilms to stick to surfaces. This approach of synthesis allows microorganism to provide alginates with well-described monomer compositions, probably permitting the manufacturing of "tailor-made" bacterial alginates (Loureiro dos Santos, 2017).

3.20.3 USAGE

Because of its biocompatibility, low toxicity, low cost, and moderate gelation, alginate is utilized in a number of ingredients and biomedical programs (Gombotz & Fong, 1998). Alginate is used inside the meals enterprise as a thickening agent, gelling agent, emulsifier, stabilizer, and texture improver. Alginate is now utilized in a number of ingredients, together with ice cream, jelly, acid milk drinks, dressings, immediately noodles, beer, and so on.

Alginic acid is utilized in pharmaceutical programs as a service to boost up pill disintegration for a quicker launch of the medicinal element; it's also utilized in cosmetics because of its capability as a thickener and moisture retainer. Alginate, for example, aids in the retention of lipstick coloration at the lip floor with the aid of using forming a gel community. Because they're structurally just like the macromolecular-primarily based totally additives inside the frame and may frequently be brought into the frame through minimally invasive administration (Dhoot et al., 2004), alginate-primarily based totally HGs are exceptionally promising applicants to be used as drug shipping structures and biomedical implants. Alginate is a great candidate for protein drug shipping due to the fact proteins may be included into alginate-primarily based totally formulations beneath exceedingly moderate conditions, minimizing denaturation, and the gels can guard them from degradation till launch (Silva & Mooney, 2010). In biomedical research, sodium alginate gels are an increasing number of getting used as a version gadget for mammalian mobileular tradition. These gels are without difficulty adaptable to 2-D or greater physiologically applicable 3D tradition structures. The loss of mammalian mobileular receptors for alginate, blended with the low protein adsorption to alginate gels, lets in those substances to function a really perfect clean slate for the incorporation of exceptionally unique and quantitative modes of mobileular adhesion. Furthermore, simple findings located *in vitro* may be without difficulty translated *in vivo* because of alginate's biocompatibility and simplicity of advent into the frame (Silva & Mooney, 2010).

3.20.4 ADVANTAGES

The maximum massive advantage of alginates is their liquid-gel behavior in aqueous answers. When monovalent ions (e.g., sodium in sodium alginate) are exchanged for divalent ions (mainly calcium), the response proceeds nearly immediately, reworking the answer right into a gel structure. The gelled mass is a copolymer made of one-of-a-kind styles of monomer devices (Puscaselu et al., 2020).

3.20.5 DISADVANTAGES

Alginate additionally has some of drawbacks, together with bad balance or low mechanical and barrier houses, incompatibility with heavy metals, and warmth remedy instability, a number of which cannot be remedied later. These houses may be progressed with the aid of using combining alginate with different biopolymers, mainly protein-primarily based totally ones, or with artificial polymers, in addition to with the aid of using converting or combining the remedies used for the duration of manufacturing (Puscaselu et al., 2020).

3.20.6 STABILITY

The quantity of water inside the relative humidity impacts the hygroscopic equilibrium of alginate. Dry sodium alginate in sealed boxes at temperatures beneath 25°C and shop in an exceedingly strong environment. Salt alginate gives strong results between pH 5 and 9. The DP, molecular weight, and viscosity of sodium alginate answer are all immediately associated with garage time and may be used to lessen the DP. The oligomeric heterozygosity's excessive DP is much less strong than alginate sodium alginate's balance. The proton-catalyzed hydrolysis of sodium alginate may be, in step with reports, relying on the hydrolysis time, pH, and temperature. At room temperature, propylene glycol alginate answers with pH 3–4 are strong; pH much less than 2 or more than 6 are unstable, even at room temperature, will extensively lessen viscosity (http://www.chinafooding.com/485/sodium_alginate_ph_value_and_stability, n.d.).

3.21 OTHER MATERIALS USED FOR NDDS

Other than polymer, there are several components that are used for NDDS. Formulation requires different solvents, permeation enhancers, antioxidants, stabilizers, buffers, tonicity modifiers, humectants, preservatives, and other additives which make the formulation completely stable and effective. Various organic and inorganic solvents are used while preparing liposomes, nanosomes, and nanoparticles. Examples like chloroform and methanol mixture are used to make liposomes by the hand-shaken method (Md. Rafiul Haque, 2021).

Niosomes preparation requires cholesterol with diethyl ether solvent (Dheeraj, 2016). Super disintegrants, humectants, sweeteners, and fillers are used in sublingual tablets or chewing gums. Examples like magnesium and calcium carbonate, aluminum silicate, clay, talc are used as fillers or texturizers. Then glycerin, lecithin, oleic acid, palmitic acid are used as softeners in various MCG. Corn syrup and sorbitol are used for aqueous sweetener (Dheeraj, 2016). Sodium bisulfite or metabisulfite are generally used as antioxidants in an acceptable concentration of 0.3%

in ocular drug delivery (Chein, 2009). Ophthalmic solutions are considered isotonic when their tonicity is equal to 0.9% w/v sodium chloride solution. Glycols, transcutol, labrasol are used in nasal drug delivery to increase solubility (CARBO, 1988). In floating DDS, there is a need for a gas-forming agent to float the dosage form. Calcium carbonate is used as a gas-forming agent in gastroretentive DDS. In the case of microspheres, glutaraldehyde solution is used to stabilize the preparation. Likewise, various additives other than polymer are also important in the formulation of NDDS.

ACKNOWLEDGMENT

Authors are thankful to M.E.T.'s Bhujbal Knowledge City, Institute of Pharmacy, Adgaon, Nashik, and Maharashtra, India for support and providing necessary facilities.

CONFLICT OF INTEREST

There is no conflict of interest.

KEYWORDS

- **chitosan**
- **copolymer**
- **cyclodextrins**
- **Eudragit**
- **natural polymer**
- **novel drug delivery system**
- **poloxamers**
- **poly(D,L-lactide-co-glycolide)**
- **polyethylene glycol**
- **polymers**
- **synthetic polymer**

REFERENCES

Aburahma, M. H., & Badr-Eldin, S. M., (2014). Compritol 888 ATO: a multifunctional lipid excipient in drug delivery systems and nanopharmaceuticals. *Expert Opinion on Drug Delivery*, *11*(12), 1865–1883.

Ahire, E., Thakkar, S., Borade, Y., & Misra, M., (2020). Nanocrystal based orally disintegrating tablets as a tool to improve dissolution rate of Vortioxetine. *Bulletin of Faculty of Pharmacy, Cairo University*, *58*(1, 2), 11–20.

Ahire, E., Thakkar, S., Darshanwad, M., & Misra, M., (2018). Parenteral nanosuspensions: A brief review from solubility enhancement to more novel and specific applications. *Acta Pharmaceutica Sinica B*, *8*(5), 733–755.

Ahmed, E. M., (2015). Hydrogel: Preparation, characterization, and applications: A review. In: *J. Adv. Res.* (Vol. 6, No. 2, pp. 105–121). Elsevier B.V. https://doi.org/10.1016/j.jare.2013.07.006.

Ajazuddin, & Saraf, S., (2010). Applications of novel drug delivery system for herbal formulations. In: *Fitoterapia* (Vol. 81, No. 7, pp. 680–689). https://doi.org/10.1016/j.fitote.2010.05.001.

Alexander, P., (2010). Organic rheological additives. *Mfg. Chem. Res. J. Pharm. Technol.*, *3*(3).

Almeida, M., Magalhães, M., Veiga, F., & Figueiras, A., (2018). Poloxamers, poloxamines and polymeric micelles: Definition, structure, and therapeutic applications in cancer. In: *J. Polym. Res.* (Vol. 25, No. 1). Springer Netherlands. https://doi.org/10.1007/s10965-017-1426-x.

Anchana Devi, Kamatchi, P., & Leela, K. (2016). Extraction, characterization, and application of gelatin from *Carcharhinus amblyrhyncho* and *Sphyraena barracuda*. *IOSR Journal of Biotechnology and Biochemistry (IOSR-JBB)*, *2*(6), 40–49. https://www.iosrjournals.org/iosr-jbb/papers/Vol2-issue6/Version-1/I02064049.pdf (accessed on 6 December 2023).

Aronson, J. K., (2010). *Meyler's Side Effects of Analgesics and Anti-inflammatory Drugs*.

Astete, C. E., & Sabliov, C. M., (2006). Synthesis and characterization of PLGA nanoparticles. In: *Journal of Biomaterials Science, Polymer Edition* (Vol. 17, No. 3, pp. 247–289). VSP BV. https://doi.org/10.1163/156856206775997322.

Baek, G., & Kim, C., (2011). Rheological properties of Carbopol containing nanoparticles. *Journal of Rheology*, *55*(2), 313–330. https://doi.org/10.1122/1.3538092.

Bajpai, P. K., Singh, I., & Madaan, J., (2014). Development and characterization of PLA-based green composites: A review. In: *Journal of Thermoplastic Composite Materials* (Vol. 27, No. 1, pp. 52–81). https://doi.org/10.1177/0892705712439571.

Berney, B. M., & Deasy, P. B., (1979). Evaluation of carbopol 934 as a suspending agent for sulphaoimidine suspensions. *International Journal of Pharmaceutics*, *3*(2–3), 73–80.

Bettencourt, A., & Almeida, A. J., (2012). Poly(methyl methacrylate) particulate carriers in drug delivery. In: *Journal of Microencapsulation* (Vol. 29, No. 4, pp. 353–367). https://doi.org/10.3109/02652048.2011.651500.

Bhavsar, M. D., & Amiji, M. M., (2008). Development of novel biodegradable polymeric nanoparticles-in-microsphere formulation for local plasmid DNA delivery in the gastrointestinal tract. *AAPS PharmSciTech*, *9*(1), 288–294. https://doi.org/10.1208/s12249-007-9021-9.

Bou-Gharios, G., Abraham, D., & De Crombrugghe, B., (2019). Type I collagen structure, synthesis, and regulation. In: *Principles of Bone Biology* (pp. 295–337). Elsevier. https://doi.org/10.1016/B978-0-12-814841-9.00013-0.

Calvarro, J., Perez-Palacios, T., & Ruiz, J., (2016). Modification of gelatin functionality for culinary applications by using transglutaminase. *International Journal of Gastronomy and Food Science, 5, 6*, 27–32. https://doi.org/10.1016/j.ijgfs.2016.11.001.

Corbo, D. C., Huang, Y. C., & Chien, Y. W., (1988). Nasal delivery of progestational steroids in ovariectomized rabbits: I. progesterone—comparison of pharmacokinetics with intravenous and oral administration. *International Journal of Pharmaceutics, 46*(1–2), 133–140.

Chein, Y. W., (2009). *Novel Drug Delivery System* (2nd edn.).

Cheng, Y., Deng, S., Chen, P., & Ruan, R., (2009). Polylactic acid (PLA) synthesis and modifications: A review. *Frontiers of Chemistry in China, 4*(3), 259–264. https://doi.org/10.1007/s11458-009-0092-x.

Dash, T. K., & Konkimalla, V. B., (2012). Polymeric modification and its implication in drug delivery: Poly-ε-caprolactone (PCL) as a model polymer. *Molecular Pharmaceutics, 9*(9), 2365–2379. https://doi.org/10.1021/mp3001952.

DeMerlis, C. C., & Schoneker, D. R., (2003). Review of the oral toxicity of polyvinyl alcohol (PVA). In: *Food and Chemical Toxicology* (Vol. 41, No. 3, pp. 319–326). Elsevier Ltd. https://doi.org/10.1016/S0278–6915(02)00258-2.

Dheeraj, B. D. (2010). *Journal Novel Drug Delivery System*. Nirali Prakashan, 5.18,14.30.

Dhoot, N. O., Tobias, C. A., Fischer, I., & Wheatley, M. A., (2004). Peptide-modified alginate surfaces as a growth permissive substrate for neurite outgrowth. *Journal of Biomedical Materials Research – Part A, 71*(2), 191–200. https://doi.org/10.1002/jbm.a.30103.

Dranca, I., & Vyazovkin, S., (2009). Thermal stability of gelatin gels: Effect of preparation conditions on the activation energy barrier to melting. *Polymer, 50*(20), 4859–4867. https://doi.org/10.1016/j.polymer.2009.06.070.

Dumortier, G., Grossiord, J. L., Agnely, F., & Chaumeil, J. C., (2006). A review of poloxamer 407 pharmaceutical and pharmacological characteristics. In: *Pharmaceutical Research* (Vol. 23, No. 12, pp. 2709–2728). Springer Science and Business Media Deutschland GmbH. https://doi.org/10.1007/s11095-006-9104-4.

Dwivedi, R., Kumar, S., Pandey, R., Mahajan, A., Nandana, D., Katti, D. S., & Mehrotra, D., (2020). Polycaprolactone as biomaterial for bone scaffolds: Review of literature. In: *Journal of Oral Biology and Craniofacial Research* (Vol. 10, No. 1, pp. 381–388). Elsevier B.V. https://doi.org/10.1016/j.jobcr.2019.10.003.

El-Setouhy, D. A., & El-Malak, N. S. A., (2010). Formulation of a novel tianeptine sodium orodispersible film. *AAPS PharmSciTech, 11*(3), 1018–1025. https://doi.org/10.1208/s12249-010-9464-2.

Espinoza, S. M., Patil, H. I., San, M. M. E., Casañas, P. R., & Ige, P. P., (2020). Poly-ε-caprolactone (PCL), a promising polymer for pharmaceutical and biomedical applications: Focus on nanomedicine in cancer. *International Journal of Polymeric Materials and Polymeric Biomaterials, 69*(2), 85–126. https://doi.org/10.1080/00914037.2018.1539990.

Faheem, A. M., & Abdelkader, D. H., (2019). Novel drug delivery systems. In: *Engineering Drug Delivery Systems* (pp. 1–16). Elsevier. https://doi.org/10.1016/B978-0-08-102548-2.00001-9.

Fattahi, F. S., Khoddami, A., & Avinc, O., (2019). Poly (lactic acid) nano-fibers as drug-delivery systems: Opportunities and challenges. In: *Nanomedicine Research Journal* (Vol. 4, No. 3, pp. 130–140). Tehran University of Medical Sciences. https://doi.org/10.22034/NMRJ.2019.03.002.

Franco, P., & De Marco, I., (2020). The use of poly(N-vinyl pyrrolidone) in the delivery of drugs: A review. In: *Polymers* (Vol. 12, No. 5). MDPI AG. https://doi.org/10.3390/POLYM12051114.

Fu, K., Pack, D. W., Klibanov, A. M., & Langer, R. (1999). *Visual Evidence of Acidic Environment Within Degrading Poly(Lactic-co-glycolic Acid) (PLGA) Microspheres.* http://rsb.info.nih.gov/ (accessed on 6 December 2023).

Gaaz, T. S., Sulong, A. B., Akhtar, M. N., Kadhum, A. A. H., Mohamad, A. B., Al-Amiery, A. A., & McPhee, D. J., (2015). Properties and applications of polyvinyl alcohol, halloysite nanotubes and their nanocomposites. In: *Molecules* (Vol. 20, No. 12, pp. 22833–22847). MDPI AG. https://doi.org/10.3390/molecules201219884.

Gombotz, W. R., & Fong, W. S., (1998). Protein release from alginate matrices. In: *Advanced Drug Delivery Reviews* (Vol. 31).

Gupta, B., Mishra, V., Gharat, S., Momin, M., & Omri, A., (2021). Cellulosic polymers for enhancing drug bioavailability in ocular drug delivery systems. In: *Pharmaceuticals* (Vol. 14, No. 11). MDPI. https://doi.org/10.3390/ph14111201.

Ha, C. S., & Gardella, J. A., (2005). Surface chemistry of biodegradable polymers for drug delivery systems. In: *Chemical Reviews* (Vol. 105, No. 11, pp. 4205–4232). https://doi.org/10.1021/cr040419y.

Hajiali, F., Tajbakhsh, S., & Shojaei, A., (2018). Fabrication and properties of polycaprolactone composites containing calcium phosphate-based ceramics and bioactive glasses in bone tissue engineering: A review. In: *Polymer Reviews* (Vol. 58, No. 1, pp. 164–207). Taylor and Francis Inc. https://doi.org/10.1080/15583724.2017.1332640.

Hiraishi, N., Yau, J. Y. Y., Loushine, R. J., Armstrong, S. R., Weller, R. N., King, N. M., Pashley, D. H., & Tay, F. R., (2007). Susceptibility of a polycaprolactone-based root canal-filling material to degradation. III. Turbidimetric evaluation of enzymatic hydrolysis. *Journal of Endodontics*, *33*(8), 952–956. https://doi.org/10.1016/j.joen.2007.05.004.

Houchin, M. L., & Topp, E. M., (2009). Physical properties of PLGA films during polymer degradation. *Journal of Applied Polymer Science*, *114*(5), 2848–2854. https://doi.org/10.1002/app.30813.

http://www.chinafooding.com/485/Sodium_alginate_pH_value_and_stability (n.d.). (accessed on 6 December 2023).

https://chembam.com/resources-for-students/the-chemistry-of/gelatin/ (n.d.). (accessed on 6 December 2023).

https://healthcare.evonik.com/ (n.d.). (accessed on 6 December 2023).

https://pubchem.ncbi.nlm.nih.gov/ (n.d.). (accessed on 6 December 2023).

https://www.pharmaexcipients.com/ (n.d.). (accessed on 6 December 2023).

Islam, S., Bhuiyan, M. A. R., & Islam, M. N., (2017). Chitin and chitosan: Structure, properties and applications in biomedical engineering. In: *Journal of Polymers and the Environment* (Vol. 25, No. 3, pp. 854–866). Springer New York LLC. https://doi.org/10.1007/s10924-016-0865-5.

Jiménez-Gómez, C. P., & Cecilia, J. A., (2020). Chitosan: A natural biopolymer with a wide and varied range of applications. In: *Molecules* (Vol. 25, No. 17). MDPI AG. https://doi.org/10.3390/molecules25173981.

Kaiser, N., Klein, D., Karanja, P., Greten, Z., & Newman, J., (2009). Inactivation of chlorhexidine gluconate on skin by incompatible alcohol hand sanitizing gels. *American Journal of Infection Control*, *37*(7), 569–573. https://doi.org/10.1016/j.ajic.2008.12.008.

Kapoor, D. N., Bhatia, A., Kaur, R., Sharma, R., Kaur, G., & Dhawan, S., (2015). PLGA: A unique polymer for drug delivery. In: *Therapeutic Delivery ()*. (Vol. 6, No. 1, pp. 41–58). Future Science Ltd. https://doi.org/10.4155/tde.14.91.

Kaseem, M., (n.d.). *Properties and Medical Applications of Polylactic Acid.* Related papers melt rheology of poly(lactic acid)/low density polyethylene polymer blends Mosab Kaseem preparation and characterization of binary and ternary blends with poly(lactic acid), polystyrene, an… Mosab Kaseem a review on recent researches on polylactic acid/carbon nanotube composites. www.expresspolymlett.com (accessed on 6 December 2023).

Kaur, S., & Dhillon, G. S., (2014). The versatile biopolymer chitosan: Potential sources, evaluation of extraction methods and applications. In: *Critical Reviews in Microbiology* (Vol. 40, No. 2, pp. 155–175). https://doi.org/10.3109/1040841X.2013.770385.

Khamanga, S. M., & Walker, R. B., (2011). Drug transport mechanisms from Carbopol/ Eudragit verapamil sustained-release tablets. *Dissolution Technologies*, *18*(3), 30–38. https://doi.org/10.14227/DT180311P30.

Khoushab, F., & Yamabhai, M., (2010). Chitin research revisited. In: *Marine Drugs* (Vol. 8, No. 7, pp. 1988–2012). MDPI AG. https://doi.org/10.3390/md8071988.

Khutoryanskiy, V. V., (2007). Hydrogen-bonded interpolymer complexes as materials for pharmaceutical applications. *International Journal of Pharmaceutics*, *334*(1–2), 15–26.

Kreuter, J., & Liehl, E., (1978). Protection induced by inactivated influenza virus vaccines with polymethylmethacrylate adjuvants. *Medical Microbiology and Immunology*, *165*, 111–117.

Kreuter, J., Mauler, R., Gruschkau, H., & Speiser, P. P., (1976). The use of new polymethylmethacrylate adjuvants for split influenza vaccines. *Pathobiology*, *44*(1), 12–19.

Kulshrestha, A. S., & Mahapatro, A., (2008). Polymers for biomedical applications. *ACS Symposium Series*, *977*, 1–7. https://doi.org/10.1021/bk-2008-0977.ch001.

Kumosa, L. S., Zetterberg, V., & Schouenborg, J., (2018). Gelatin promotes rapid restoration of the blood-brain barrier after acute brain injury. *Acta Biomaterialia*, *65*, 137–149. https://doi.org/10.1016/j.actbio.2017.10.020.

Kurakula, M., & Rao, G. S. N. K., (2020). Pharmaceutical assessment of polyvinylpyrrolidone (PVP): As excipient from conventional to controlled delivery systems with a spotlight on COVID-19 inhibition. In: *Journal of Drug Delivery Science and Technology* (Vol. 60). Editions de Sante. https://doi.org/10.1016/j.jddst.2020.102046.

Labet, M., & Thielemans, W., (2009). Synthesis of polycaprolactone: A review. *Chemical Society Reviews*, *38*(12), 3484–3504. https://doi.org/10.1039/b820162p.

Li, L., & LaBarbera, D. V., (2017). 3D high-content screening of organoids for drug discovery. In: *Comprehensive Medicinal Chemistry III* (Vols. 2–8, pp. 388–415). Elsevier Inc. https://doi.org/10.1016/B978-0-12-409547-2.12329-7.

Limpan, N., Prodpran, T., Benjakul, S., & Prasarpran, S., (2012). Influences of degree of hydrolysis and molecular weight of poly(vinyl alcohol) (PVA) on properties of fish myofibrillar protein/PVA blend films. *Food Hydrocolloids*, *29*(1), 226–233. https://doi.org/10.1016/j.foodhyd.2012.03.007.

Liu, F., Wang, B., Xing, Y., Zhang, K., & Jiang, W., (2020). Effect of polyvinyl alcohol on the rheological properties of cement mortar. *Molecules*, *25*(3). https://doi.org/10.3390/molecules25030754.

Liu, M., Guo, B., Du, M., & Jia, D., (2007). Drying induced aggregation of halloysite nanotubes in polyvinyl alcohol/halloysite nanotubes solution and its effect on properties of composite film. *Applied Physics A: Materials Science and Processing, 88*(2), 391–395. https://doi.org/10.1007/s00339-007-3995-8.

Loureiro, D. S. L. A., (2017). Natural polymeric biomaterials: Processing and properties. In: *Reference Module in Materials Science and Materials Engineering.* Elsevier. https://doi.org/10.1016/b978-0-12-803581-8.02253-0.

Lv, L. C., Huang, Q. Y., Ding, W., Xiao, X. H., Zhang, H. Y., & Xiong, L. X., (2019). Fish gelatin: The novel potential applications. In: *Journal of Functional Foods* (Vol. 63). Elsevier Ltd. https://doi.org/10.1016/j.jff.2019.103581.

Makadia, H. K., & Siegel, S. J., (2011). Poly lactic-co-glycolic acid (PLGA) as biodegradable controlled drug delivery carrier. *Polymers, 3*(3), 1377–1397. https://doi.org/10.3390/polym3031377.

Mandal, P., & Shunmugam, R., (2020). Polycaprolactone: A biodegradable polymer with its application in the field of self-assembly study. In: *Journal of Macromolecular Science, Part A: Pure and Applied Chemistry* (Vol. 58, No. 2, pp. 111–129). Bellwether Publishing, Ltd. https://doi.org/10.1080/10601325.2020.1831392.

Maria, T. M. C., De Carvalho, R. A., Sobral, P. J. A., Habitante, A. M. B. Q., & Solorza-Feria, J., (2008). The effect of the degree of hydrolysis of the PVA and the plasticizer concentration on the color, opacity, and thermal and mechanical properties of films based on PVA and gelatin blends. *Journal of Food Engineering, 87*(2), 191–199. https://doi.org/10.1016/j.jfoodeng.2007.11.026.

Mariod, A. A.., & Fadul, H., 2013. Gelatin, source, extraction and industrial applications. *Acta Scientiarum Polonorum Technologia Alimentaria, 12*(2), 135–147.

Matyjaszewski, K., (2012). Atom transfer radical polymerization (ATRP): Current status and future perspectives. *Macromolecules, 45*(10), 4015–4039. https://doi.org/10.1021/ma3001719.

Mishra, P. R., Al Shaal, L., Müller, R. H., & Keck, C. M., (2009). Production and characterization of Hesperetin nanosuspensions for dermal delivery. *International Journal of Pharmaceutics, 371*(1–2), 182–189.

Misra, A., & Shahiwala, A., (2020). Novel drug delivery technologies: Innovative strategies for drug re-positioning. In: *Novel Drug Delivery Technologies: Innovative Strategies for Drug Re-Positioning.* Springer Singapore. https://doi.org/10.1007/978-981-13-3642-3.

Moers-Carpi, M. M., & Sherwood, S., (2013). Polycaprolactone for the correction of nasolabial folds: A 24-month, prospective, randomized, controlled clinical trial. *Dermatologic Surgery, 39*(3 PART 1), 457–463. https://doi.org/10.1111/dsu.12054.

Mogoşanu, G. D., & Grumezescu, A. M., (2014). Natural and synthetic polymers for wounds and burns dressing. *International Journal of Pharmaceutics, 463*(2), 127–136. https://doi.org/10.1016/j.ijpharm.2013.12.015.

Mohamed, R. M., & Yusoh, K., (2015). A review on the recent research of polycaprolactone (PCL). *Advanced Materials Research, 1134*, 249–255. https://doi.org/10.4028/www.scientific.net/amr.1134.249.

Mohammadi-Samani, S., & Taghipour, B., (2015). PLGA micro and nanoparticles in delivery of peptides and proteins; Problems and approaches. *Pharmaceutical Development and Technology, 20*(4), 385–393. https://doi.org/10.3109/10837450.2014.882940.

Nair, L. S., & Laurencin, C. T., (2007). Biodegradable polymers as biomaterials. In: *Progress in Polymer Science (Oxford)* (Vol. 32, No. 8, 9, pp. 762–798). https://doi.org/10.1016/j.progpolymsci.2007.05.017.

Nurkeeva, Z. S., Khutoryanskiy, V. V., Mun, G. A., Sherbakova, M. V., Ivaschenko, A. T., & Aitkhozhina, N. A., (2004). Polycomplexes of poly (acrylic acid) with streptomycin sulfate and their antibacterial activity. *European Journal of Pharmaceutics and Biopharmaceutics*, *57*(2), 245–249.

Olatunji, O., (2015). Natural polymers: Industry techniques and applications. In: *Natural Polymers: Industry Techniques and Applications*. Springer International Publishing. https://doi.org/10.1007/978-3-319-26414-1.

Omathanu, P., & Ramesh, P., (2001). Polymers in drug delivery. *Curr. Opin. Chem. Biol*, *5*(4), 447–451.

Oprea, M., & Voicu, S. I., (2020). Recent advances in composites based on cellulose derivatives for biomedical applications. In: *Carbohydrate Polymers* (Vol. 247). Elsevier Ltd. https://doi.org/10.1016/j.carbpol.2020.116683.

Ouwerx, C., Velings, N., Mestdagh, M. M., & Axelos, M. A. V., (1998). Physico-chemical properties and rheology of alginate gel beads formed with various divalent cations. In: *Polymer Gels and Networks* (Vol. 6).

Pavot, V., Berthet, M., Rességuier, J., Legaz, S., Handké, N., Gilbert, S. C., Paul, S., & Verrier, B., (2014). Poly(lactic acid) and poly(lactic-co-glycolic acid) particles as versatile carrier platforms for vaccine delivery. In: *Nanomedicine* (Vol. 9, No. 17, pp. 2703–2718). Future Medicine Ltd. https://doi.org/10.2217/nnm.14.156.

Pornsak, S., (n.d.). *Chemistry of Pectin and Its Pharmaceutical Uses: A Review Pornsak Sriamornsak Pornsak Sriamornsak, 207.*

Prabhakar, P., & Puranik, P. K., (2010). Carbopol polymers: A versatile polymer for pharmaceutical applications. *Res. J. Pharm. and Tech.*, *3*, 672–675.

Puscaselu, R. G., Lobiuc, A., Dimian, M., & Covasa, M., (2020). Alginate: From food industry to biomedical applications and management of metabolic disorders. In: *Polymers* (Vol. 12, No. 10, pp. 1–30). MDPI AG. https://doi.org/10.3390/polym12102417.

Qiu, K., & Netravali, A. N., (2013). A composting study of membrane-like polyvinyl alcohol-based resins and nanocomposites. *Journal of Polymers and the Environment*, *21*(3), 658–674. https://doi.org/10.1007/s10924-013-0584-0.

Qiu, K., & Netravali, A. N., (2013). Halloysite nanotube reinforced biodegradable nanocomposites using noncrosslinked and malonic acid crosslinked polyvinyl alcohol. *Polymer Composites*, *34*(5), 799–809. https://doi.org/10.1002/pc.22482.

Ruhé, P. Q., Hedberg-Dirk, E. L., Padron, N. T., Spauwen, P. H., Jansen, J. A.. & Mikos, A. G., (2006). Porous poly (DL-lactic-co-glycolic acid)/calcium phosphate cement composite for reconstruction of bone defects. *Tissue Engineering*, *12*(4), 789–800.

Ramos, M., Valdés, A., Beltrán, A., & Garrigós, M., (2016). Gelatin-based films and coatings for food packaging applications. *Coatings*, *6*(4), 41. https://doi.org/10.3390/coatings6040041.

Raymond, C. R., (2006). *Handbook of Pharmaceutical Excipients,* 5th edn.

Razzak, M. T., & Darwis, D., (2001). Irradiation of polyvinyl alcohol and polyvinyl pyrrolidone blended hydrogel for wound dressing. In: *Radiation Physics and Chemistry* (Vol. 62).

Samadi, N., Abbadessa, A., Di Stefano, A., Van, N. C. F., Vermonden, T., Rahimian, S., Teunissen, E. A., et al., (2013). The effect of lauryl capping group on protein release and degradation of poly(D,L-lactic-co-glycolic acid) particles. *Journal of Controlled Release*, *172*(2), 436–443. https://doi.org/10.1016/j.jconrel.2013.05.034.

Shahidi, F., & Abuzaytoun, R., (2005). Chitin, chitosan, and co-products: chemistry, production, applications, and health effects. *Advances in Food and Nutrition Research, 49*(4526), 49003–49008.

Shokri, J., & Adibki, K., (2013). Application of cellulose and cellulose derivatives in pharmaceutical industries. In: *Cellulose – Medical, Pharmaceutical and Electronic Applications*. Intech. https://doi.org/10.5772/55178.

Silva, E. A., & Mooney, D. J., (2010). Effects of VEGF temporal and spatial presentation on angiogenesis. *Biomaterials, 31*(6), 1235–1241. https://doi.org/10.1016/j.biomaterials.2009.10.052.

Singhvi, M., & Gokhale, D., (2013). Biomass to biodegradable polymer (PLA). In: *RSC Advances* (Vol. 3, No. 33, pp. 13558–13568). https://doi.org/10.1039/c3ra41592a.

Song, R., Murphy, M., Li, C., Ting, K., Soo, C., & Zheng, Z., (2018). Current development of biodegradable polymeric materials for biomedical applications. In: *Drug Design, Development and Therapy* (Vol. 12, pp. 3117–3145). Dove Medical Press Ltd. https://doi.org/10.2147/DDDT.S165440.

Sukhbir, S., Neelam, Sandeep, A., & Yash, P. S., (2015). An overview of multifaceted significance of eudragit polymers in drug delivery systems. *Asian J. Pharm. Clin. Res., 8*(5), 1–6.

Sundarraj, A., & Raj, S., (2012). *A Review on Pectin: Chemistry due to General Properties of Pectin and Its Pharmaceutical Uses Cite This Paper Open Access, 1*. https://doi.org/10.4172/scientificreports.550.

Szymańska, E., & Winnicka, K., (2015). Stability of chitosan – A challenge for pharmaceutical and biomedical applications. In: *Marine Drugs* (Vol. 13, No. 4, pp. 1819–1846). MDPI AG. https://doi.org/10.3390/md13041819.

Talele, S. G., Ahire, E. D., Talele, G. S., & Derle, D. V., (2021). An innovative approach as self-emulsifying drug delivery system for phytoconstituents. In: *Enhancing the Therapeutic Efficacy of Herbal Formulations* (pp. 69–84). IGI Global.

Tang, Y., Zhou, D., & Zhang, J., (2013). Novel polyvinyl alcohol/styrene butadiene rubber latex/carboxymethyl cellulose nanocomposites reinforced with modified halloysite nanotubes. *Journal of Nanomaterials*. https://doi.org/10.1155/2013/542421.

Thakral, S., Thakral, N. K., & Majumdar, D. K., (2013). Eudragit®: A technology evaluation. In: *Expert Opinion on Drug Delivery* (Vol. 10, No. 1, pp. 131–149). https://doi.org/10.1517/17425247.2013.736962.

Thakur, B. R., Singh, R. K., & Handa, A. K., (1997). Chemistry and uses of pectin – A review. In: *Critical Reviews in Food Science and Nutrition* (Vol. 37, No. 1, pp. 47–73). Taylor and Francis Inc. https://doi.org/10.1080/10408399709527767.

Tojo, K., (2005). *Mathematical Models of Transdermal and Topical Drug Delivery* (2nd edn.). Biocom Systems Inc, Japan.

Toncheva, A., Spasova, M., Paneva, D., Manolova, N., & Rashkov, I., (2014). Polylactide (PLA)-based electrospun fibrous materials containing ionic drugs as wound dressing materials: A review. In: *International Journal of Polymeric Materials and Polymeric Biomaterials* (Vol. 63, No. 13, pp. 657–671). Taylor and Francis Inc. https://doi.org/10.1080/00914037.2013.854240.

Vitthal, S. K., & Charles, S., (2016). Use of Polymers and Thickeners in Semisolid and Liquid Formulations Chapter 5. *Essential Chemistry for Formulators of Semisolid and Liquid Dosages. science Direct*, 43–69.

Woodruff, M. A., & Hutmacher, D. W., (2010). The return of a forgotten polymer – polycaprolactone in the 21st century. In: *Progress in Polymer Science (Oxford)* (Vol. 35, No. 10, pp. 1217–1256). Elsevier Ltd. https://doi.org/10.1016/j.progpolymsci.2010.04.002.

Xu, Y., Kim, C. S., Saylor, D. M., & Koo, D., (2017). Polymer degradation and drug delivery in PLGA-based drug–polymer applications: A review of experiments and theories. In: *Journal of Biomedical Materials Research – Part B Applied Biomaterials* (Vol. 105, No. 6, pp. 1692–1716). John Wiley and Sons Inc. https://doi.org/10.1002/jbm.b.33648.

Yang, Z., Hemar, Y., Hilliou, L., Gilbert, E. P., McGillivray, D. J., Williams, M. A. K., & Chaieb, S., (2016). Nonlinear behavior of gelatin networks reveals a hierarchical structure. *Biomacromolecules*, *17*(2), 590–600. https://doi.org/10.1021/acs.biomac.5b01538.

Zarrintaj, P., Ramsey, J. D., Samadi, A., Atoufi, Z., Yazdi, M. K., Ganjali, M. R., Amirabad, L. M., et al., (2020). Poloxamer: A versatile tri-block copolymer for biomedical applications. In: *Acta Biomaterialia* (Vol. 110, pp. 37–67). Acta Materialia Inc. https://doi.org/10.1016/j.actbio.2020.04.028.

PART II

Novel Drug Delivery System in Central Nervous System Diseases

CHAPTER 4

Novel Drug Delivery Systems for the Treatment of Chronic Alzheimer's

ABHIJEET KULKARNI,[1] VISHAL GULECHA,[1] PRAVIN JHA,[1] AMAR ZALTE,[1] RAMDAS DOLAS,[1] SUSHIL NARKHEDE,[1] SUMIT DEORE,[1] AKSHADA KOTHULE,[2] VIVEK WAGH,[2] MAYUR MORE,[2] and VISHAL PANDE[3]

[1]*School of Pharmaceutical Sciences, Sandip University, Nashik, Maharashtra, India*

[2]*Department of Pharmaceutics, Sanjivani College of Pharmaceutical Education and Research, Kopargaon, Maharashtra, India*

[3]*N. N. Sattha College of Pharmacy, Ahmednagar, Maharashtra, India*

ABSTRACT

Alzheimer's disease (AD) is a brain disorder, characterized by progressive neuronal loss, categorizing it as one of the leading causes of death due to neurodegenerative disorders. According to current statistics given by the National Institute of Aging, it is the 7th leading cause of death in the United States and the prime reason for dementia. Impaired cognitive function and compromised behavioral capabilities significantly affect the routine lifestyle of senior patients. This chapter comprehensively covers the neuropathogenesis of AD, current treatment modalities, and drug delivery approaches, including nanotechnology applications.

4.1 INTRODUCTION

First mention of Alzheimer's disease (AD), in the year 1907, was credited to German neuro-physician and psychiatrist Alois Alzheimer who first

Novel Drug Delivery Systems in the Management of Chronic Diseases.
Raj K. Keservani, Umesh D. Laddha, Eknath D. Ahire, and Sanjay J. Kshirsagar (Eds.)

reported the case of "presenile dementia" which was later called AD (Fan et al., 2020). It is one of the most prevalent neurodegenerative diseases characterized by gradual impairment of cognitive as well as behavioral functions primarily affecting memory, language, and thinking. The quality of life of the AD-affected patient slowly gets severely compromised as neuronal damage progresses and eventually leads to full dependency on others to perform simple tasks (Kumar et al., 2021; Ahire et al., 2020). According to the National Institute of Aging (https://www.nia.nih.gov/), AD is the prime culprit and most common cause of dementia. In a recent annual report released by the Alzheimer's Association (2022 Alzheimer's Disease Facts and Figures), one in three seniors' deaths can be ascribed to AD or related dementia, which is more than breast and prostate cancer combined. By the year 2050, the prevalence of AD and related dementia is projected to affect 12.7 senior people aged $\geq$ 65. The occurrence of AD in early age (early-onset type) between 30s and mid-60s is very sporadic, however, symptoms start appearing in mid-60s (late-onset type) in most individuals. Worldwide, around 50 million patients' lives are being affected by AD and the prevalence continues to grow, posing a great healthcare challenge of the 21st century (Fan et al., 2020).

4.2 PATHOGENESIS

Normal healthy average adult brain weighs around 1,370 g (male) and 1,200 g (female) with corresponding volume of 1,260 cm^3 and 1,130 cm^3, respectively (Wikipedia). Basic structural and functional of unit of such healthy brain is neuron and average brain has around 100 billion long, branched neurons which are connected to other neurons by synapses, which are approximately 100 trillion. These synaptic connections functions to carry chemical signals across the different parts of brain rapidly which forms the basis of thoughts, emotions, locomotor activities and other cognitive functions. It is essential to have healthy neuron function to have healthy brain function. However, in AD, neuronal damage/destruction and progressive neurodegeneration occurs with the age leading to associated chronic inflammation, brain changes or reduced cerebral volume (brain atrophy). The basic comparison between healthy and AD affected brain is shown in Figure 4.1 which clearly depicts significant brain volume reduction along with atrophied "gyri" (folds in the brain) and sulci (grooves in the brain) which are largely responsible for higher cognitive functions viz., consciousness, language, and memories (Silva et al., 2008).

The fundamental hypothesis that has been proposed in the pathogenesis of AD is beta amyloid cascade and tau protein hyper-phosphorylation.

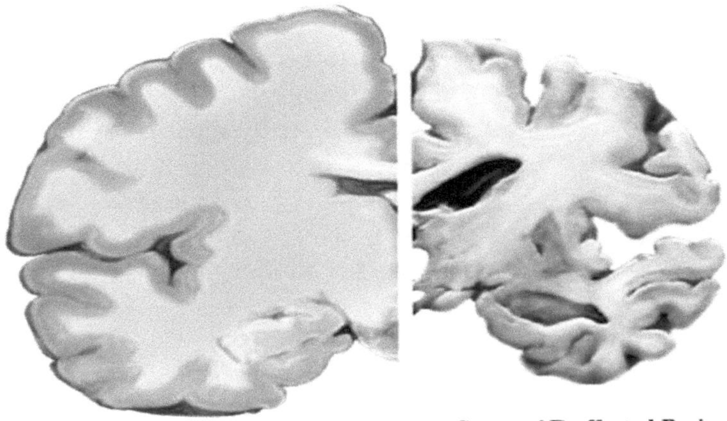

FIGURE 4.1 Comparison between healthy and AD affected brain.
Source: Adapted from: National Institutes of Health.

4.2.1 BETA-AMYLOID PLAQUES AND TAU HYPER-PHOSPHORYLATION HYPOTHESIS

Brains of the AD affected elderly patients is characterized by accumulation of fragment into clumps (plaques) of beta-amyloid protein outside of the neuron which leads to the severe neuronal damage disturbing the inter-neuronal communications.

Tau is microtubule related phosphoprotein found in neuronal axons whose function is to maintain neuronal synapse and neuronal signaling. Hyper phosphorylation of tau leads to its defective functioning causing reduced synapse number and tau aggregation. Further, accumulation of this small oligomeric chain of abnormal protein in circular fashion (called "tau tangles") interferes with transport of essential nutrient molecules affecting normal neuronal function and further leading to neuronal death. This progressive brain changes and associated neuronal destructions lead to inflammation and reduced brain volume. It is thought that microglia (immune cells of brain) attempt to clear off the debris of damaged and dying neurons to restore the brain function until it can keep a pace with rate of destruction which slowly becomes unmanageable. Worsening the AD condition further brain's ability to metabolize glucose get compromised leading to severe complications.

Current research is focused to target or dissolve the beta-amyloid protein and tau aggregation (Fan et al., 2020).

4.2.2 *GAMUT OF AD PROGRESSION*

AD progresses with the series of stages, initiating as without symptom (preclinical stage) to mild cognitive impairment (MCI) to severe AD related dementia (profound clinical stage) which finally leads to physical disability. This dementia caused by AD can further be classified as mild, moderate, and severe. Figure 4.2 illustrates the gamut of AD. The duration of each stage of this series of gradual brain changes is not uniform and can be dependent on various factors like genetic make-up, sex and age, etc.

4.2.2.1 PRECLINICAL AD

This is the first stage of AD progression where brain changes that are not noticeable even to the person affected as the brain adjust for changes to lead a normal life. There is no presence of any signs/symptoms related memory loss however, quantifiable amount of characteristic proteins (like beta-amyloid) start appearing in the plasma, cerebrospinal fluid or can be detectable in positron emission tomography (PET) scans. Though early detection tools for AD are available there is desperate need of simple accurate and precise techniques for early diagnosis.

4.2.2.2 MILD COGNITIVE IMPAIRMENT (MCI)

At MCI stage as the name implies mild signs of affected cognitive function such as thinking, memorizing or language issues, start appearing however person still can perform his daily activities. Though the symptoms are subtle one but can be perceived by the person or his/her close ones. At this stage progressive neuronal damage can be clearly diagnosed by biomarker analysis as the brain is not more able to compensate for the neuronal loss (Fonseca-Santos et al., 2015).

4.2.2.3 MILD AD DEMENTIA

Successive phase after the MCI leads to the dementia. Almost 30% of the MCI affected individuals precipitate into dementia within 5 years. At mild AD stage affected individual start to need assistance or more time to perform some of the simple tasks such as numerical calculation, payment of bills.

4.2.2.4 MODERATE AD DEMENTIA

Usually, the longest phase of AD dementia where person affected find difficulties to perform activities which involves multiple steps like taking bath or dressing oneself. Individual experiences confusion and agitated state of mind at times faces behavioral changes. Few are unable to control the passing of urine or feces from the body (incontinence). Recognition of faces, relatives also become problematic.

4.2.2.5 SEVERE AD DEMENTIA

Since this stage is characterized by severe neurodegeneration, many vital parts of brain get affected which are responsible for verbal communication, locomotors activities. Symptoms of severe AD Dementia may include following but not limited to – repeating same question, forgetting learned skill or information, fail to stay aware of time-day-dates, numeracy, organization, planning skills get affected, vision, color identification and spatial relationship get affected, issues with verbal and written communication, poor judgment, wandering (difficulty in retracing steps), withdrawal from society, behavioral or mood changes eventually requiring constant around-the-clock surveillance.

Patient usually becomes bed-ridden increasing the risk of various infections. Severe chronic inflammation builds up inside the whole body, which may lead to organ failure. In case if the brain areas which control the swallowing function get affected, food particle accidently may get entered into the lung tract causing aspiration pneumonia which is prime reason that leads to death of AD affected parsons.

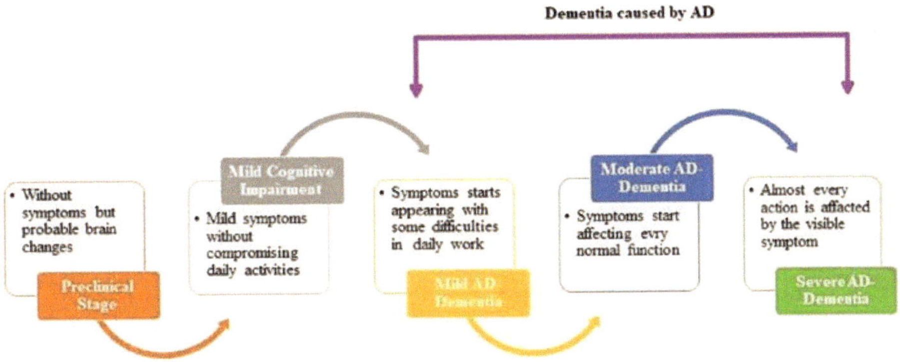

FIGURE 4.2 Gamut of AD progression.

4.3 ETIOLOGY

The etiology or clear mechanism of AD is not clearly known. However, probable factors that are responsible for cerebral changes (progressive neuronal loss) involves abnormality in vascular health and related deprived oxygen supply to brain, stress, and free radicle hypothesis, infections, intoxications, nutritional or vitamin deficiency, cerebral tumors, and others (Breijyeh & Karaman, 2020). These factors can be primarily categorized as

4.3.1 AGE-RELATED

There is difference in the normal age-related changes in the reduction in the brain volume and AD affected brain, which is characterized by brain shrinking, inflammation, blood vessel damage, which may be detrimental to neurons and other brain cells. It has been established that there is significant correlation exist between the age above 65 with AD related dementia.

4.3.2 GENE RELATED

Family history of AD or altered gene can be predominant cause of AD especially for AD occurring at 30 to mid-60s (early onset type). First degree relative-FDR (parents, siblings, children) and those who have more than one FDR with AD are more likely to develop disease. Individual with genetic abnormality of Down syndrome known as Trisomy 21 is characterized by extra copy (total 3 copies) of chromosomes (instead of 2) is significantly correlated with early occurrence of AD. It is believed that this extra chromosome carries gene that is responsible for the expression of APP-Amyloid Precursor Protein which increases the synthesis of beta-amyloid protein in the brain (Richter et al., 2010). Observation reported by National Down Syndrome Society states that almost 50% of the individuals with Down syndrome have developed AD after their 60s and 30% after their 50s (Zhou et al., 2011). Out of many genes responsible for AD, apolipoprotein E (APOE) involved in the lipid metabolism, located on chromosome 19, specifically with its e4 type has significant correlation with late onset type AD (Pimenova & Goate, 2020). Every individual is inherited with one of the 3 types (alleles) of APOE genes viz., e2, e3, e4 from each parent (Marei et al., 2021). These forms total 6 APOE pairs of permutation-combinations. Probable risk of developing AD follows sequence of e4 > e3 > e2 (Robinson et al., 2017). Those inherits one copy of e4 possesses has 3 times more or those who possesses two copies e4 has 8–12 times more risk of developing AD as compared to e3. As e3 form considered as neutral with respect to risk of causing AD (Giau et al., 2015). Data of AD diagnosed patients from US also reconfirm the fact that 56% of them were having one copy and 90% were having two copies of APOE e4 gene. Few scientists have also tried to establish the link of ATP-binding cassette transporter (ABCA7) with AD in Blacks (Murphy et al., 2013). Presence of this gene, ABCA7 doubles the risk of AD as compared with the deficient (Galimberti & Scarpini, 2011).

4.3.3 GENE MUTATION

Abnormal changes in the genes (chemical pairs) sequence are termed as gene mutation (Xia et al., 2015). Most of the early onset type cases of AD can be ascribed autosomal dominant disorder related to mutations specifically in three genes viz., AAP (Chromosome 21) (Kennedy et al., 1994), Presenilin-1 (Chromosome 14) and Presenilin-2 (Chromosome 1) (Breijyeh & Karaman, 2020). Mutations on these genes surely lead to AD at early age of 30s to 50s (Gordon et al., 2018).

4.3.4 HEALTH AND LIFESTYLE RELATED

Presence of clinical conditions such as heart disease, high blood pressure, diabetes, obesity along with exposure to pollutants and other environmental factors are also being found to increase the risk of AD.

4.4 DIAGNOSIS OF AD

Tools and techniques that can measure the brain changes (physiological and pathological) will be able to diagnose the early signs of AD (Ahmad et al., 2017). Biomarkers are the bio-chemicals/genes/molecules that are derived out of biological processes and found in blood or body fluids which may be characteristic of particular disease or condition. For example, presence of salient biomarkers (beta amyloid or tau protein) in the brain fluid CSF or plasma is indicative of AD related brain changes. DIAD-Dominantly Inherited AD, condition characterized by genetic mutation that leads to AD where symptoms for beta-amyloid protein start appearing in fluids before two decades, glucose metabolism began to decrease 18 years before, brain size reduction starts 13 years before the actual clinical occurrence of AD (Lu et al., 2018). Another such biomarker, (NF-L-neurofilament-light) polypeptide chain, alarms about the forthcoming AD almost 22 years ahead of time. Along with this PET scans clearly indicates position of abnormal protein accumulation in the brain. Although enough tools are available to identify brain changes, a greater number of in-depth studies are required to optimize it for AD in cost effective manner (Owens et al., 2021).

4.5 CURRENT TREATMENT MODALITIES

Losses of acetylcholine (Ach) neurons have been the established hypothesis of AD in early disease progression Current approved molecules by the FDA majorly attempts to provide symptomatic relief like donepezil (Aricept®), rivastigmine (Exelon®), galantamine (Reminyl®/Nivalin®/Razadyne®) which are from the class of anti-cholinesterase the enzyme responsible for degradation of the Ach. Normal cholinergic function of the brain is required for the regulation of higher order cognitive functions (Yiannopoulou & Papageorgiou, 2013). In addition to that memantine (Namenda®XR) N-methyl-D-aspartate (NMDA) antagonist used as alternative therapeutic agent to treat moderate or severe AD symptoms. It is hypothesized to protect neurons from toxic effects from over activation of some neurotransmitter like glutamate (Vaz & Silvestre, 2020).

Recently, instead of mere symptom modifying strategies attention of the researcher have focused on addressing the fundamental biology of beta-amyloid cascade hypothesis which is responsible for AD. This disease modifying strategies aims to reduce generation, block aggregation or promote dissolution/clearance of beta amyloid plaques.

All the existing drugs used are for mild AD symptoms and that to have adverse effects and none are successful in arresting the progression of AD. Last year in June 2021 monoclonal antibody (MAB) based, especially beta amyloid directed drug, Aducanumab (Aduhelm™), was approved by FDA to Biogen company to treat MCI or mild AD related dementia. Though the molecule was approved through accelerated mode of approval process and drug lacks long term study data (for early access to serious patients) there is hope that drug will prove to be successful even at post approval detailed clinical trials. Many other MAB (Gantenerumab, Lecanemab, and Donanemab) targeted towards beta amyloid are being evaluated in clinical trials. To undergo treatment with aducanumab one has to have confirmed MCI, positive PET scans for plaques, no history of stroke or brain hemorrhage or should not be on blood thinners or above 85 ages. The pricing of the IV infusion MAB is very costly, which is approximately 5,000$ per month. Another novel MAB – Saracatinib being investigated clinically has been shown to be successful in reversing the memory loss and nerve function in animal models. Sargramostim is a new granulocyte-macrophage colony-stimulating factor (GM-CSF) which helps in activating innate immune system of microglia cells in the brain. It has shown reduction in the amyloid load in transgenic mice model for AD. Few researchers hypothesized that tau hypothesis is closely related to cognitive decline and developing tau vaccine-based immunotherapy directed towards tau tangles in the brain (Keskin et al., 2019).

4.6 NOVEL DRUG DELIVERY FOR AD

As it is evident from the above treatment modalities that most of the therapies for AD are naturally directed towards the brain. Some of the approved FDA medicaments are administered through oral route and few (rivastigmine) are being formulated a transdermal patch. It is always important to understand that how much fraction of the orally administered drug is getting bioavailable at the brain site of action. Challenges posed by oral route involve oral absorption, liver metabolism, protein binding, distribution, and of course most difficult barrier would be the blood-brain-barrier (BBB). As all these therapies are getting exposed to whole body untoward side effects is natural outcome. To avoid these complications novel nano-carrier systems would come to a rescue which addresses most of the issues with conventional therapies (Kumar et al., 2020). They offer multi-functionalization, ability to carry high drug payload, ability to cross BBB, low hepatic clearance, low, if any protein binding, minimizing dose requirement thus preventing unpleasant obvious adverse effects caused duet to conventional therapies. As most of the unmet needs and related transport issues are addressed by the nanonization, nanodrug delivery would be the future and popular choice for CNS diseases including AD (Ahire et al., 2018). Figure 4.3 highlights some of the common nanocarrier systems being applied for biomedical applications.

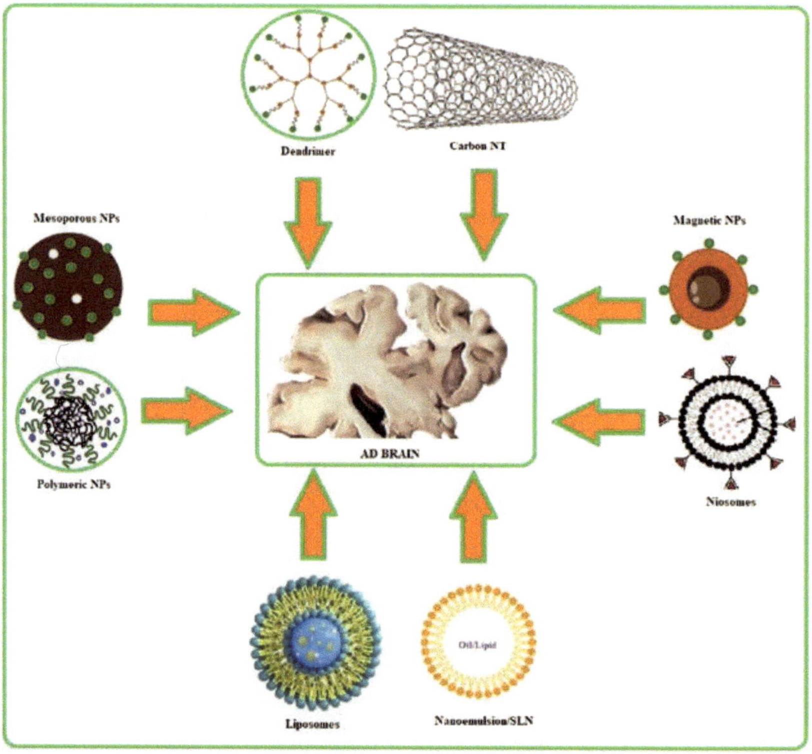

FIGURE 4.3 Few popular nano-carriers used in drug delivery.

In addition to the above benefits, nanocarrier based therapies are more attractive as they can be manipulated with respect to their overall desired stability (physicochemical or biological), possibility to transport any type of drug (lipophilic or hydrophilic), flexibility with route of administration (parenteral, inhalational or oral). Moreover, they can be targeted to specific areas of the brain by functionalization with various ligand molecules like peptide/aptamers (Apts)/antibodies (Jokar et al., 2019). Owing to larger surface-to-volume ratio, multiple numbers of ligand molecules can modify surface properties increasing the target specificity of the nano-conjugates. The stealthing process by coating PEG or polysorbate increases the circulation time and prevent opsonization and clearance of nanocarrier system by reticuloendothelial system (RES) on systemic exposure. All these properties make nano-drug delivery best choice to target AD affected brain. Table 4.1 highlights few of the recent applications of novel drug delivery system (NDDS) in management of AD.

TABLE 4.1 Few Representatives of Novel Nanodrug Delivery Approaches for the Management AD

Drug Delivery System	Drug	Method/ Technique	Particulates	Animal Species Used	References
Dendrimer	Small interfering RNA (siRNA)	Self-assembly and conjugation.	Conjugation of NL4 peptide and apolipoprotein A-I (ApoA-I) onto dendrimer leads to dendrimer assembly. *In vitro* results demonstrated that siRNA could effectively target both BBB and down-regulate beta-site APP cleaving enzyme (BACE1) gene expression. Also, the ANNP/siRNA showed low cytotoxicity.	Nude mice	Zhang et al. (2018)
Niosomes	Rivastigmine N-acetyl cysteine.	Ethanol injection method.	Sustained drug release pattern was shown by the drug loaded niosomes. AChE and DPPH enzyme inhibition assays showed a better effect than the free drug solutions. A better drug profile and greater distribution was shown by *in-vivo* pharmacokinetic and organ bio-distribution studies in the brain.	Male Wistar rats	Kulkarni et al. (2021)
Nanoparticles	Poly(lactic-co-glycolic acid)	Conjugation method	The nanoparticle efficiently increases its half-life in blood circulation by preventing engulfment via phagocytes and enhances its brain distribution. Nec-1s in the engulfed NPs modulated DAMs to beneficial microglia, rescuing cognition and synapse loss by reducing Aβ level and neuroinflammation.	Transgenic mice	Zhang et al. (2020)
Liposomes	Rivastigmine	Ammonium sulfate gradient loading method.	CPP-Lp enhanced the permeability across the BBB by *in vitro* model of murine brain microvascular endothelial cells. Liposomes showed the ability to improve drug distribution and adequate retention in CNS regions than that of IV administration. CPP-Lp improve the brain delivery and enhance pharmacodynamics which respect to BBB penetration.	Male Sprague-Dawley rats	Yang et al. (2013)

TABLE 4.1 *(Continued)*

Drug Delivery System	Drug	Method/ Technique	Particulates	Animal Species Used	References
Nanocapsules	BSA, potassium ferrocyanide, manganese chloride	Facile biomimetic mineralization.	HMPWCs enhanced cognitive performance without significant side effects. They can attenuate the cognitive impairments of Tau-related AD-like rats without significant side effects.	Male Sprague-Dawley rats	Cai et al. (2020)
Liposomes	Transferrin Penetration	Thin lipid film hydration method and post-insertion technique.	Immunofluorescence studies of brain sections of mice illustrated the superior ability of dual functionalized liposomes to accumulate in the brain and transfect neurons.	C57BL/6J mice	Rodrigues et al. (2018)
Niosomes	Rivastigmine	Lipid film hydration method.	3^2Full factorial design based niosomal *in situ* gel was prepared. The niosomes developed has good perfusion compared to pure drug in the brain.	Male gender rats and Sprague Dawley rat strain	Sarswathi & Motilal (2021)
Liposomes (transdermal)	Rivastigmine	Thin layer hydration technique.	Optimized liposome showed two-fold increase in both rivastigmine permeation through the skin. Optimized liposome demonstrated good properties to provide this ideal pharmacokinetic profile.	Rat	Salimi et al. (2020)
Liposomes	Centella asiatica extract.	Solvent evaporation method and ionotropic gelation method.	*In vitro* intestinal study proved that the permeation rate increases due to the encapsulation of chitosan-phospholipid complex. Chitosan coated liposomes were stable due to compact chitosan coating layer.	A duodenal part from goat intestine.	Hebbar et al. (2020)

TABLE 4.1 *(Continued)*

Drug Delivery System	Drug	Method/ Technique	Particulates	Animal Species Used	References
Nanoparticles	Rivastigmine	Nano-precipitation method.	Rivastigmine loaded PLGA NPs were compared with nasal liposomal formulation by scopolamine-induced amnesia and colchicine induced cognitive dysfunction animal models. This PKPD analysis showed liposomal formulation offers better PK parameters with high correlation to Ach enzyme inhibition.	Male Wistar rats	Rompicherla et al. (2021)
Liposomes	*Aphanamixis polystachya* leaf extract.	Solvent injection method.	Significant improvement in the memory function, locomotor activity and ambulatory performance of dementia induced mice was observed for the liposomal batches. *In-vivo* studies for liposomal drug delivery of *A. polystachya* leaf extract showed significantly better performance compared to extract.	Swiss albino mice	Shariare et al. (2020)
Nano-emulsion gel	Tacrine hydrochloride	Spontaneous emulsification	PK studies showed greater extent of absorption from NEGT4 in comparison to capsule and hydrogel. Biochemical assessment showed superior anti-amnesic activity of NEGT4, decreased lipid peroxidation and acetylcholinesterase activity.	Male Wistar rats	Setya et al. (2019)
Liposomes	Lactoferrin, DDAB, DOPE	Thin layer hydration technique.	Compared with conventional liposomes and CLs, PCL and Lf-PCLs showed an improved performance in the uptake efficiency and cytotoxicity. PCL and Lf-PCL-8 were observed to show much improved characteristics of the localization in the brain as compared to conventional liposomes.	Kunming mice and new-born Sprague–Dawley rats.	Chen et al. (2010)
Niosomes	Rivastigmine	Film hydration method.	Incorporation of rivastigmine into bilayer assemblies of niosomes compromises a good candidate for new drug delivery system.	Not applicable	Pardakhty et al. (2018)

4.6.1 DENDRIMERS

Tree-like structure of branched chain of polymers is called dendrimers (Gautam et al., 2015) (Greek word: Dendron means tree; meros means part) where the internal core of the dendrimer is essentially the same with concentric symmetric extended spheroidal structures leading to three-dimensional shapes. Ramification of the core (root) molecule generally happens like a growing tree giving out branches towards the periphery (Divergent method). Usually, the central core starts with two similar chemical functional moieties. Repeating units of other molecules further lead to chain extension or branching which further results in concentric radial layers that are closely packed. The gap that gets generated within this heavily branched structure can be occupied by the drug molecules which get entrapped in dendrimer structures. So, a large amount of dose can be loaded with these structures. Usually, the divergent method allows loosely held central core to sequentially connect or branch further until the steric hindrance limits further reactions to extend the branches. On the contrary, in the convergent method structure is created starting with the ending functional groups towards the core where the central molecule is placed at the end.

Beta amyloid hypothesis has been greatly correlated with the formation of amyloid protein plaques and the development of AD. Strategies researchers are working towards are either to inhibit the formation of oligomeric peptide or to prevent its aggregate structure formation.

Zhang et al. attempted the challenge of BBB and poor intra-brain cell selectivity by a novel dual-targeted DDS. Two targeting ligands (NL4 and Apolipoprotein A-I) were attached to the dendrimer to deliver siRNA to brain nerve cells so that it will inhibit Aβ protein formation and downregulation of BACE1 (enzyme responsible for beta amyloid). The prepared Apo A-I and NL4 dual-modified nanoparticle (ANNP) was around 80 nm with a stable spherical structure. It was observed that cell uptake was significantly enhanced with almost 2.5-fold better penetration power. These dual-modified dendrimeric NPs were found to be helpful in selective gene knockdown capacity with lower cytotoxicity. So ApoA-I and NL4 modified ANNP/siRNA would be a good nanocarrier option to target AD-affected brain.

4.6.2 LIPOSOMES

Liposomes are the globular vesicular systems consisting usually of bilayer spherical structures of phospholipids creating internal aqueous core. The lipids used for the preparation of liposomes are biocompatible and biodegradable similar to one present in biological system making it one of the most acceptable system nano-vesicular system of first generation. The examples of lipids used are cholesterol, phosphotidyl choline, glycerophospholipids, sphingomylein, which are also natural components of the bio-membrane which make liposomes more permeable (Keservani & Gautam, 2020), compatible and stable *in vivo*. Various formulation methods (ultrasonication, high-pressure homogenization, reverse phase evaporation) are applied based on which various arrangements of lamellae can be possible. Based on size and number of lamellae liposomes can be categorized as small unilamellar vesicles with one bilayer (SUVs-100 nm), large unilamellar vesicles with one bilayer (LUVs – more than 100 nm). Multilamellar vesicles (MLVs – in micrometer) are characterized by multiple concentric layers of lipid bilayers. Further amendments in the basic structure of liposomes have resulted into novel vesicular structures like phytosomes, ethosomes, transferosomes, niosomes (Keservani et al., 2020; Keservani & Gautam, 2022; Xu et al., 2021).

Due to amphiphilic nature of the liposomes, this nanosystem is considered ideal to deliver drug to brain as it resembles the biosystem, as it carries both (hydrophilic and hydrophobic molecules), is least toxic, offers greater permeability. To improve the circulation time, size reduction and pegylation of the liposomes have also been done. Generally, it has been reported that negatively charged surface of liposomes help activate RES system clear offs the liposome rapidly whereas conferring cationic charge with positively charged lipids can be used as transfection vehicles for the DNA material. When negatively charges phosphate functionality of nucleic acid or genetic material reacts with cationic materials to form complex structure is called lipoplexes (Ahmad et al., 2017). Examples of such positively charged lipid include DOPE (dioleoyl-phosphatidylethanolamine), DOTAP (dioleoyl-trimethyl ammonium propane). Cationic liposomes possess greater endocytosis and endosomic internalization compared to normal liposomes. The exact mechanism of greater BBB penetration of liposomes is presumed to be due to smaller size (< 100 nm) and endocytic transport.

Yang et al. (2013) developed the cell penetrating peptide (CPP) modified rivastigmine loaded liposome (CPP-Lp) which enhanced brain permeability and CNS distribution when administered intranasally compared to i.v. administration.

Rodrigues et al. (2020) developed transferrin-CPP-dual functionalized liposome for gene delivery targeted to the brain. Tf-CPP-liposomes enhanced *in vitro* BBB permeability model and transfection efficiency of primary neurons.

Salimi et al. (2021) developed rivastigmine loaded liposomal formulation as a transdermal drug delivery to maintain therapeutic concentration in AD patients. *Ex vivo* permeation study, compared with *in vivo* pharmacokinetic parameters. Prepared formulation showed higher permeation as compared to control indicating topical liposomal formulation as an alternative. Some of the prominent applications of liposomal formulation are highlighted in Table 4.1.

4.6.3 NIOSOMES

Niosomes are the microscopic lamellar vesicular system formed out of self-assembly of non-ionic surfactant (alkyl polyglycerol ether with cholesterol) upon hydration. Structurally, bilayer nature of noisome is similar to the liposome except stability conferred by the non-ionic surfactant and smaller size (10 to 100 nm). It also can carry hydrophilic molecules in the inner core whereas hydrophobic molecule entrapped in lamellar structures.

Kulkarni et al. (2021) have formulated N-acetyl cysteine (NAC) functionalized rivastigmine loaded niosomes for better pharmacokinetic properties. The prepared system was validated using Box-Behnken design for optimization and administered with nasal route. Greater nasal permeations resulted in the higher distribution in the brain indicating success of the niosomal formulation circumventing the BBB. Nasal *in situ* niosomal gel loaded with rivastigmine was prepared film hydration method applying factorial design. *In vivo* perfusion study in rats indicated prolonged release of drug for 8 h with higher localization in the brain (Sarswathi & Motilal, 2021).

4.6.4 NANOPARTICLES OR NANOCAPSULE

Polymer based nanoparticles (Singh & Keservani, 2017) are the systems where drug is either dispersed as matrix or encapsulated into the polymer chains to modify the drug release rate. Drug is either dispersed or loaded as a solution into the polymer with approximate size range of 50 nm to 200 nm. Various polymers synthetic, semisynthetic or natural can be utilized to prepare the stable, biocompatible, biodegradable polymeric nanoparticles (PNPs) or nanocapsules (Keservani et al., 2018). Examples of polymeric material include – polylactide-co-glycolides (PLGA), polyanhydrides, polycaprolactone (PCL), polycyanoacrylates, polyglycolides (PGA), etc. Many natural polymers like chitosan and its derivatives have been employed in nose to brain delivery of NPs (Keservani et al., 2019; Alexander & Saraf, 2018;

Lee & Minko, 2021). Overcoming the BBB and to avoid opsonization or increase the circulation time in the plasma various surface modification have been implemented like coating with PEG or poloxamers or other molecules. Recently PLGA nanoparticle functionalized with conjugated with CD47 extracellular domain and BBB CPP and Nec-1 which is microglia activating agent. Prepared system specifically targets the microglia as expected and reduces the beta amyloid burden, inflammation, and reduces associated synaptic loss and cognition deficits improving the AD (Zhang et al., 2020). Recently two different nano-formulation (PLGA NPs and Liposomes) were compared to establish better performance in acute (scopolamine-induced amnesia) and chronic (colchicine induced cognitive dysfunction) animal models of AD, wherein the liposomal formulation proved to offer better pharmacokinetic profile compared to PLGA NPs and better Ach enzyme inhibition activity which is in sync with pharmacokinetic parameters (Rompicherla et al., 2021). Hollow Manganese Prussian White nanocapsules (HMPWCs) and bovine serum albumin (BSA) with biomimetic mineralization attenuates tau related neuro inflammation and inhibit hyperphosphorylation of tau protein. This hypothesis is put forward as an alternative strategy to combat tau related AD associated neurodegeneration (Cai et al., 2020).

4.6.5 NANOEMULSION (NE)

Oral tacrine hydrochloride treatment suffers from many disadvantages hepa,totoxicity, extensive first pass metabolism, and the need for frequent dosing. However, it can be used as reversible inhibition of Acetylcholinesterase enzyme make it very useful candidate for AD management. To avoid the oral route challenges tacrine loaded nanoemulsion (NE) was prepared for transdermal application which was further converted into nano-emulsion based gel (NEG). Substantial improvement is observed in the scopolamine-induced amnesia rat model with minimum hepatotoxicity (Setya et al., 2019).

4.7 CURRENT CHALLENGES AND FUTURE NEEDS

Current AD therapies still face the challenge of the successful and sufficient transport across the BBB. As conventional routes suffer with side effects and dose losses alternative routes are being evaluated for the therapeutics. Tacrine was discontinued in US (2013) owing to its liver toxicity whereas rivastigmine being administered with comparative ease by transdermal route.

Effective and stable DDSs are highly needed to overcome the gap of limited FDA approved molecules on the patient bedside. Intranasal DDS with olfactory route is attracting seminal attention of the researchers for AD targeting. Many clinical trials are being undertaken to see the success of nanocarriers through intranasal route as limited correlation can be established with animal nasal route application (Cunha et al., 2021). Many investigations with nanotechnology show promising future with proteins, peptides, monoclonal antibodies, nucleic acid delivery and even nano-vaccine in AD and many are being tested in clinical trials as it was witnessed with accelerated FDA approval Aducanumab. However, safety with long duration studies (detailed post approval clinical trials) needs to be established and related cost of the therapy makes it difficult for the common elderly patients to avail it as part of routine treatment. In conclusion, with all the promises of NDDS, dream of affordable therapy for AD patients remains matter of time when sufficient clinical safety and efficacy would deliver the molecule from bench-side to bedside for AD patient (Ahire et al., 2021a, b).

4.8 CONCLUSION

Alzheimer's disease (AD) is the biggest societal challenge being faced mostly in countries dense with elderly population. In the coming days, the population and prevalence of individuals affected with AD are going to increase substantially. There is an enormous gap between the available avenues and modalities of treatment, with a handful of molecules approved by the FDA. Most of all, the treatment caters to the need of alleviating symptoms with attested side effects. There is great scope for the formulation scientist to deliver a novel delivery system to the AD-affected target site without compromising on safety and efficacy. A newer range of therapeutics with biologicals is getting limelight in the clinical trials, much needed to be done with respect to the toxicity elucidation. Early diagnosis, though helps in the management of AD; however, a search for the best therapeutics is still ongoing, and authors are optimistic that in the near future, transformation from paper to prescription will surely happen for AD.

ACKNOWLEDGMENT

The authors are thankful to the Hon. Management of Sandip University, Dean of School of Pharmaceutical Sciences, and all the faculties for their kind support in accomplishing this manuscript.

DECLARATION OF INTEREST

The authors declare no conflict of interest.

KEYWORDS

- **Alzheimer's disease**
- **dementia**
- **gene mutation**
- **genetic mutation**
- **nanoemulsion**
- **nanoparticles**
- **neurons**

REFERENCES

Ahire, E. D., Sonawane, V. N., Surana, K. R., & Talele, G. S., (2021). Drug discovery, drug-likeness screening, and bioavailability: Development of drug-likeness rule for natural products. In: *Applied Pharmaceutical Practice and Nutraceuticals* (pp. 191–208). Apple Academic Press.

Ahire, E. D., Talele, S. G., & Shah, H. S., (2020). Nanoparticles as a promising technology in microbial pharmaceutics. In: *Applied Pharmaceutical Science and Microbiology* (pp. 133–158). Apple Academic Press.

Ahire, E., Thakkar, S., Borade, Y., & Misra, M., (2020). Nanocrystal based orally disintegrating tablets as a tool to improve dissolution rate of Vortioxetine. *Bulletin of Faculty of Pharmacy, Cairo University*, *58*(1, 2), 11–20.

Ahire, E., Thakkar, S., Darshanwad, M., & Misra, M., (2018). Parenteral nanosuspensions: A brief review from solubility enhancement to more novel and specific applications. *Acta Pharmaceutica Sinica B*, *8*(5), 733–755.

Ahmad, J., Akhter, S., Rizwanullah, Md., Khan, M. A., Pigeon, L., Addo, R. T., et al., (2017). Nanotechnology based theranostic approaches in Alzheimer's disease management: Current status, and future perspective. *Current Alzheimer Res.*, *14*(11), 1164–1181.

Alexander, A., & Saraf, S., (2018). Nose-to-brain drug delivery approach: A key to easily accessing the brain for the treatment of Alzheimer's disease. *Neural Regen Res.*, *13*(12), 2102–2104.

Breijyeh, Z., & Karaman, R., (2020). Comprehensive review on Alzheimer's disease: Causes and treatment. *Molecules*, *25*(24), 5789.

Cai, X., Zhang, K., Xie, X., Zhu, X., Feng, J., Jin, Z., et al., (2020). Self-assembly hollow manganese Prussian white nanocapsules attenuate Tau-related neuropathology and cognitive decline. *Biomaterials, 231*, 119678.

Chen, H., Tang, L., Qin, Y., Yin, Y., Tang, J., Tang, W., et al., (2010). Lactoferrin-modified procationic liposomes as a novel drug carrier for brain delivery. *Eur. J. Pharm. Sci., 40*(2), 94–102.

Cunha, S., Forbes, B., Lobo, J. M. S., & Silva, A. C., (2021). Improving drug delivery for Alzheimer's disease through nose-to-brain delivery using nanoemulsions, nanostructured lipid carriers (NLC) and in situ hydrogels. *Int. J. Nanomedicine, 16*, 4373–4390.

Dos, B., Rodrigues, S., Oue, H., Banerjee, A., Kanekiyo, T., & Singh, J., (2018). Dual functionalized liposome-mediated gene delivery across triple co-culture blood-brain barrier model and specific *in vivo* neuronal transfection. *J. Control Release, 286*, 264–278.

Fan, L., Mao, C., Hu, X., Zhang, S., Yang, Z., Hu, Z., et al., (2020). New insights into the pathogenesis of Alzheimer's disease. *Front Neurol., 10*, 1312.

Fonseca-Santos, B., Gremião, M. P. D., & Chorilli, M., (2015). Nanotechnology-based drug delivery systems for the treatment of Alzheimer's disease. *Int J Nanomedicine, 10*, 4981–5003.

Galimberti, D., & Scarpini, E., (2011). Disease-modifying treatments for Alzheimer's disease. *Ther. Adv. Neurol. Disord., 4*(4), 203–216.

Gautam, S. P., Keservani, R. K., Gautam, T., Gupta, A. K., & Sharma, A. K., (2015). An alternative approach for acetylation of amine-terminated polyamidoamine (PAMAM) dendrimer. *ARS Pharmaceutica, 56*(3), 155–159.

Giau, V. V., Bagyinszky, E., An, S. S. A., & Kim, S. Y., (2015). Role of apolipoprotein E in neurodegenerative diseases. *Neuropsychiatr. Dis. Treat., 11*, 1723–1737.

Gordon, B. A., Blazey, T. M., Su, Y., Hari-Raj, A., Dincer, A., Flores, S., et al., (2018). Spatial patterns of neuroimaging biomarker change in individuals from families with autosomal dominant Alzheimer's disease: A longitudinal study. *Lancet Neurol., 17*(3), 241–250.

Hebbar, S., Ravi, G. S., Shetty, A., & Dubey, A., (2020). Chitosan-coated liposomes of *Centella asiatica* extract: An *in vitro* formulation design for oral delivery. *Asian Journal of Pharmaceutics, 14*(4), 677–688.

Jokar, S., Khazaei, S., Behnammanesh, H., Shamloo, A., Erfani, M., & Beiki, D., (2019). Recent advances in the design and applications of amyloid-β peptide aggregation inhibitors for Alzheimer's disease therapy. *Biophys. Rev., 11*, 901–925.

Kennedy, A. M., Brown, J., & Rossor, M., (1994). The genetics of Alzheimer's disease. *Baillieres Clin. Neurol., 3*(2), 217–240.

Keservani, R. K., & Gautam, S. P., (2020). Formulation and evaluation of baclofen liposome vesicles using lecithin. *ARS Pharmaceutica, 61*(3), 175–180.

Keservani, R. K., & Gautam, S. P., (2022). Skeletal muscle relaxant activity of different formulations of span 60 niosomes. *ARS Pharmaceutica, 63*(91), 32–44.

Keservani, R. K., Sharma, A. K., & Kesharwani, R. K., (2018). Nanoparticulate nanocarriers in drug delivery. In: *Nanobiomaterials* (pp. 3–22). Apple Academic Press. ISBN: 9781771885911.

Keservani, R. K., Sharma, A. K., & Kesharwani, R. K., (2019). *Nanocarriers for Brain Targeting: Principles and Applications*. CRC Press. ISBN:9781771887304.

Keservani, R. K., Sharma, A. K., & Ramteke, S., (2010). Novel vesicular approach for topical delivery of baclofen via niosomes. *Lat. Am. J. Pharm., 29*, 1364–1370.

Keskin, A. O., Durmaz, N., Uncu, G., Erzurumluoglu, E., Yıldırım, Z., & Tuncer, N., et al., (2019). Future treatment of Alzheimer's disease. *Geriatric Medicine and Gerontology* (1st edn., Vol. 1, pp. 1–20). Intech open, London.

Kulkarni, P., Rawtani, D., & Barot, T., (2021). Design, development, and *in-vitro/in-vivo* evaluation of intranasally delivered rivastigmine and N-acetyl cysteine loaded bifunctional niosomes for applications in combinative treatment of Alzheimer's disease. *Eur. J. Pharm. Biopharm., 163*, 1–15.

Kumar, A., Chaudhary, R. K., Singh, R., Singh, S. P., Wang, S. Y., Hoe, Z. Y., et al., (2020). Nanotheranostic applications for detection and targeting neurodegenerative diseases. *Front Neurosci., 14*, 305.

Kumar, A., Sidhu, J., Goyal, A., & Tsao, J. W., (2021). *Alzheimer Disease* (pp. 1–27). Treasure Island (FL): StatPearls.

Lee, D., & Minko, T., (2021). Nanotherapeutics for nose-to-brain drug delivery: An approach to bypass the blood-brain barrier. *Pharmaceutics, 13*(12), 2049.

Lu, J., Fu, J., Zhong, Y., Yang, Q., Huang, J., Li, J., et al., (2018). Association between ABCA1 gene polymorphisms and the therapeutic response to donepezil therapy in Han Chinese patients with Alzheimer's disease. *Brain Res Bull., 140*, 1–4.

Marei, H. E., Althani, A., Afifi, N., Hasan, A., Caceci, T., Pozzoli, G., et al., (2021). Generation of gene edited hiPSC from familial Alzheimer's disease patient carrying N141I missense mutation in presenilin 2. *Stem Cell Res, 56*, 102552.

Murphy, S. R., Chang, C. C. Y., Dogbevia, G., Bryleva, E. Y., Bowen, Z., Hasan, M. T., et al., (2013). Acat1 knockdown gene therapy decreases amyloid-β in a mouse model of Alzheimer's disease. *Mol. Ther., 21*(8), 1497–1506.

Owens, L. V., Benedetto, A., Dawson, N., Gaffney, C. J., & Parkin, E. T., (2021). Gene therapy-mediated enhancement of protective protein expression for the treatment of Alzheimer's disease. *Brain Res., 1753*, 147264.

Pimenova, A. A., & Goate, A. M., (2020). Novel presenilin 1 and 2 double knock-out cell line for *in vitro* validation of PSEN1 and PSEN2 mutations. *Neurobiol. Dis., 138*, 104785.

Richter, L., Munter, L. M., Ness, J., Hildebrand, P. W., Dasari, M., Unterreitmeier, S., et al., (2010). Amyloid beta 42 peptide (Abeta42)-lowering compounds directly bind to Abeta and interfere with amyloid precursor protein (APP) transmembrane dimerization. *Proc. Natl. Acad. Sci. U S A, 107*(33), 14597–14602.

Robinson, M., Lee, B. Y., & Hane, F. T., (2017). Recent progress in Alzheimer's disease research, part 2: Genetics and epidemiology. *J. Alzheimers Dis., 57*(2), 317–330.

Rompicherla, S. K. L., Arumugam, K., Bojja, S. L., Kumar, N., & Rao, C. M., (2021). Pharmacokinetic and pharmacodynamic evaluation of nasal liposome and nanoparticle-based rivastigmine formulations in acute and chronic models of Alzheimer's disease. *Naunyn-Schmiedebergs Arch of Pharmacol., 394*(8), 1737–1755.

Salimi, A., Gobadian, H., & Sharif, M. B., (2021). Dermal pharmacokinetics of rivastigmine-loaded liposomes: An ex vivo–*in vivo* correlation study. *J. Liposome Res., 31*(3), 246–254.

Saraswathi, T. S., & Mothilal, M., (2021). Development of rivastigmine-loaded self-assembled nanostructures of nonionic surfactants for brain delivery. *International Journal of Applied Pharmaceutics, 13*(5), 205–215.

Setya, S., Madaan, T., Razdan, B. K., Farswan, M., & Talegaonkar, S., (2019). Design and development of novel transdermal nanoemulgel for Alzheimer's disease: Pharmacokinetic, pharmacodynamic and biochemical investigations. *Curr. Drug Deliv., 16*(10), 902–912.

Silva, G. A., (2008). Nanotechnology approaches to crossing the blood-brain barrier and drug delivery to the CNS. *BMC Neurosci.*, *9*(Suppl 3), S4.

Singh, V. K., & Keservani, R. K., (2017). Application of nanoparticles as a drug delivery system. In: *Novel Approaches for Drug Delivery* (pp. 364–389). IGI Global. ISBN: 9781522507512.

Vaz, M., & Silvestre, S., (2020). Alzheimer's disease: Recent treatment strategies. *Eur. J. Pharmacol.*, *887*, 173554.

What Is Alzheimer's Disease? National Institute on Aging: https://www.nia.nih.gov/health/what-alzheimers-disease (accessed on 6 December 2023).

Xia, D., Watanabe, H., Wu, B., Lee, S. H., Li, Y., Tsvetkov, E., et al., (2015). Presenilin-1 knockin mice reveal loss-of-function mechanism for familial Alzheimer's disease. *Neuron.*, *85*(5), 967–981.

Xu, Y., Zhao, M., Zhou, D., Zheng, T., & Zhang, H., (2021). The application of multifunctional nanomaterials in Alzheimer's disease: A potential theranostics strategy. *Biomed Pharmacother.*, *137*, 111360.

Yang, Z. Z., Zhang, Y. Q., Wang, Z. Z., Wu, K., Lou, J. N., & Qi, X. R., (2013). Enhanced brain distribution and pharmacodynamics of rivastigmine by liposomes following intranasal administration. *Int. J. Pharm.*, *452*(1, 2), 344–354.

Yiannopoulou, K. G., & Papageorgiou, S. G., (2013). Current and future treatments for Alzheimer's disease. *Ther. Adv. Neurol. Disord.*, *6*(1), 19–33.

Zhang, C., Gu, Z., Shen, L., Liu, X., & Lin, H., (2018). A dual targeting drug delivery system for penetrating blood brain barrier and selectively delivering siRNA to neurons for Alzheimer's disease treatment. *Curr Pharm Biotechnol.*, *18*(14), 1124–1131.

Zhang, L., Liu, X. G., Liu, D. Q., Yu, X. L., Zhang, L. X., Zhu, J., et al., (2020). A conditionally releasable "do not eat me" CD47 signal facilitates microglia-targeted drug delivery for the treatment of Alzheimer's disease. *Adv. Funct. Mater.*, *30*(24), 1910691.

Zhou, Z. D., Chan, C. H. S., Ma, Q. H., Xu, X. H., Xiao, Z. C., & Tan, E. K., (2011). The roles of amyloid precursor protein (APP) in neurogenesis, implications to pathogenesis and therapy of Alzheimer disease. *Cell Adh. Migr.*, *5*(4), 280–292.

CHAPTER 5

Parkinson's Targeting: A Novel Drug Delivery Approach

SAPANA P. AHIRRAO, DEEPAK S. BHAMBERE, SANJAY J. KSHIRSAGAR, and JAYESH D. KADAM

MET's Institute of Pharmacy, Bhujbal Knowledge City, Adgaon, Nashik, Maharashtra, India

ABSTRACT

Treatment of chronic neurodegenerative disease, Parkinson's disease (PD), is one of the major challenges to scientists due to the presence of the blood-brain barrier (BBB). Considering the limited permeation of different anti-Parkinson drugs across the brain and the associated motor and non-motor complications of the disease, scientists have worked on novel drug delivery systems (NDDSs) to develop an effective, brain-targeted therapeutic approach in a variety of central nervous system (CNS) diseases. Conventional DDSs used for PD are associated with limited brain targeting and poor bioavailability. Recent advancements in drug delivery technologies using microparticles (MPs), nanoparticles (NPs), and hydrogels (HGs) have achieved brain targeting in the treatment of various CNS diseases. In this review, the applications of hydrogel, nanoparticles, liposomes, and micelle-based DDSs, as well as the polymers used in the development of these novel therapeutic approaches and their significant potential in the treatment of Parkinson's disease, are discussed in detail. Considering the nose-to-brain transport of various drugs as an effective means of brain targeting, researchers can design an effective therapeutic approach in the treatment of chronic CNS diseases using these novel drug delivery approaches.

Novel Drug Delivery Systems in the Management of Chronic Diseases.
Raj K. Keservani, Umesh D. Laddha, Eknath D. Ahire, and Sanjay J. Kshirsagar (Eds.)

5.1 INTRODUCTION

Parkinson's disease (PD) is the most common neurodegenerative disease after Alzheimer's disease (AD). PD is a neurodegenerative disorder of the central nervous system (CNS), which impairs motor skills, cognitive processes, and other functions. Clinical symptoms of PD include both motor and non-motor symptoms. Novel neurotherapeutics have been found but remain inactive due to limited permeation across the blood-brain barrier (BBB). The brain is one of the least available organs for drug delivery due to the presence of the BBB (Keservani et al., 2019). The BBB limits the proliferation of large or hydrophilic molecules in the treatment of most CNS and psychiatric disorders. Benjamí Oller-Salvia et al. studied the structure of the BBB and investigated the role of molecular vectors known as BBB shuttles for direct drug targeting as elucidated in detail in Figure 5.1 (Abulrob et al., 2005; Anden et al., 1965).

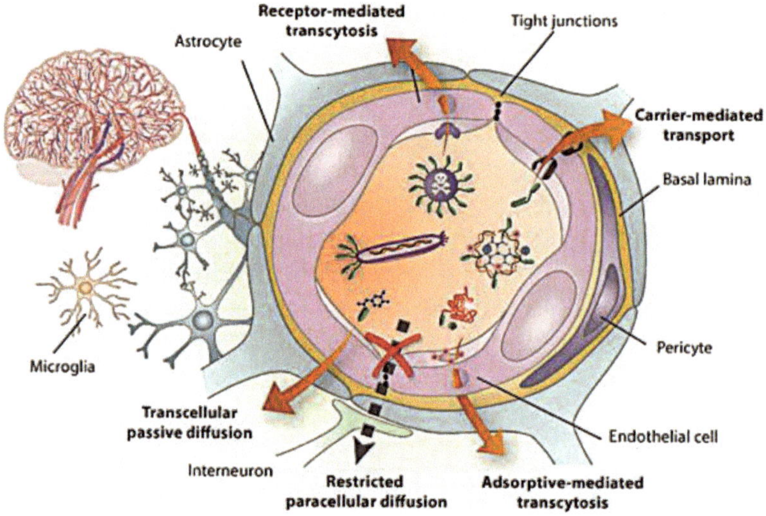

FIGURE 5.1 BBB shuttle-mediated transport across the BBB.

Source: Reprinted with permission from Oller-Salvia et al. 2016. © Royal Society of Chemistry.

Parkinson's disease; one of the chronic neurodegenerative diseases affecting various motor and non-motor functions and brain targeting of many antiparkinsonian therapies is a promising challenge to the scientist. Delivery of drug to brain remains a barrier to the development of neurological therapies that lead to long-term drug development and high failure rates in clinical trials (Antonini, et al., 2018).

Many anti-Parkinson's drugs are available such as benzatropine, ropinirole, tolcapone, selegiline, amantadine, levodopa. Levodopa (also known as L-dopa) is a commonly prescribed anti-Parkinson's containing carbidopa to effectively control the symptoms. Only 5–10% of oral L-DOPA is able to cross the BBB and the remainder is normally digested in dopamine by enzymes, namely decarboxylases, causing a variety of side effects, such as nausea, dyskinesia, and stiffness. In order to reduce the side effects of the drug such as L-DOPA associated with systemic overload, advanced delivery systems should be developed to treat PD (Attilio et al., 2021).

5.2 NOVEL DRUG DELIVERY SYSTEM (NDDS) FOR TARGETING PARKINSON'S

By considering the low efficacy of therapeutic drugs due to low brain bioavailability and associated side effects, nanotechnology appears promising in developing DDSs of targeting and controlled-release characteristics. Many nanoparticle-based strategies, including liposomes, nanospheres, and cationic albumin nanoparticles, are under development to deliver drugs across the BBB. Targeting increases efficacy, reduces side effects, and reduces systemic drug exposure. Organized nanoparticle structures allow the incorporation of various targeting moieties to enhance drug delivery to target sites, reduce off-target toxicity, and facilitate cellular uptake of therapeutics. Administration using nano-sized carrier is an effective way to solve the challenging treatment of neurodegenerative diseases (NDs) and direct CNS targeting in various CNS disorders (Banks, 2015).

5.3 CLASSIFICATION OF NOVEL DRUG DELIVERY SYSTEMS (NDDSS)

Novel drug delivery system (NDDS) (Keservani et al., 2017) is classified in different ways by various researchers based on parameters like particle size, surface charge, type of polymers used, and physicochemical composition. Among the DDSs, microparticles (MPs), nanoparticles (NPs), and hydrogels (HGs) appear to be the most effective in providing neuroprotection compared to liposomes and micelles. MPs and NPs are particulate carrier systems in the micrometer and nanometer size range. MPs are generally used for long-term drug delivery, while NPs are commonly used as small molecule carriers for targeted and intracellular delivery. On the other hand, HGs are three-dimensional polymer networks that absorb a large amount of water, which becomes their main component (Benjamí Oller-Salvia et al., 2016) (Figure 5.2).

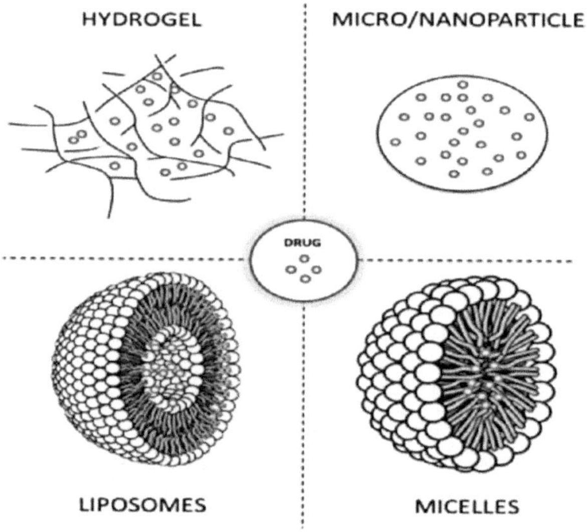

FIGURE 5.2 Novel drug delivery systems for Parkinson's disease treatment.
Source: Reprinted from Del Rey et al., 2018. https://creativecommons.org/licenses/by/4.0/

5.3.1 HYDROGELS (HGS)

Hydrogels (HGs) are polymeric substances that are connected in physical or chemical ways. High water (usually 70–99%) provides physical similarity to tissues and can give the HGs excellent biocompatibility and the capability to easily encapsulate hydrophilic drugs. HGs vary in size, composition, and performance, and these combined factors determine how they are used in drug delivery. HGs have the beneficial properties of being able to store large amounts of water within their structure without melting. This high-water content makes them more compatible with living tissue but unfortunately allows low-molecular weight substances to disperse more quickly. Cross-linking agents can be used to reduce these distribution losses and provide structural strength, but this can increase the irritation of hydrogel collisions with surrounding tissues. The minimum number of crosslinks needed to form a three-dimensional matrix is approximately 1.5 crosslinks per chain, and this yields the maximum expansion possible without separation of the chains into a true solution (Bergman et al., 2015).

5.3.1.1 CLASSIFICATION OF HYDROGEL DELIVERY SYSTEMS

Different types of hydrogels based on various parameters are listed in Table 5.1.

TABLE 5.1 Types of Hydrogels

Classification Criteria	Types of Hydrogels
Based on source	Natural and synthetic origins
According to polymeric composition	Homopolymeric hydrogels
	Copolymeric hydrogels
	Multipolymer interpenetrating polymeric hydrogel
Based on physicochemical composition	Amorphous
	Crystalline
Based on network electrical charge	• Non-ionic • Ionic • Amphoteric electrolyte • Zwitterion

NDs, including PD and AD, seriously affect the health and quality of life of middle-aged and elderly people. Implantation of therapeutic drugs or functional cells into HGs for efficient delivery into organisms is a modern therapeutic option for the treatment of ND. Clinically, L-dopa is the most effective drug for the treatment of PD, but it has a short half-life, poor bioavailability in vivo, and high doses, and long-term administration will lead to bioresistance, motor complications, movement disorders, and other serious side effects. The best results of L-dopa therapy were reported using self-assembling HGs to deliver L-dopa for the treatment of PD (Wang et al., 2010; Bertler et al., 1959).

5.3.1.2 HYDROGEL AS A CARRIER IN TREATMENT OF PARKINSON'S DISEASE (PD)

Hydrogel is considered as one of the effective carriers for various NDDS in the treatment of NDs. Various therapeutic agents and stem cell-based transplantation have been successfully targeted in the CNS region using different hydrogel-based DDSs; some of the approaches are listed below (Blesa et al., 2012):

- Functional cell delivery;
- Brain tumor therapy;

- Hydrogels constructed with therapeutic agents;
- Therapeutic agents;
- Hydrogel-based tissue engineering approaches.

Advanced immunoregulatory strategies including recombinant proteins, immune suppression, gene therapy or cell therapy cannot be easily targeted by systemic administration because of damaged tissues. However, CNS targeting using hydrogel-based approach reported by various researchers (Qing et al., 2007).

5.3.2 NANOPARTICLES

Polymeric nanoparticles (PNPs) are colloidal drug carriers, composed of drugs encapsulated either in solid-state or solution or adsorbed to polymers or noncovalently or chemically attached to the surface. PNPs are particles within a range of size from 1 to 1,000 nm and can be loaded with active compounds bonded internally or labeled over a polymeric core. The term "nanoparticle" represents both nanocapsules and the nanosphere, which are classified by morphological structure. PNP is stable, has a high level of drug loading, and provides a controlled release of the drug from nanoparticles (Keservani et al., 2018; Sharma et al., 2018) and the surface can be easily modified by attaching the appropriate ligands allowing them to deliver more drugs to the desired area of the body with greater efficiency. Therefore, nanoparticle DDS possess great potential for CNS targeting of various drugs. In order to overcome the pharmacokinetic limitations of conventional therapies and have specific applications such as drug protection from damage, continuous drug release, permeation across BBB and also deliver drug to specific cells to target particular intracellular pathways, nanotechnology-based formulation became a specific and effective drug-delivery system in the treatment of various complex CNS disorders (Ahire et al., 2020a, b, Talele et al., 2021).

5.3.2.1 POLYMERS USED IN NANOPARTICLE FORMULATION

Various researchers have reported the use of different natural and synthetic polymers in nanoparticle formulation; some of these are listed in Table 5.2.

TABLE 5.2 Different Natural and Synthetic Polymers in Nanoparticle Formulation

Synthetic polymers used in nanoparticle formulation.	Poly(alkyl cyanoacrylate)
	Poly(lactic-co-glycolic acid)
	Poly-ϵ-caprolactone
	Polyamidoamine dendrimers
Natural polymers used nanoparticles formulation.	Chitosan
	Alginate
	Pectin
	Sodium Alginate

5.3.2.2 BIODEGRADABLE AND NON-BIODEGRADABLE POLYMERS

Polymeric nanoparticles (PNPs); an effective DDS in the field of nanomedicine that is extensively investigated by various researchers in the treatment of cancer, neurodegenerative disorders, and cardiovascular diseases. Various biodegradable and non-biodegradable polymers used in the formulation of nanoparticle formulations exhibiting variation in pharmacokinetic profile of entrapped drug, toxicity, and overall performance of formulation (Bonifati, 2010).

5.3.2.3 NANOPARTICLES USING NONBIODEGRADABLE POLYMERS

Initial uses of PNPs for nano-based applications were based on non-biodegradable polymers. For non-biodegradable nanoparticles, systems need to be designed such that particles exhibit rapid and efficient clearance through carriers such as feces or urine, do not accumulate or distribute in tissues at a toxic level because they cannot be easily degraded and excreted or are physically removed (Ahirrao et al., 2022).

5.3.2.4 NANOPARTICLES USING BIODEGRADABLE POLYMERS

Biodegradable polymeric particles became of interest owing to reduced toxicity concerns, ability to influence specific drug release kinetic patterns, and increased. Biodegradable polymers include PLA, PLGA, poly(ε-caprolactone) (PCL), and poly(amino acids) in addition to natural polymers (Braak et al., 2003). Nanoparticles are classically divided by their structural composition, mentioned in Table 5.3.

TABLE 5.3 Types of Nanoparticles – Examples and Their Applications

General Class	Examples	Applications
Polymeric nanoparticles (Jonathan Baskin et al.)	• Polymeric nanosphere • Polymeric nanocapsules • Polymeric micelle • Dendrimer	Circumvent first-pass metabolism when absorbed via the lymphatic system encapsulate drugs and provide sustained protection from degradation in the circulation.
Inorganic nanoparticles (Paul et al.)	Gold, magnetic iron oxide, synthetic allotropes of carbon, carbon nanotubes.	Non-toxic, hydrophilic, biocompatible, and highly stable compared to organic material have electrochemical properties that are size and shape dependent.

Diagrammatic representation of different nanoparticle classes mentioned in Figure 5.3.

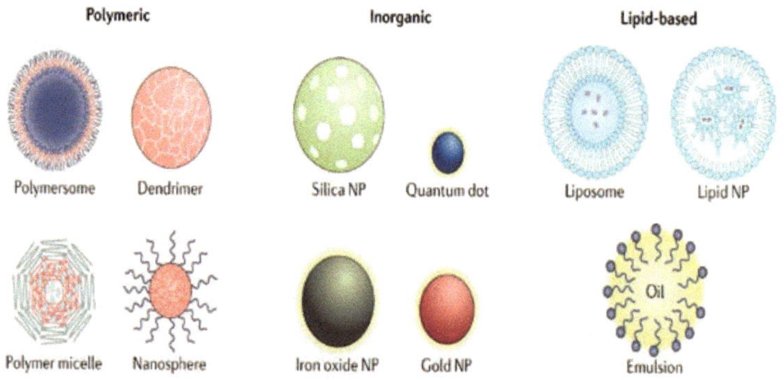

FIGURE 5.3 Various nanoparticle drug delivery system.

Source: Reprinted from Yang et al., 2021. https://creativecommons.org/licenses/by/4.0/

5.3.2.5 POLYMERIC NANOPARTICLES (PNPS)

Polymeric NPs can be synthesized from natural or synthetic materials, just as monomers or preformed polymers are synthesized using various techniques such as emulsification (solvent displacement or diffusion), nanoprecipitation, ionic gelation, and microfluidics, all leading to different end products. The most common forms of polymeric NPs are nanocapsules (cavities surrounded by a polymeric membrane or shell) and nanospheres (solid matrix systems) (Ahire et al., 2018). Within these two main categories, NPs are further divided into subclasses such as polymersomes, micelles, and dendrimers. Polymeric Micelles, which are also generally responsive block copolymers, synthesize themselves to form a hydrophilic conventional nanosphere and hydrophobic coating that works to protect physicochemical properties of drugs and improve systemic circulation time. Polymeric micelles can load a variety of therapeutic drugs, including small molecules and proteins. Dendrimers (Gautam et al., 2015) possess complex three-dimensional structures in which mass, size, shape, texture, and surface chemistry can be highly controlled. Charged polymers can be used to make non-dendrimer NPs as well. Polyelectrolytes are just one example: these polymers have a replicating electrolyte group, which gives them different pH strength. Polyelectrolytes are incorporated into many NP structures to improve structures such as bioavailability and mucosal transport (Bras et al., 2012; Brockmann et al., 2013; Caffrey et al., 2007; Cardo et al., 2012).

5.3.2.6 INORGANIC NPS

Inorganic NPs are well designed and can be arranged in a variety of sizes, structures, and geometries. Inorganic NPs have unique electrical, magnetic, and optical properties, due to the properties of the basic elements themselves. Some of the nanotechnology-based platforms used to control PD are mentioned in Figure 5.4 (Chaudhuri et al., 2006; Chung et al., 2018).

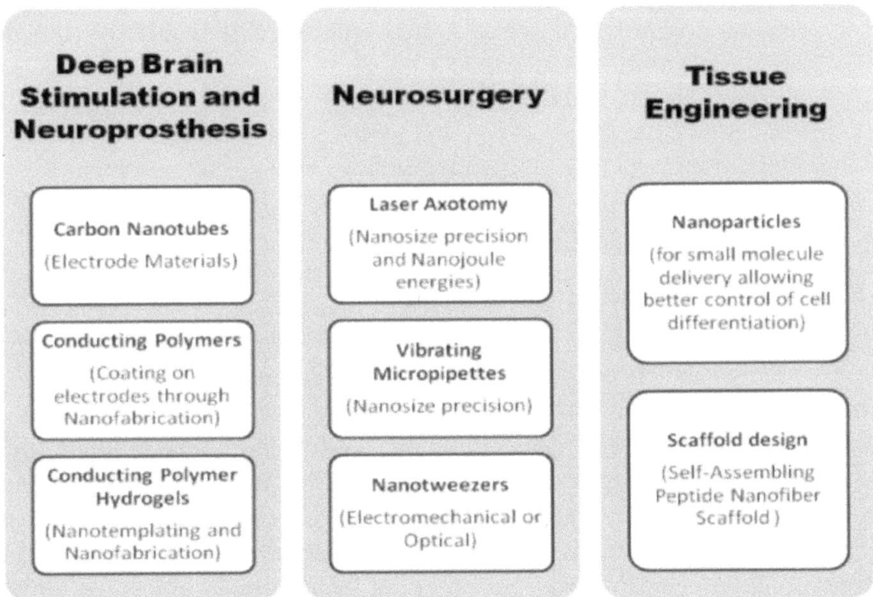

FIGURE 5.4 Nanotechnology-based platforms for the management of Parkinson's disease.

Source: Reprinted from Adhikary et al., 2015. https://creativecommons.org/licenses/by/4.0/

Functional elements of nanoparticles investigated in preclinical trials of drug delivery in PD to improve the pharmacokinetic properties of conventional treatments. Severe symptoms of PD are controlled by nanotechnology-based platforms by some of the following mechanisms (Cotzias, 1968):

1. **Improving the Release Profile of Drugs:** Release profile of dopamine and then other dopaminergic medications, such as levodopa, dopamine agonists and monoamine oxidase B inhibitors reported using preclinical animal models.
2. **Immunoevasion:** Circulation times of drugs improved by using polyethylene glycol (PEG) nanoparticles resulting in immunoevasion by steric hindrance.
3. **Mucosal Entry:** Direct targeting to the brain via olfactory and trigeminal neural pathways by circumventing the BBB by using mucoadhesive polymers, such as chitosan or by conjugating nanoparticles with ligands capable of binding to the nasoepithelial surface, such as lectins.
4. **Blood-Brain-Barrier (BBB) Entry:** Specific delivery of drug to brain remains a barrier to the development of CNS therapies that lead to long-term drug development and high failure rates in clinical trials. Due to the small size and lipophilicity of lipid-based vehicles can enter the BBB and the BBB receptors are directed using the local activation of nanoparticles with ligands (Crucho & Barros, 2017).

5.3.3 LIPOSOMES

Liposomes are currently the most widely used system for the delivery of targeted drugs to the brain. Liposomes are nano-sized vesicles composed of one or more bilayers of phospholipid blocking an aqueous droplet. Liposomes (Keservani et al., 2020) become a useful DDS due to its attractive features such as small size, ability to load both hydrophilic or lipophilic drugs and reduced toxicity compared to other conventional DDSs. The components of lipid are usually phosphatidylcholine (PC), which is found in egg or soybeans lecithin. Given its uniquely closed structure, liposomes can trap hydrophilic agents in aqueous core and hydrophobic molecules in the lipid membrane (Thombre et al., 2022).

5.3.3.1 ADVANTAGES OF LIPOSOMES

Liposomes overcome the shortcomings of traditional formulations; some of them are listed below:

- Poor aqueous solubility;
- Low bioavailability;
- Nonspecific distribution in the body.

5.3.3.2 RECENT ADVANCEMENT IN LIPOSOMAL-BASED STRATEGIES

Liposomes can be used as an active biological carrier and widely used as a DDS to improve drug efficiency and eliminate drug-related toxins or unwanted side effects. Now there are days left for various local operations to have a direct site action and to improve the fullness of the many treatments throughout the BBB. Some of the advanced types of liposome-based systems are listed in subsection (Daneman & Prat, 2015; Del et al., 2018).

5.3.3.2.1 Long-Circulating Liposome

Liposomes are considered one of the important nanotechnology-based dosage forms, which have been investigated by various researchers. Liposomes are widely used for drug delivery and drug targeting by incorporating a variety of therapeutic drugs, biologicals, and new biotechnology-derived agents. Though conventional liposomes possess a wide variety of applications like being biocompatible, encapsulating both hydrophilic and hydrophobic drugs, and preventing degradation of drugs compared to free drugs, they are associated with some limitations. Conventional liposomes have a short blood circulation time due to rapid recognition and removal from the blood by cells of the mononuclear phagocytic system (MPS), particularly those in the liver and spleen, which often eliminates the intended beneficial effects and can pose a considerable risk of toxicity to these cells. Thus, in the treatment of CNS disorders, there is a strong need for novel liposome formulations that should possess site-specific action and prolong circulation time in the bloodstream (Drin et al., 2003; Elezaby et al., 2003).

Significant research findings of liposome-based formulations in CNS targeting have been studied by various researchers listed in Table 5.4.

TABLE 5.4 Liposome-based Formulation in CNS Targeting

Formulation Details	Research Finding	Researchers
Docetaxel (DCX) and glutathione reduced gold nanoparticle.	Site specific delivery of DCX for brain targeted cancer therapy.	Roshan et al. (2021)
Trojan horse liposomes.	For plasmid DNA gene therapy of the brain.	William (2020)
Geniposide-loaded liposomes	Improved brain targeting of geniposide for cerebral ischemia reperfusion injury (CIRI) therapy.	Jinyan et al. (2021)
Grape pomace extracts loaded liposomes.	Promising results in *in-vitro* model of Parkinson's disease due to antioxidant properties of extract.	Attilio et al. (2021)

5.3.4 MICELLES

Polymeric micelles are self-assembled, nano-sized (5–100 nm), colloidal particles consisting of amphiphilic copolymers which form spontaneously in an aqueous environment. Above the so-called 'critical micelle concentration' (CMC), the single chains of polymers associate to minimize the contact of the hydrophobic segment with the aqueous media, leading to the formation of a core-shell micellar structure. Polymeric micelles are among the most promising delivery systems in nanomedicine. Polymeric micelles are spherically shaped NPs made of hydrophobic and hydrophilic units in aqueous solutions. Hydrophobic drugs/compounds that can slowly penetrate the entire BBB can be incorporated into micellar nanoparticles to improve therapeutic efficacy. Block-copolymer micelles are spherical supermolecular assemblies of amphiphilic or copolymers charged separately in a wet environment. The size of most polymeric micelles is within a range of 10–100 nm, which ensures that it can be selectively rotated in the target area using enhanced accessibility and final impact. Micellar nanocarriers have emerged as a flexible and powerful nanotherapeutic compound (Enas, 2015; Gabathuler, 2010).

Polymeric micelles have been used for drug delivery applications because of their unique properties. These properties include variability in the selection of hydrophobic regions, nanoscale magnitude, and increased solubility and stability of anticancer agents. Controlling the ratio of the number of hydrophilic/hydrophobic units and selecting different hydrophobic units can adjust the physicochemical properties of Polymeric micelles (Guarnieri et al., 2013).

As per literature survey, in the treatment of various chronic diseases like brain tumor, CNS disorders and NDs, drug loaded NPs successfully achieved the CNS targeting due to its nanometer size range. By considering the solubility of drug, nature of disease to be treated and drug targeting, suitable formulation method will be selected to nanoparticles (Gustavsson et al., 2011).

5.3.4.1 PHARMACEUTICAL APPLICATIONS OF POLYMERIC MICELLES

Polymeric micelles are mainly designed to refine three fundamental parameters in drug performance:

1. **Solubilization of Hydrophobic or Water-Soluble Drugs**: Recent studies involving polymeric micelles have focused on the effect of micelles on solubilization enhancement, dissolution rate, increased aqueous solubility, and oral bioavailability of poorly water-soluble drugs. The results showed that micelles promise to increase the water solubility of many drugs by providing a hydrophobic reservoir where drugs can differentiate by chemical interactions, physical closure, or ionic interactions, depending on the chemical structure and physicochemical properties of the polymer.
2. **Controlled or Continuous Release of the Drug:** The ability of micelles to show sustained release (SR) requires specific properties: micelles must be stable to dilution due to their high thermodynamic stability (low CMC) or to a high kinetic stability based on highly viscous, low chain mobility core properties.
3. **Identifying a Specific Type of Cell or Organ:** Active targeting is aimed at improving drug delivery and minimizing side effects by increasing the targeted intake of interest.

   There are two common ways to identify active targeting:

   i. Utilizing biologically specific interactions; and
   ii. Taking advantage of locally applied signals to me (external or internal);

Some of the active targeting achieved through polymeric micelles as listed.

1. **Molecular Targeting of Polymeric Micelles:** Ligand-Mediated Targeting: Since some diseases, especially cancer, cause excessive stress on cell surface receptors, attaching certain ligands to terminals exposed to hydrophilic block water can be used to produce active micelles targeted. The latter is expected to improve the accumulation of nanocarriers in interested tissues and at the same time increase selective absorption by receptor-mediated endocytosis. Ligands are widely used by monoclonal antibodies (mAbs), carbohydrate moieties, aptamers (Apts), and peptides (Jankovic, 2002).
2. **Carbohydrate-Based Identification:** Carbohydrate molecules have also been used to activate micelles. This helps to identify asialoglycoprotein receptor (ASGPR) excessively expressed in hepatocellular carcinoma (HCC) for imaging or controlled release of loaded drugs in high dosage. Micelle formulation with galactose moieties on the face was prepared with poly(hydroxyethyl) aspartamide ethylene diamine-poly-lactic acid copolymer.
3. **mAb-based Targeting:** Among all the specific ligands, the immune system has provided extensive opportunities in terms of target diversity and communication specificity. Micelles targeted antibodies, immunomicelles, can be prepared by attaching chemical antibodies or antibody fragments to a free surface exposed to water hydrophilic block of polymer that forms micelle. Torchilin et al. developed phospholipid-based tumor-based immunomicelles by chemical interactions of mAbs 2C5 or 2G4 in the micellar region. PTX-loaded PTX immunomicelles showed higher drug concentrations in Lewis lung carcinoma tumors in mice that resulted in higher therapeutic efficacy compared with free PTX or PTX loaded in empty micelles (Jesus et al., 2019).
4. **Permeation of Nanoparticles Across BBB:** Treatment of NDs such as PD, Huntington's, and AD is a major challenge in the medical field as these diseases include irreversible neuronal cell death. Drug transport across the BBB is a major problem as it is highly heterogeneous and is able to penetrate in the same way as active molecules. The transported PNPs cross the endothelium carrier in the form of a network, receptor-mediated, and adsorptive-mediated shown in Figure 5.5.

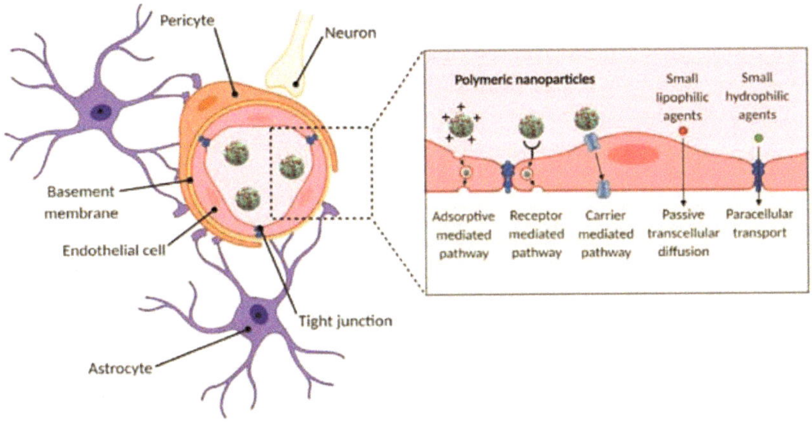

FIGURE 5.5 Transport of nanoparticles across BBB.

Source: Reprinted from Zhang et al., 2021. https://creativecommons.org/licenses/by/4.0/

5. **Structure of BBB Healthy and Disease Structure:** Continuous endothelial membrane, BBB; separating circulating blood from the brain and protecting the brain from external damage. In neurodegenerative disease, BBB depletion causes toxic molecules in the blood, cells, and microbial agents to enter the brain and be associated with inflammatory and immune responses. Structure of a stable BBB and healthy BBB shown in Figure 5.6.

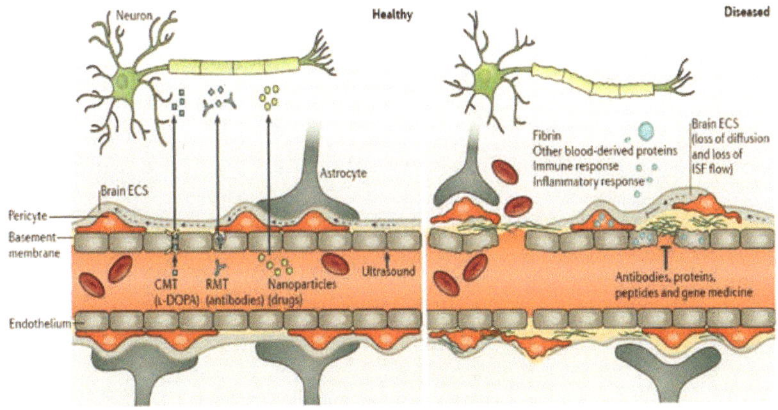

FIGURE 5.6 Structure of BBB in healthy and diseases state.

Source: Reprinted with permission from Sweeney et al., 2018. © Springer Nature.

In a healthy blood-brain barrier (BBB), BBB-breaking strategies and delivery of neuropharmaceuticals to the brain are dependent on carrier-mediated transport (CMT), receptor-mediated transcytosis (RMT), nanoparticles, and/or temporary opening of the BBB (e.g., using concentrated ultrasound).

Under pathological conditions (right), disturbed BBB causes debris and cells taken from the blood to accumulate in the perivascular enlarged areas. This accumulation inhibits the normal distribution of molecules throughout the CNS by concentration-gradient-driven propulsion throughout brain extracellular spaces (ECSs) and disrupts the formation of internal fluid (ISF) and ISF flow, which inhibits the immune system, medicinal, proteins, peptides, genetics, medications, and other medications to successfully achieve their neuronal goals (Sweeney et al., 2018).

5.3.5 CNS TARGETING USING NPS-BASED COMPOUNDS

Challenges of drug delivery caused by BBB and considering the increasing number of patients with neurodegenerative disorders, effective CNS regulation is required to control the complications of this disease. Nanotechnology-based DDSs offer a variety of benefits to overcome pharmacokinetic limitations of conventional therapies by providing effective CNS guidance. Nanoparticle delivery systems pass through the BBB and reach neurons in a variety of ways, such as paracellular pathway, transcellular lipophilic pathway, cell mediated transport, and receptor-mediated endocytosis; details shown in Figure 5.7. The nanotechnology-based approach provides controlled drug release, reduces drug toxicity, and assists in the transmission of neurotransmitters through the BBB (Jinyan et al., 2021).

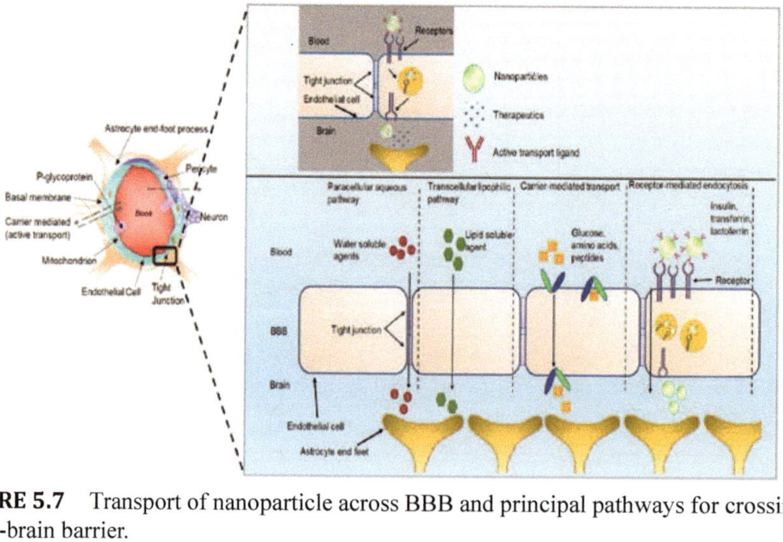

FIGURE 5.7 Transport of nanoparticle across BBB and principal pathways for crossing the blood-brain barrier.

Source: Reprinted from Duwa et al., 2021. © 2021 Springer Nature.

5.3.5.1 GENERAL FEATURES OF NANOPARTICLES FOR BRAIN TARGETING

Development of an optimized drug delivery for brain targeting is major challenge to formulation scientist due to complex barrier; BBB. Various factors are taken into consideration while developing the optimized nanotechnology-based formulation for brain targeting which are listed below:

1. **Drug-Related Factors:** Solubility, lipophilicity, permeability, molecular size, molecular weight, protein binding should be taken into consideration during the formulation and development of nanotechnology-based formulations for brain targeting. Various research studies reported that molecules with high lipophilicity and low molecular size can passively diffuse down their concentration gradient across the BBB. Small molecules with a molecular mass under 400 Da undergo significant free diffusion through the BBB. Lipophilicity is measured in terms of log P, the higher the log P, the higher the BBB permeability (Asha et al., 2020).

   It has been reported that all large molecule drugs having a molecular weight more than 1 Kilo Daltons, including recombinant proteins, enzymes, or gene-based medicines, do not pass through the BBB, and approximately 98% of small molecules minimally cross the BBB. Therefore, for a therapeutic/molecule to enter into the brain, it must cross the BBB by binding with specific protein transporters and/or receptors located at brain capillary endothelium cells.

2. **Disease-Related Items:** The nature of the disease and the severity play an important role in the development of nanotechnology-based construction in the direction of the brain. Researchers who successfully use a variety of nano-drug delivery systems target a variety of CNS-related diseases/disorders (Alzheimer's disease, Parkinson's disease, brain cancer). The nature of the disease should be considered as the severity of the brain disease/disorder begins from the first stage with mild or absent symptoms to the later stage with severe symptoms. The most important obstacle to successfully managing CNS diseases is proper diagnosis; as the onset of symptoms does not change and the lack of appropriate CNS diagnostic tools or symptoms (Jonathan et al., 2020).

3. **Patient-Related Features:** Age, gender, weight, and the immune system of the patient play an important role as all of these factors affect the parameters of different pharmacokinetics and pharmacodynamics.

5.3.5.2 RECENT ADVANCES IN PARKINSON'S TREATMENTS

Control of both motor and non-motor symptoms of Parkinson's is the main challenge to ensure successful treatment. The etiology of PD is less well known. Most patients are classified as cases of Parkinson's idiopathic syndrome, that is, that occur 'automatically' or for an unknown reason. Traumatic disorders in PD have been extensively investigated, leading to better diagnosis and the establishment of proven standardized measurements and treatment options.

5.3.6 DRUG DELIVERY SYSTEMS (DDSS) FOR NEUROTROPHIC FACTOR THERAPY

Neurotrophic traits, as well as glial cell-derived neurotrophic factor (GDNF) in particular, have been considered one of the most promising molecules for PD. Neuroprotection has been detected after intravenous administration of GDNF nanoformulations by crossing the BBB through receptor-mediated delivery. From a recent study, it is noted that the detection of neurotrophic factor in the brain through nasal congestion is enhanced when GDNF is covered in liposomal formation. Studies conducted using animal models of PD (mice and monkeys) revealed improvements in motor function and restoration of dopaminergic function by better regulation of specific brain regions.

5.3.7 DRUG DELIVERY SYSTEMS (DDSS) FOR STEM CELL THERAPY

Many promising embryonic stem cells (ESCs) and pluripotent stem cells for neural grafting have been considered as possible treatment options for PD. Stem cell therapy may provide significant symptomatic benefit for those with severe motor symptoms, but it will not be of little benefit to patients when the main problems are not motor.

It may be possible that stem cell-based regenerative therapy is employed to restore the function of already lost dopaminergic neurons, while new pathogens may be used to prevent further neuronal death. Stem cell-based treatment appears to be the most promising method for the development of effective regenerative therapy that can be used in a large number of patients. A few experimental drugs based on PD are listed in Table 5.5.

TABLE 5.5 Experimental Cell-based Treatments for Parkinson's Disease

Cell Types Trialed in Humans	Examples
Autografts	Adrenal medullary cells, carotid body cells, mesenchymal stem cells.
Allografts	Human fetal ventral mesencephalon, retinal pigment epithelial cells.
Xenografts	Porcine ventral mesencephalon.

5.4 CONCLUSION

By considering the severe symptoms of PD, lack of complete cure, and permeation of drugs across the BBB, significant contributions have been made by various researchers in the development of optimized nanotechnology-based formulations for the treatment of PD. PNPs are currently being engineered using various natural and synthetic polymers through suitable fabrication techniques. However, the development of an optimized nanoformulation for CNS targeting is the most challenging task for researchers, considering the physicochemical properties of the active pharmaceutical ingredient, compatibility of the selected polymer, IVIVC (in vitro, in vivo correlation), and significant challenges in preclinical and clinical studies.

ACKNOWLEDGMENT

Authors want to thank the METs, Institute of Pharmacy, Nashik, Maharashtra 422003, India.

CONFLICT OF INTEREST

Nil

KEYWORDS

- **drug delivery**
- **hydrogels**
- **liposomes**
- **micelles**
- **nanoparticles**
- **neurotrophic factor therapy**
- **Parkinson's disease**

REFERENCES

Abulrob, A., Zhang, J., Tanha, J., MacKenzie, R., & Stanimirovic, D., (2005). Single domain antibodies as blood–brain barrier delivery vectors. In: *International Congress Series* (Vol. 1277, pp. 212–223). Elsevier.

Adhikary, Rishi & Sandbhor Gaikwad, Puja & Banerjee, Rinti. (2015). Nanotechnology platforms in Parkinson's Disease. ADMET & DMPK. 3. 10.5599/admet.3.3.189.

Ahire, E. D., Talele, S. G., & Shah, H. S., (2020). Nanoparticles as a promising technology in microbial pharmaceutics. In: *Applied Pharmaceutical Science and Microbiology* (pp. 133–158). Apple Academic Press.

Ahire, E., Thakkar, S., Borade, Y., & Misra, M., (2020). Nanocrystal based orally disintegrating tablets as a tool to improve dissolution rate of Vortioxetine. *Bulletin of Faculty of Pharmacy, Cairo University*, *58*(1, 2), 11–20.

Ahire, E., Thakkar, S., Darshanwad, M., & Misra, M., (2018). Parenteral nanosuspensions: A brief review from solubility enhancement to more novel and specific applications. *Acta Pharmaceutica Sinica B*, *8*(5), 733–755.

Ahirrao, S. P., Sonawane, M. H., Bhambere, D. S., Udavant, P. B., Ahire, E. D., & Kanade, R., (2022). Cocrystal formulation: A novel approach to enhance solubility and dissolution of etodolac. *Biosciences Biotechnology Research Asia*, *19*(1), 111.

Andén, N. E., Dahlström, A., Fuxe, K., & Larsson, K., (1965). Further evidence for the presence of nigro-neostriatal dopamine neurons in the rat. *American Journal of Anatomy*, *116*(1), 329–333.

Antonini, A., Moro, E., Godeiro, C., & Reichmann, H., (2018). Medical and surgical management of advanced Parkinson's disease. *Movement Disorders*, *33*(6), 900–908.

Asha, S. K. M., Mahendran, B., Reddy, K. V. V. S. N., & Jawahar, N., (2020). A comprehensive review of nano drug delivery system in the treatment of CNS disorders. *Journal of Drug Delivery Science and Technology*, *57*, 101628.

Attilio, M., Matteo, B., Andrea, D., Chiara, L., Giada, G., Andrea, P., & Gianni, C., (2021). Liposomes loaded with polyphenol-rich grape pomace extracts protect from neurodegeneration in a rotenone-based *in vitro* model of Parkinson's disease. *Biomater. Sci., 9*, 8171–8818.

Banks, W. A., (2015). Peptides and the blood–brain barrier. *Peptides, 72,* 16–19.

Barker, R. A., Drouin-Ouellet, J., & Parmar, M., (2015). Cell-based therapies for Parkinson's disease—past insights and future potential. *Nature Reviews Neurology, 11*(9), 492–503.

Bergman, H., Wichmann, T., & DeLong, M. R., (1990). Reversal of experimental parkinsonism by lesions of the subthalamic nucleus. *Science, 249*(4975), 1436–1438.

Bertler, Å., & Rosengren, E., (1959). Occurrence and distribution of dopamine in brain and other tissues. *Experientia, 15*(1), 10, 11.

Blesa, J., Phani, S., Jackson-Lewis, V., & Przedborski, S., (2012). Classic and new animal models of Parkinson's disease. *Journal of Biomedicine and Biotechnology, 2012.*

Blesa, J., Trigo-Damas, I., Dileone, M., Del, R. N. L. G., Hernandez, L. F., & Obeso, J. A., (2017). Compensatory mechanisms in Parkinson's disease: Circuits adaptations and role in disease modification. *Experimental Neurology, 298,* 148–161.

Bonifati, V., (2010). Shaking the genome: New studies reveal genetic risk for Parkinson's disease. *The Lancet Neurology, 9*(2), 136–138.

Braak, H., Del Tredici, K., Rüb, U., De Vos, R. A., Steur, E. N. J., & Braak, E., (2003). Staging of brain pathology related to sporadic Parkinson's disease. *Neurobiology of Aging, 24*(2), 197–211.

Bras, J., Verloes, A., Schneider, S. A., Mole, S. E., & Guerreiro, R. J., (2012). Mutation of the parkinsonism gene ATP13A2 causes neuronal ceroid-lipofuscinosis. *Human Molecular Genetics, 21*(12), 2646–2650.

Brockmann, K., Schulte, C., Hauser, A. K., Lichtner, P., Huber, H., Maetzler, W., Berg, D., & Gasser, T., (2013). SNCA: Major genetic modifier of age at onset of Parkinson's disease. *Movement Disorders, 28*(9), 1217–1221.

Caffrey, T. M., & Wade-Martins, R., (2007). Functional MAPT haplotypes: Bridging the gap between genotype and neuropathology. *Neurobiology of Disease, 27*(1), 1–10.

Cardo, L. F., Coto, E., De Mena, L., Ribacoba, R., Lorenzo-Betancor, O., Pastor, P., Samaranch, L., et al., (2012). A search for SNCA 3′ UTR variants identified SNP rs356165 as a determinant of disease risk and onset age in Parkinson's disease. *Journal of Molecular Neuroscience, 47*(3), 425–430.

Chaudhuri, K. R., Healy, D. G., & Schapira, A. H., (2006). Non-motor symptoms of Parkinson's disease: Diagnosis and management. *The Lancet Neurology, 5*(3), 235–245.

Chung, T. H., Hsu, S. C., Wu, S. H., Hsiao, J. K., Lin, C. P., Yao, M., & Huang, D. M., (2018). Dextran-coated iron oxide nanoparticle-improved therapeutic effects of human mesenchymal stem cells in a mouse model of Parkinson's disease. *Nanoscale, 10*(6), 2998–3007.

Cotzias, G. C., (1968). L-dopa for parkinsonism. *The New England Journal of Medicine, 278*(11), 630.

Crucho, C. I., & Barros, M. T., (2017). Polymeric nanoparticles: A study on the preparation variables and characterization methods. *Materials Science and Engineering: C, 80,* 771–784.

Daneman, R., & Prat, A., (2015). The blood–brain barrier. *Cold Spring Harbor Perspectives in Biology, 7*(1), a020412.

Débora, B. V., & Lionel, F. G. (2016). *Getting Into the Brain: Liposome-Based Strategies for Effective Drug Delivery Across the Blood–Brain Barrier, 18*(11), 5381–5414.

Del Rey, N. L. G., Quiroga-Varela, A., Garbayo, E., & Carballo-Carbajal, I., (2018). Advances in Parkinson's disease: 200 years later. *Frontiers in Neuroanatomy, 12,* Article 113.

Drin, G., Cottin, S., Blanc, E., Rees, A. R., & Temsamani, J., (2003). Studies on the internalization mechanism of cationic cell-penetrating peptides. *Journal of Biological Chemistry, 278*(33), 31192–31201.

Duwa, R., Jeong, JH. & Yook, S. Development of immunotherapy and nanoparticles-based strategies for the treatment of Parkinson's disease. J. Pharm. Investig. 51, 465–481 (2021). https://doi.org/10.1007/s40005-021-00521-3

Elezaby, R. S., Gad, H. A., Metwally, A. A., Geneidi, A. S., & Awad, G. A., (2017). Self-assembled amphiphilic core-shell nanocarriers in line with the modern strategies for brain delivery. *Journal of Controlled Release, 261,* 43–61.

Enas, M. A., (2015). Hydrogel: Preparation, characterization, and applications: A review. *Journal of Advanced Research, 6*(2), 105–121.

Gabathuler, R., (2010). Approaches to transport therapeutic drugs across the blood–brain barrier to treat brain diseases. *Neurobiology of Disease, 37*(1), 48–57.

Gautam, S. P., Keservani, R. K., Gautam, T., Gupta, A. K., & Sharma, A. K., (2015). An alternative approach for acetylation of amine-terminated polyamidoamine (PAMAM) dendrimer. *ARS Pharmaceutica, 56*(3), 155–159.

Guarnieri, D., Falanga, A., Muscetti, O., Tarallo, R., Fusco, S., Galdiero, M., Galdiero, S., & Netti, P. A., (2013). Shuttle-mediated nanoparticle delivery to the blood–brain barrier. *Small, 9*(6), 853–862.

Gustavsson, A., Svensson, M., Jacobi, F., Allgulander, C., Alonso, J., Beghi, E., Dodel, R., et al., (2011). Cost of disorders of the brain in Europe 2010. *European Neuropsychopharmacology, 21*(10), 718–779.

Jankovic, J., (2002). Levodopa strengths and weaknesses. *Neurology, 58*(suppl 1), S19–S32.

Jesus, S., Schmutz, M., Som, C., Borchard, G., Wick, P., & Borges, O., (2019). Hazard assessment of polymeric nanobiomaterials for drug delivery: What can we learn from literature so far. *Frontiers in Bioengineering and Biotechnology, 7,* 261.

Jinyan, W., Yu, L., Songyu, L., Yulu, Z., Yan, X., & Dan, L., (2021). Geniposide-loaded liposomes for brain targeting: Development, Evaluation, and *in vivo* studies. *AAPS PharmSciTech, 22.*

Jonathan, B., June, E. J., & Simon, J. G. L., (2020). Nanoparticles for drug delivery in Parkinson's disease. *Journal of Neurology.*

Keservani, R. K., & Gautam, S. P., (2022). Skeletal muscle relaxant activity of different formulation of span 60 niosome. *ARS Pharmaceutica, 63*(1), 32–44.

Keservani, R. K., Sharma, A. K., & Kesharwani, R. K., (2016). *Novel Approaches for Drug Delivery.* IGI Global. ISBN13: 9781522507543.

Keservani, R. K., Sharma, A. K., & Kesharwani, R. K., (2018). Nanoparticulate nanocarriers in drug delivery. In: *Nanobiomaterials* (pp. 3–22). Apple Academic Press. ISBN: 9781771885911.

Keservani, R. K., Sharma, A. K., & Kesharwani, R. K., (2019). *Nanocarriers for Brain Targeting: Principles and Applications.* CRC Press. ISBN:9781771887304.

Malakoutikhah, M., Guixer, B., Arranz-Gibert, P., Teixidó, M., & Giralt, E., (2014). 'A la carte' peptide shuttles: Tools to increase their passage across the blood–brain barrier. *ChemMedChem, 9*(7), 1594–1601.

Malakoutikhah, M., Teixido, M., & Giralt, E., (2011). Shuttle-mediated drug delivery to the brain. *Angewandte Chemie International Edition, 50*(35), 7998–8014.

Melanie, D. S., Abhay, P. S., & Berislav, V. Z. (2018). *Blood–Brain Barrier Breakdown in Alzheimer Disease and Other Neurodegenerative Disorders. 4*(3), 133–150.

Mensch, J., Oyarzabal, J., Mackie, C., & Augustijns, P., (2009). *In vivo, in vitro* and in silico methods for small molecule transfer across the BBB. *Journal of Pharmaceutical Sciences, 98*(12), 4429–4468.

Michael, J. M., Margaret, M. B., Rebecca, M. H., & Marissa, E. W., (2021). *Engineering Precision Nanoparticles for Drug Delivery, 20,* 101–124.

Milowska, K., Szwed, A., Mutrynowska, M., Gomez-Ramirez, R., De La Mata, F. J., Gabryelak, T., & Bryszewska, M., (2015). Carbosilane dendrimers inhibit α-synuclein fibrillation and prevent cells from rotenone-induced damage. *International Journal of Pharmaceutics, 484*(1, 2), 268–275.

Mitragotri, S., Burke, P. A., & Langer, R., (2014). Overcoming the challenges in administering biopharmaceuticals: Formulation and delivery strategies. *Nature Reviews Drug Discovery, 13*(9), 655–672.

Murray, C. J., Vos, T., Lozano, R., Naghavi, M., Flaxman, A. D., Michaud, C., Ezzati, M., et al., (2012). Disability-adjusted life years (DALYs) for 291 diseases and injuries in 21 regions, 1990–2010: A systematic analysis for the Global Burden of Disease Study 2010. *The Lancet, 380*(9859), 2197–2223.

Niewoehner, J., Bohrmann, B., Collin, L., Urich, E., Sade, H., Maier, P., Rueger, P., et al., (2014). Increased brain penetration and potency of a therapeutic antibody using a monovalent molecular shuttle. *Neuron, 81*(1), 49–60.

Obermeier, B., Daneman, R., & Ransohoff, R. M., (2013). Development, maintenance, and disruption of the blood-brain barrier. *Nature Medicine, 19*(12), 1584–1596.

Oller-Salvia, B., Sánchez-Navarro, M., Ernest, G., & Meritxell, T., (2016). Blood–brain barrier shuttle peptides: An emerging paradigm for brain delivery. *Chem. Soc. Rev., 45,* 4690.

Pattni, B. S., Chupin, V. V., & Torchilin, V. P., (2015). New developments in liposomal drug delivery. *Chemical Reviews, 115*(19), 10938–10966.

Paul, W., & Sharma, C. P., (2010). *Inorganic Nanoparticles for Targeted Drug Delivery* (204–235). Woodhead Publishing Limited.

Qian, Z. M., Li, H., Sun, H., & Ho, K., (2002). Targeted drug delivery via the transferrin receptor-mediated endocytosis pathway. *Pharmacological Reviews, 54*(4), 561–587.

Qin, Y., Chen, H., Zhang, Q., Wang, X., Yuan, W., Kuai, R., Tang, J., et al., (2011). Liposome formulated with TAT-modified cholesterol for improving brain delivery and therapeutic efficacy on brain glioma in animals. *International Journal of Pharmaceutics, 420*(2), 304–312.

Qing, L., Xinxin, S., Xianglin, D., Qiong, G., Bolei, Y., Ying, L., & Wei, J., (2022). Recent trends in the development of hydrogel therapeutics for the treatment of central nervous system disorders. *NPG Asia Materials, 14,* 14.

Rekas, A., Lo, V., Gadd, G. E., Cappai, R., & Yun, S. I., (2009). PAMAM dendrimers as potential agents against fibrillation of α-synuclein, a Parkinson's disease-related protein. *Macromolecular Bioscience, 9*(3), 230–238.

Roco, M. C., (2003). Nanotechnology: Convergence with modern biology and medicine. *Current Opinion in Biotechnology, 14*(3), 337–346.

Roshan, S., Sonalia, A. J., Matte, K. V., Ankita, S. B., & Narendra, (2021). Gold liposomes for brain-targeted drug delivery: Formulation and brain distribution kinetics. *Materials Science and Engineering: C, 120,* 111652.

Saraiva, C., Praça, C., Ferreira, R., Santos, T., Ferreira, L., & Bernardino, L., (2016). Nanoparticle-mediated brain drug delivery: Overcoming blood–brain barrier to treat neurodegenerative diseases. *Journal of Controlled Release*, *235*, 34–47.

Sharma, A. K., Keservani, R. K., & Keseharwani, R. K., (2018). *Nanobiomaterials: Applications in Drug Delivery*. CRC Press. ISBN:9781771885911.

Singh, R., & Lillard, Jr. J. W., (2009). Nanoparticle-based targeted drug delivery. *Experimental and Molecular Pathology*, *86*(3), 215–223.

Soddu, E., Rassu, G., Giunchedi, P., Sarmento, B., & Gavini, E., (2015). From naturally-occurring neurotoxic agents to CNS shuttles for drug delivery. *European Journal of Pharmaceutical Sciences*, *74*, 63–76.

Sweeney, M., Sagare, A. & Zlokovic, B. Blood–brain barrier breakdown in Alzheimer disease and other neurodegenerative disorders. Nat Rev Neurol 14, 133–150 (2018). https://doi.org/10.1038/nrneurol.2017.188

Talele, S. G., Ahire, E. D., Talele, G. S., & Derle, D. V., (2021). An innovative approach as self-emulsifying drug delivery system for phytoconstituents. In: *Enhancing the Therapeutic Efficacy of Herbal Formulations* (pp. 69–84). IGI Global.

Thomas, B. S., (2018). *Parkinson's Disease Pathogenesis and Clinical Aspects* (83–93). Codon Publications.

Thombre, N. A., Niphade, P. S., Ahire, E. D., & Kshirsagar, S. J., (2022). Formulation development and evaluation of microemulsion based lornoxicam gel. *Biosciences Biotechnology Research Asia*, *19*(1), 69–80.

Wang, J. T., et al., (2021). Enhanced delivery of neuroactive drugs via nasal delivery with a self-healing supramolecular gel. *Adv. Sci.*, *8*, e2101058.

Weisen, Z., Ami, M., Ziqiu, T., Lars, E., & Nicolas, H. V., (2021). Development of polymeric nanoparticles for blood–brain barrier transfer—Strategies and challenges. *Adv. Sci.*, *8*, 1–32.

William, M. P., (2020). Brain delivery of nanomedicines: Trojan horse liposomes for plasmid DNA gene therapy of the brain. *Front. Med. Technol.*

Wolak, D. J., & Thorne, R. G., (2013). Diffusion of macromolecules in the brain: Implications for drug delivery. *Molecular Pharmaceutics*, *10*(5), 1492–1504.

Yang, J., Jia, C., Yang, J. (2021). Designing Nanoparticle-based Drug Delivery Systems for Precision Medicine. International Journal of Medical Sciences, 18(13), 2943-2949. https://doi.org/10.7150/ijms.60874.

Zhang, W., Mehta, A., Tong, Z., Esser, L., Voelcker, N. H., Development of Polymeric Nanoparticles for Blood–Brain Barrier Transfer—Strategies and Challenges. Adv. Sci. 2021, 8, 2003937. https://doi.org/10.1002/advs.202003937

Zhu, Y., Liu, C., & Pang, Z., (2019). Dendrimer-based drug delivery systems for brain targeting. *Biomolecules*, *9*(12), 790.

CHAPTER 6

Nose-to-Brain Delivery as a Novel Drug Delivery System

SMITA P. KAKAD, SANJAY J. KSHIRSAGAR, CHETAN SONAWANE, MILIND SHEWALE, and RUSHIKESH SHINDE

Department of Pharmaceutics, MET's Institute of Pharmacy, Nashik, Maharashtra, India

ABSTRACT

The blood-brain barrier and blood-cerebrospinal fluid barrier are major barriers to delivering drugs into the central nervous system (CNS). They prevent the entry of most substances. To enhance drug delivery systems (DDSs) for the CNS, intrathecal or intraparenchymal methods are preferred. These methods are non-invasive, but they involve a risk of infection. Scientists are focusing on the nose-to-brain delivery system, which is an invasive method for drug administration that crosses the blood-brain barrier. The nose-to-brain delivery involves the transport of drugs from the nasal cavity to the brain through two major pathways: the olfactory nerve pathway and the trigeminal nerve pathway located at the upper part of the nasal cavity. The advantage of nose-to-brain delivery is that it bypasses the blood-brain barrier, avoids first-pass metabolism, and facilitates effective drug delivery for CNS targeting. Along with these advantages, there are certain limitations associated with it, such as low bioavailability, mucociliary clearance, enzymatic degradation, and active efflux transporters, which can affect the therapeutic efficacy of the drug molecule. To counteract this problem, various approaches have been developed for the development of nasal formulations, such as nanoparticles, polymeric nanoparticles (PNPs), lipid nanoparticles (niosomes, liposomes), microspheres, nano-emulsions, nanosuspensions,

Novel Drug Delivery Systems in the Management of Chronic Diseases.
Raj K. Keservani, Umesh D. Laddha, Eknath D. Ahire, and Sanjay J. Kshirsagar (Eds.)

nanogels, etc. These novel drug delivery systems are very effective for CNS diseases such as migraines, Alzheimer's disease (AD), neuro-AIDS, and brain tumors for effective targeting.

6.1 INTRODUCTION

The blood-brain barrier is a microvasculature of the central nervous system (CNS) because it separates the brain from other parts of the body (Bonferroni et al., 2019). The blood-brain barrier is a highly successful protecting mechanism that protects the brain and CNS (Veronesi et al., 2020). The blood-brain barrier prevents the entry of toxic, external agents, pathogens, and neurotherapeutics into the brain or CNS (Kulkarni et al., 2015). The blood-brain barrier exists due to the special properties of brain endothelial cells located on the walls of blood vessels. Brain endothelial cells differ from endothelial cells of non-neuronal tissue because they are held together by tight junctions which prevent the entry of substances but on the other hand, the brain endothelial cells contain a certain type of transport mechanism in both directions for both influx and efflux of substances. The efflux transporter passively eliminates the lipophilic toxins whereas influx transporters which are specific deliver the nutrients to the brain across the BBB (Bonferroni et al., 2019; Daneman et al., 2015). The BBB and BCF prevent distribution of drugs from blood to the central nervous system (Pardeshi et al., 2012). In past outbreaks, studies show that more than 97% of potential therapies targeting CNS disorders are restricted by the blood-brain barrier (Veronesi et al., 2020). Therefore, local delivery strategies such as intracerebroventricular or intraparenchymal injection, intracranial delivery with mini pumps, catheter infusion are developed for direct delivery of drugs to the brain. But all of these techniques are risky and harmful (Dhuria et al., 2010). The nasal route has attracted the broad attention of researchers. Traditionally, it is used for the treatment of local diseases. Later on, research targeted this route for the treatment of many diseases due to its non-invasive, safety, reliability, and higher absorption (Kulkarni et al., 2015; Pardeshi et al., 2012). Researchers focus on the nasal route for brain targeting, i.e., nose to brain delivery of drug moieties and explore its advantages over conventional routes such as bypassing the BBB, avoidance of first-pass metabolism, safety, convenience of administration, and painless (Colombo et al., 2011; Chen et al., 2008). The drug delivery via intranasal route has various advantages over conventional systematic drug delivery systems (DDS) as it directly delivers the drug into the brain (Keservani et al., 2019; Agrawal et al., 2018). Drug delivery from the nose (Gujarathi et al., 2022) to the brain is mediated through the olfactory mucous membrane pathway and trigeminal nerve pathway located at the upper part of the nasal cavity (Kulkarni et al., 2015). The fastest route of drug travel from the nasal cavity to the brain is the olfactory mucous pathway (Zhao et al., 2012). Nose to brain delivery is mediated through neuroepithelial which is the only part of the CNS that is directly in contact with the external environment (Pardeshi et al., 2013). The above pathways of drug transport from the nose to the brain are proven in animal models, but it is not confirmed that the same transports occur in human beings and also other limitations such as poor drug permeation through the mucosa and mucociliary clearance associated with intranasal delivery (Ali et al., 2010). Therefore, novel strategies such as novel carrier systems, colloidal carriers, mucoadhesive devices, controlled drug systems, and prodrugs are preferred to overcome the above limitations (Agrawal et al., 2018).

6.1.1 INTRANASAL DELIVERY

6.1.1.1 NASAL CAVITY

For understanding the nose to brain delivery, it is essential to know about the morphology and physiology of the nose. Structurally, the midline septum of the nose divides it into two nasal cavities (Hirlekar et al., 2018). The length, volume, and surface area of the nasal cavity are 12 cm, 15–20 ml, and 150 cm square, respectively. Each cavity comprises three different regions, namely the vestibule, the respiratory region, and the olfactory region (Semwal et al., 2003). The nasal vestibule, located inside the nostrils, has a surface area of about 0.6 cm2. Three nasal turbinates make up the respiratory region, and they are important for creating greater contact between breathed air and the mucosal surface. The olfactory region of the nasal cavity is located at the top of the cavity in humans and makes up 10% of the total surface area (Pardeshi et al., 2013). The nose plays an important role in function breathing and olfaction, which is responsible for filtration, heating, and humidification of inhaled air before it reaches the lower respiratory tract. The hairs of the nasal vestibule or mucus layer of the respiratory area in the nasal cavity trap the inhaled particles and pathogens. The only routes by which the brain is connected to the outside environment are the olfactory and trigeminal pathways (Mistry et al., 2009).

6.1.1.2 RESPIRATORY EPITHELIUM AND MUCOCILIARY CLEARANCE MECHANISM

It contains four types of cells namely ciliated and non-ciliated columnar, basal and goblet cells that are responsible for the exchange of water ions between cells and the motility of cilia by active transport. These cells prevent the mucosa from drying by trapping moisture and promoting mucociliary clearance (Pardeshi et al., 2013; Mistry et al., 2009). The cilia fold half of their original length (7 mm) during the recovery stroke (Semwal et al., 2003). The mucus in nasal secretions is comprised of a mixture of 95% water and lactoferrins.

The very efficient defense mechanism in humans that protects the lungs against inhaled particulates, droplets, and microorganisms is mucociliary clearance. The mucus moves from the anterior side to the posterior side in the nasal cavity (Mistry et al., 2009; Illum, 2006).

6.1.1.3 OLFACTORY REGION

The olfactory epithelium is made up of three types of cells: olfactory neural cells, sustentacular or supporting cells, and basal cells, which are positioned at the upper part of the nasal cavity. It is made up of neuroepithelium, which is the only component of the CNS that is physically connected to the outside world, and basal cells, which are precursors of sustentacular or supporting cells that provide mechanical support to other cells. It contains olfactory receptor cells that are sensitive to odor molecules, which allows to identify food, mates, predators, and provide sensual pleasures and warn against danger. This region is about 2 cm^2 gray and contains millions of primary receptor cells with cilia (Figure 6.1) (Semwal et al., 2003; Rao et al., 1997).

6.2 PATHWAYS

The drug transport from nasal cavity to brain by cellular transport mechanism as mentioned in Figure 6.2 for nose to brain delivery the drug generally transfers from respiratory epithelium and olfactory region through olfactory nerve cells and trigeminal nerve cells. That deliver the drug from nose to different parts of brain including frontal cortex of olfactory bulb, cerebrum, and cerebrum. The pathways by which the drug is transfer from nose to brain as mentioned in Figure 6.2 (Agrawal et al., 2018).

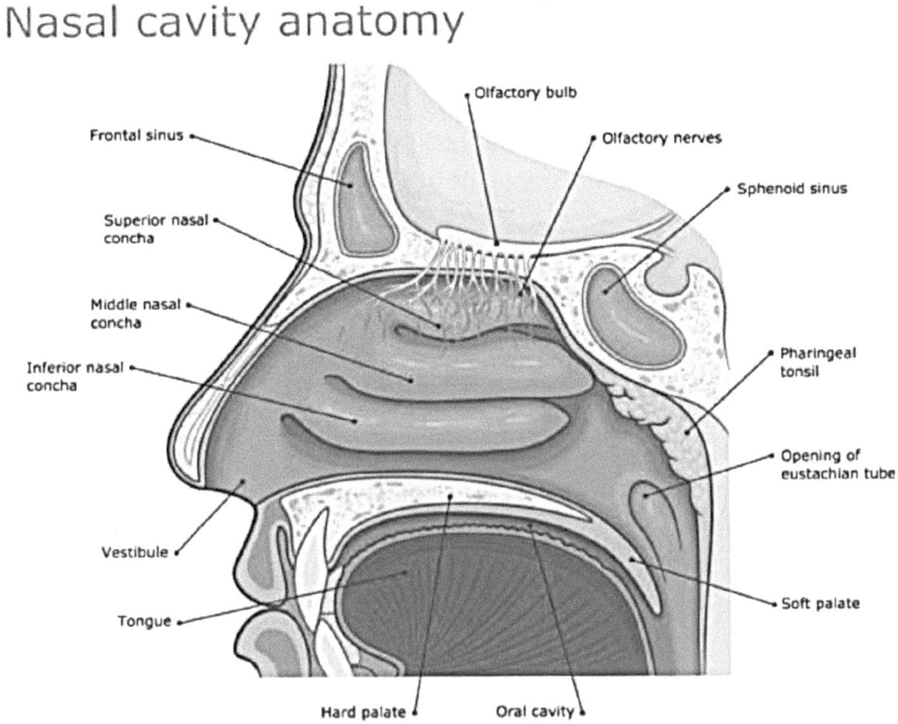

FIGURE 6.1 Anatomy of nasal cavity.

Source: Reprinted from Star Health. https://www.starhealth.in/blog/nose-anatomy

6.2.1 OLFACTORY NERVE PATHWAY

As the name indicates the drug is travel to the brain from olfactory region of nasal cavity by olfactory nerve pathway. It occurs by two cellular transport mechanisms one is intraneuronal or intracellular and other is extra neuronal or extracellular pathway.

6.2.1.1 INTRACELLULAR PATHWAY/INTRANEURONAL PATHWAY

It follows axonal transport by which uptake of drug into olfactory sensory neuron occurs by endocytosis and reaches in the olfactory bulb or drug transfer through the sustentacular of the lamina propria following trans-cytosis. It is May slow transport pathway it needs hours to days to reach the drug in different parts of brain as mentioned in Figure 6.2.

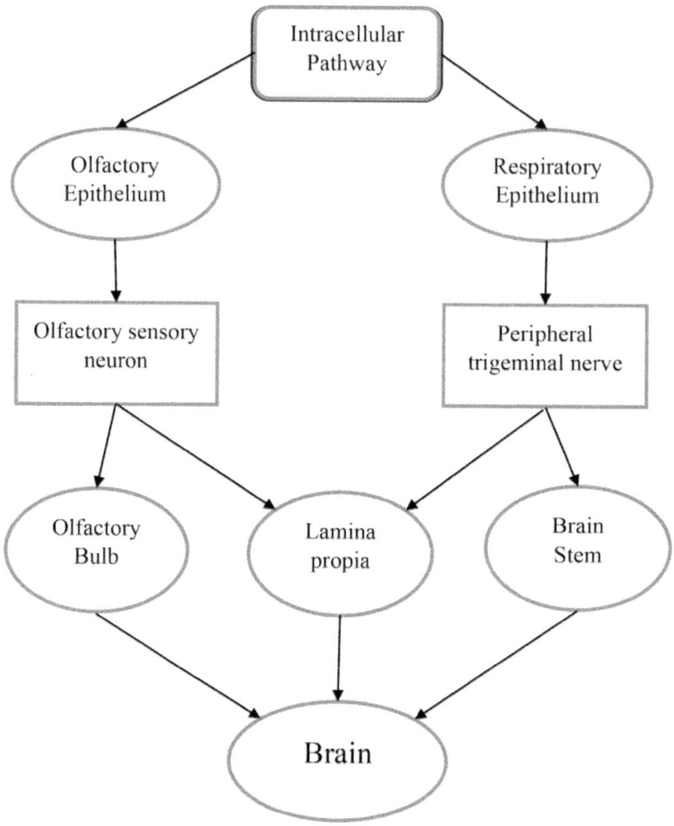

FIGURE 6.2 Intraneuronal pathway.

6.2.1.2 EXTRA NEURONAL PATHWAY

In this pathway, drugs are transferred from the gap present between the two neuronal cells. The drug is transported from the olfactory region to the olfactory bulb, and from the olfactory bulb, it enters other brain regions by diffusion. There are tight junctions on ensheathing cells between the two neurons. The drug passes through them, and additionally, the regrowth of the olfactory nerve fibers provides a path for the transformation of the drug to the brain. Through this pathway, the drug is directly delivered to the CSF. It is the fastest pathway that transfers drugs to the brain within minutes. On the contrary, particles of the same size are transported via the axon of the neuronal cell to the olfactory bulb, and from there, they reach the cerebral hemisphere and then the cerebrum cerebellum. Therefore, it is a very potential pathway in which drug transport occurs from the nose to the brain (Hirlekar et al., 2018; Talegaonkar et al., 2004; Pardeshi et al., 2013) (Figure 6.3).

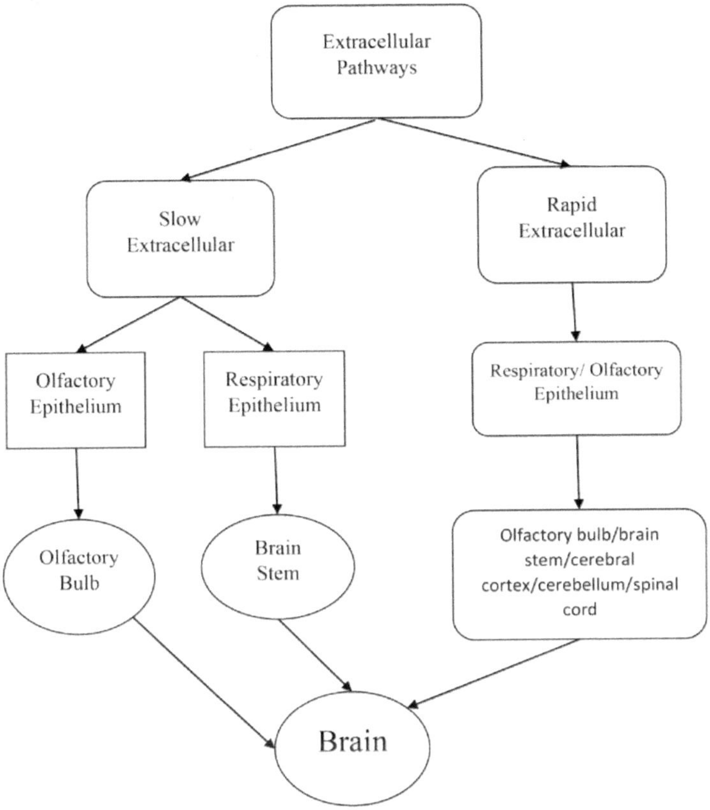

FIGURE 6.3 Extracellular pathway.

6.2.2 *TRIGEMINAL NERVE PATHWAY*

It is important pathway that connects nose and brain. It consists small branch of trigeminal nerves that transverse from olfactory and respiratory region to the brain stem. The ophthalmic and maxillary branches of trigeminal nerve present in nasal mucosa which act mediator for brain delivery of drug via nose. In this pathway the uptake of drug occurs by endocytosis from the respiratory epithelium to the trigeminal nerve which further transport the drug to brain stem (Hirlekar et al., 2018; Pardeshi et al., 2013; Van Woensel et al., 2013).

6.2.2.1 CEREBROSPINAL PATHWAY

This pathway includes subarachnoid space containing cerebrospinal fluid, perineural space along the olfactory and trigeminal nerves and nasal lymphatics. This pathway connects this space and provide passage to transfer drug from nose cavity to CSF and from CSF it is distributed to parts of brain The drug is move from nasal cavity to CSF and then into perivascular spaces where it is distributed through the brain. Based on molecular weight, lipophilicity, and degree of ionization, drug entry into CSF depends. The CSF has rapid turnover, so it affects molecules than smaller one.

Due to presence of CSF-brain barrier there is non-equivalent concentration between CSF and brain (Hirlekar et al., 2018; Misra et al., 2003; Lochhead et al., 2012).

6.2.2.2 PERIVASCULAR PATHWAY

This pathway involves distribution of drug to other distant parts of brain from the olfactory bulb and the brainstem. It takes place via intracellular pathway which involves drug from second order neuron which form synapse with peripheral olfactory sensory neurons. The extracellular pathway involves distribution or drug by to brain by bulk flow and local diffusion in the perivascular spaces. The expansion and contraction of perivascular space occur due to cardiac cycle which plays key role for fluid flow within perivascular spaces (Hirlekar et al., 2018; Lochhead et al., 2012).

6.3 ADVANTAGES

1. It bypasses the drug across blood brain barrier.
2. It reduces systemic side effect of drug by reducing systemic exposure.
3. It avoids first pass metabolism and improve bioavailability.
4. It avoids degradation of drug by gastrointestinal (GI) tract which is common in orally administered drug.
5. It is convenient, safe, and non-invasive method offer's self-administration and shows high patient compliance.
6. It required low dose to give desired therapeutic effect, therefore side-effects are reduced.
7. For protein and peptide delivery is act as alternative route to parenteral route.
8. It did not require any special modification for delivery of drug to brain which is required in other delivery system across BBB.

9. The drugs having low molecular weight when given by intranasal route shows high bioavailability (Hirlekar et al., 2018; Frey et al., 1997; Talegaonkar et al., 2004).

6.4 LIMITATIONS

1. Mucociliary Clearance: Due to mucociliary clearance drug is rapidly eliminated from nasal cavity.
2. The drug having high molecular weight have low permeability across nasal mucosa.
3. Nasal mucosa contains certain type of enzyme which causes degradation of some drugs.
4. Some drugs are irritable to nasal cavity.
5. Alter the absorption of drug by pathological conditions such as cold or allergy.
6. It has very less surface area for delivery of drug as compared to gastrointestinal tract.
7. If absorption of drug take place by intracellular axonal transport, then it is very slow and needs hours to reach olfactory bulb and days to reach the parts of brain.
8. Mechanical loss of drug take place due to use of unsuitable technique for administration of drug (Hirlekar et al., 2018; Pardeshi et al., 2013; Lochhead et al., 2012) (Table 6.1).

TABLE 6.1 Drugs Patented for Nose to Brain Delivery

Drug	Pka	Molecular Weight	Therapeutic Application	Drug Delivery System	References
Modafinil	4.06	273.35 g/mol	Sleep disorder/ narcolepsy	Lipid microemulsion	Greco et al. (2011)
Deferoxamine	7.92	560.684 g/mol	Alzheimer's disease	Liquid spray, powder spray	Frey et al. (2010)
Retinoic acid	5	300.43 g/mole	Anosmia	–	Yee et al. (2003)
Diltiazem	8.18	414.519 g/ mole	Eating disorder/ obesity	Liquid, powder	Adams et al. (2008)
Polynucleotide	–	–	Neurodegenerative disorder, tumor, nerve damage.	–	Reinhard et al. (2003)
Endogenous compound (neurotransmitter, neuropeptide)	–	–	Neurological disorder	Sustained-release micro-construct comprising a surface-eroding, non-bursting polymer.	Kubek et al. (2007)

6.5 BARRIER INVOLVE IN NASAL DRUG DELIVERY

6.5.1 LOW BIOAVAILABILITY

Generally, polar drugs have low bioavailability. Bioavailability of low molecular weight drugs is 10%, and for peptides like insulin and calcitonin, it is not more than 1%. The most essential factor responsible for the absorption of polar drug particles, particularly high molecular weight polar drugs like peptides and proteins with low membrane permeability (Illum, 2002). Drugs can pass the epithelial cell membrane in one of three ways: transcellularly by simple concentration gradients, receptor-mediated or vesicularly via receptors, or paracellularly via tight junctions between cells. Polar medicines with molecular weights less than 1,000 Da are more likely to get across the membrane via the latter method (Jadhav et al., 2007). Although tight junctions are dynamic structures that can open and close to a degree when required, their average channel size is on the order of tens of nanometers, limiting the transport of bigger molecules (McMartin et al., 1987). Endocytotic transport allows larger peptides and proteins to pass through the nasal membrane, but only in small numbers. Co-administration of absorption-enhancing excipients with such polar medicines can considerably improve nasal absorption.

6.5.2 MUCOCILIARY CLEARANCE

Another important cause for limited membrane transport is the rapid removal of the administered preparation from the cavity of nasal due to the mucociliary clearance process. This is especially true in this instance. When the medication is not quickly absorbed over the nasal mucosa, the half-life for both liquid and powder formulations that are not bioadhesive has been demonstrated. Clearance times range from 15 to 30 minutes (Illum, 1999). The use of mucoadhesive compounds and semisolid formulations may help to reduce mucociliary clearance rates and so overcome this barrier. Increased viscosity semisolid gels improve nasal residence period and increase CNS influx by up to twofold (Kublik et al., 1998).

6.5.3 ENZYMATIC DEGRADATION

Another aspect that contributes to the reduced bioavailability of peptides and proteins across the nasal mucosa, but is commonly neglected, is the possibility of enzymatic degradation of the molecule in the nasal lumen. During passage across the epithelial barrier or in the nasal cavity, exopeptidases, such as mono- and diamino peptidases, can cleave peptides at their N termini at both of these locations. Endopeptidases such as serine and cysteine, which can target internal peptide bonds, are found at the N and C termini. The use of enzyme inhibitors and/or enzyme saturation may be necessary techniques to overcome this major hurdle (Morimoto et al., 1995).

6.5.4 *ACTIVE EFFLUX TRANSPORTERS (AETS)*

These transporters may identify their substrates and actively discharge them into the bloodstream, as they are found in the biological membrane between bloodstream and the CNS and in cellular membranes (Dalpiaz et al., 2018). P-glycoprotein (ABCB1 or Pgp or MDR1), multidrug resistance-associated proteins (ABCC1 or MRP1, ABCC4 or MRP4, ABCC5, or MRP5), and breast cancer (BC) resistance protein are the active transporters currently known to efflux antiviral drugs (ABCG2 or BCRP) (Janneh et al., 2007). AET inhibitors may be investigated, however taking them together has substantial negative effects.

Nasal administration of antiviral drugs is thus proposed to address the aforementioned issues, but new devices, formulations (polymeric micro- and nanoparticles, thermoreversible gels, nano-emulsions, solid lipid microparticles (SLMs)), mucoadhesive agents (chitosan, polyvinylpyrrolidone), absorption enhancers (chitosan, papaverine) are required to target drugs for CNS and may be used AET inhibitors (Krishan et al., 2014; Paul et al., 2000). Furthermore, numerous prodrugs can block or bypass the AET systems, making them potential substrates for novel nasal formulations capable of targeting drug particles into CNS macrophages. Components of nasal preparation are mentioned in Table 6.2.

6.6 TARGETING DISEASE FOR NOSE TO BRAIN DELIVERY

6.6.1 *MIGRAINE*

Headache can be considered as a crippling and general neurovascular affliction (Chattopadhyay et al., 2021), ordinarily arranged with a migraine with ≥2 elements of throb, one-sided position, moderate to unembellished strength and crumbling by the planned actual clamor, related with ≥1 sign of sickness and/regurgitating, photophobia, and phonophobia (Arnold, 2018). Yet, every one of the side effects are emotional as well as various. Also, one out of three subjects of headache are victims of an extraordinary kind of central neurologic visual condition called aura (Robbins, 2021; Ahmed,

TABLE 6.2 Components of Nasal Preparation

Class of Compound	Examples	Effect of Each Class	References
Surfactant	Laureth-9, polysorbate 80, dioctyl sulfosuccinate.	Disrupt membranes	Machida et al. (1993)
Bile salts and derivatives	Sodium glycocholate, sodium deoxycholate.	Disrupt membranes, enzyme inhibition.	Illum (2000)
Phospholipids	DDPC, dimyristoyl phosphatidylglycerol.	Reduce nasal clearance.	Jacobs et al. (1993)
Polymer	Chitosan, poly-L-arginine chitosan, hyaluronan chitosan, Carbopol.	Reduce nasal clearance, open tight junctions.	Beukelaar et al. (1999)
Lipid	Oleic acid, linoleic acid.	Disrupt membranes	Abe et al. (1995)
Enzyme inhibitors	Amastatia, bestatin, tripsin, boroleucin.	Enzyme inhibition.	Morimoto et al. (1995)
Mucoadhesive agent	Carbopol, starch microspheres.	Open tight junctions, delayed the nasal residence time.	Erdo et al. (2018)
Miscellaneous	Cyclodextrins, modified chitosan.	Open tight junctions.	Davis et al. (2003)

2016). Throughout the spell, the veins in the mind expand and then constrict by triggering nerve summits near the overstated veins. There lies the primary justification for pain or uneasiness (Abbas et al., 2014). The duration of the inflammation varies from 2 hr. and, surprisingly, up to three dimensional and it may move starting with one half then onto the next (Oakley et al., 2014). Several innate and environmental influences are considered as two fundamental explanations for headache. As a fundamental dimension innate cases lead to two-thirds of headache cases. Fluctuation in orientation also causes an adjustment in the disease population. Females are more vulnerable to the headache than the males, especially the difference rises after puberty. The research revealed 3 out of 4 patients with headache are females (Zarcone et al., 2017) (Table 6.3).

6.6.2 *ALZHEIMER DISORDER*

AD, a neurodegenerative disorder and a critical reason for dementia throughout the world, generally affects older adults but is sometimes also found in young age (early stage AD) (Agrawal et al., 2018).

Alzheimer's disease is a form of dementia and is considered the sixth leading cause of death among older adults in the USA. The terms dementia and Alzheimer's are not the same, as dementia is only considered memory loss, while Alzheimer's is a collection of symptoms such as cognitive decline, thinking and behavior impairments, mental memory impairment, mood swings, difficulty in communication, reading, writing, and other daily activities (Mohammadi et al., 2011). Therefore, it not only affects the individual experiencing it but also the patient's family. The exact cause of the disease is still unknown, but it is believed to be linked to genetic factors, dietary habits, lifestyle changes, mental stress, etc. (Adi Stefano et al., 2011). The histopathology behind it is a multifactorial disorder regulated by various pathophysiologies. The primary hypotheses behind AD include: (i) amyloid efflux theory; (ii) tau theory; (iii) cholinergic theory; (iv) excitotoxicity theory; (v) mitochondrial dysfunction theory; and (vi) other factors (such as genetic factors, inflammatory responses, free radicals or oxidative stress, nitric oxide or toxins, environmental factors, etc.) (Brambilla et al., 2012). The current medical treatment aims to alleviate the symptoms of AD in the early stages. Nose-to-brain delivery is a novel approach for drug targeting to the central nervous system. However, preclinical trials do not guarantee therapeutic success (Table 6.4).

6.6.3 PARKINSON DISEASE

Parkinson's disease (PD) is portrayed as the most predominant neurodegenerative illness being similar to the Alzheimer's infection. PD is characterized by severe dropping of dopamine-related neurons present in substantia nigra as well as cytoplasmic incorporations. Various therapeutic specialists are available to treat initial as well as later complications of PD. However, transport of neurotherapeutics into the brain has been quite difficult for specialists, as a result of the presence of blood-brain barrier (BBB) (Braak et al., 2006). In the last many years, nasal drug delivery has acquired widespread attention. Intranasal administration as a method for targeting neurotherapeutics to the central nervous system bypassing blood-brain barrier, shows several benefits for treating neurodegenerative disorders. This route for transport of neurotherapeutics offers the benefits of easy administration, bypassing first pass metabolism, and painlessness (Pardeshi et al., 2012) (Table 6.5).

TABLE 6.3 Approved Nasal Formulation for the Treatment of Migraine

Brand Name	Company Name	Dose	Dose	References
IMITREX (Sumitriptan)	Glaxosmithkline (USA)	10 mg dose – two nasal sprays.	Maximum observed plasma concentration (C_{max}).	Djupesland et al. (2013)
MIGRANAL (dihydroergotamine (DHE) mesylate)	Bausch health companies (USA)	2 mg – two-time nasal spray.	The mean bioavailability of nasal administration was 32% relative to injectable administration.	Silberstein et al. (2020)
ZOMIG (zolmitriptan)	Amneal pharmaceuticals, USA.	Starting dose 2.5 mg. if headache relief is not achieved than 5 mg single spray.	The mean bioavailability following nasal administration of ZOMIG is 102% relative to an oral tablet form.	Yates et al. (2005)
TOSYMRA (Sumatripton)	Upsher-smith laboratories (USA)	10 mg single spray.	The mean bioavailability following nasal administration 87.7% relative to a 4 mg SC injection and 58.4% relative to 6.	Munjal et al. (2017)

TABLE 6.4 Nanotechnological Development of Anti-AD Drug Administered Through the Intranasal Route

Drug	Nanocarriers/Method	Polymer	Advantage	References
Curcumin	Lipid nanocarrier/hot solvent diffusion method.	Phospholipid 80, PEG 600 stearate.	Increase drug permeability.	Vaz et al. (2017)
Curcumin	Lipid nanoemulsion/ spontaneous emulsification method.	Labrafac lipophile, hyaluronic acid.	Increase drug concentration in brain.	Nasr (2016)
Insulin	Nanogel/ionization radiation.	Poly(N-vinyl pyrrolidone).	Significant insulin activation and protect insulin from enzymatic degradation.	Picone et al. (2016)
Rivastigmine	Nanoparticle/ionic gelation	Chitosan	Improves bioavailability.	Frazil et al. (2012)
Risperidone	Nanoemulsion/simple mixing.	Tween 80 and Capmul MCM	Enhance drug concentration in brain.	Kumar et al. (2008)
Galantamine	Nanoparticle/ionic gelation method	Chitosan	Improves memory and brain activity.	Hanafy et al. (2016)
Galantamine	Flexible liposome/thin film homogenization method.	Propylene glycol.	Improves drug bioavailability.	Bhattacharya et al. (2015)
Tarenflurbil	Nanoparticle/emulsification solvent diffusion method.	PLGA	Significant increase drug brain concentration.	Swami et al. (2015)
Piperine	Nanoparticle/ionic gelation	Chitosan	Improves drug absorption in brain.	Illum et al. (1994)

TABLE 6.5 Nanotechnological Development of Anti-PD Drug Administered Through the Intranasal Route

Drug	Nanocarriers/Method	Polymer	Advantage	References
Bromocriptine	Nanoparticle/ionic gelation.	Chitosan	Increase brain uptake.	Md et al. (2013)
Ropinirole	Nanoparticle/ionic gelation.	Chitosan	Significant target efficiency.	Jafarieh et al. (2015)
Levodopa	Nanoparticle/ionic gelation.	Chitosan	Avoid degradation of levodopa in peripheral circulation.	Sharma et al. (2014)
Levodopa	Nanoparticle/double emulsion.	PLGA	Increase bioavailability.	Gambaryan et al. (2014)
Dopamine	Nanoparticle/double emulsion.	Borneol and lactoferrin.	Reduce side effects. And maximum therapeutic efficacy.	Tang et al. (2019)
Rotigotine	Nanoparticle/ nanoprecipitation.	Modified PEG-PLGA	Increase brain uptake.	Bi et al. (2016)
Ropinirole	Solid lipid nanoparticle/ emulsion-solvent diffusion.	Stearylamine	Enhanced stability.	Bi et al. (2016)

6.6.4 NEURO-AIDS

HIV (human immunodeficiency virus) is a life-threatening disease that causes AIDS. Basically, HIV attacks the immune system by destroying a specific white blood cell called CD4 positive (CD4+) (Sagar et al., 2014). However, AIDS-related disorders may infect the CNS, which leads to nerve damage that also causes pain and weakness. So, in that case, the use of antiretroviral drugs for the treatment of neuro-AIDS is important. Drug delivery from the nose to the brain requires less time through the olfactory and trigeminal areas to achieve drug absorption (Sagar et al., 2014; Marzolini et al., 2013) (Table 6.6).

6.6.5 BRAIN TUMOR

Glioblastoma multiforme (GBM) is the most common and aggressive kind of brain cancer in humans. Glioblastoma (GBM) is the most fatal kind of brain tumor, being characterized by the rise and invasion of the surrounding tissue (Bruinsmann et al., 2019). The complex pathophysiology, the development of resistance by tumor cells, and the inability of the medicine to effectively cross the blood-brain and blood-tumor barriers lead to poor prognosis for GBM patients, with a median survival time of only one to two years

TABLE 6.6 Nanotechnological Development of Anti-Neuro-AIDS Drug Administered Through the Intranasal Route

Drug	Nanocarriers/ Method	Polymer	Advantage	References
Efavirenz	Nanoparticle/ nanoprecipitation.	Eudragit (®) RS 100	Increase drug absorption.	Seremeta et al. (2013)
Efavirenz	Polymeric micelles.	Poly(ethylene oxide)–poly(propylene oxide).	Increased bioavailability of drug in CNS.	Chiappetta et al. (2019)
Saquinavir mesylate.	Nanoemulsion/ ultrasonication	Chitosan	Enhanced drug permeability.	Maria et al. (2019)
Didanosine	Nanoparticle/ionic gelation	Chitosan with tripolyphosphate	Increase drug concentration in brain.	Al-Ghananeem et al. (2010)
Prodrug of azidothymidine (U-AZT)	Nanoparticle/ nanoprecipitation.	Chitosan	Great uptake of U-AZT in CSF	Dalpiaz et al. (2019)
Efavirenz	SLNs/HPH	Glyceryl mono stearate, glyceryl tristearate.	Increase drug absorption in brain.	Gupta et al. (2017)
Ritonavir	Nanosuspension/ HPH	Glyceryl palmitostearate	Increase drug bioavailability of drug.	Mulam et al. (2021)

(Izycka-Swieszewska et al., 2018). The present standard treatment for brain tumor is surgery. Although extensive studies have been done over the last years to develop a good therapeutic approach for GBM treatment. Nose-to-brain delivery offers an attractive, noninvasive strategy to enhance drug penetration or transport novel drug/gene carriers into the brain. Despite the fact that the exact mechanism of intranasal administration is unknown. The trigeminal and olfactory pathways are found to play an important role in bypassing the standard barriers of brain targeting (Morsy et al., 2015) (Table 6.7).

6.7 APPROACHES FOR NOSE TO BRAIN DELIVERY

The major drawback in delivering drugs through the nose to the brain is low solubility, lesser permeability, and higher degradation of drugs. To overcome this problem, various novel approaches are being developed for nose to brain delivery. Various novel formulation approaches are explored to enhance nasal drug absorption by various means such as drug solubility and permeability can be easily increased by using suitable polymers, surfactants, pH change,

TABLE 6.7 Nanotechnological Development of Anti-Neuro AIDS Drug Administered Through the Intranasal Route

Drug	Nanocarriers/Method	Polymer	Advantage	References
Bevacizumab	Nanoparticle/emulsification evaporation.	PLGA	–	Sousa et al. (2019)
Carboplatin	Nanoparticle/double emulsion solvent evaporation.	PCL	Improve drug absorption in brain.	Alex et al. (2016)
Temozolomide	Nanoparticle/emulsion solvent evaporation.	PLGA	Enhance cellular uptake of drug.	Chu et al. (2018)
Curcumin	Lipid-based NP/ high pressure homogenization.	–	Good drug permeability.	Madane et al. (2016)
Kaempferol	Nanoemulsion/ high pressure homogenization	Chitosan	Increase drug permeation through biological membranes.	Colombo et al. (2018)
Salirasib	Nanoparticle/solvent evaporation.	PEG-PLGA	Significant therapeutic activity.	Sekardag et al. (2017)
Methotrexate	Nanodispersion/solvent evaporation	PLA	Low mucociliary clearance	Jain et al. (2016)
Camptothecin	Nanomicelle/emulsion evaporation.	PCL	Enhanced drug permeability.	Taki et al. (2012)

etc. (Pandey et al., 2020). Different nasal formulations for better absorption of drugs through the nose to the brain are discussed in succeeding sections.

6.8 APPROACHES FOR NASAL FORMULATIONS

6.8.1 FORMULATION BASE APPROACHES

6.8.1.1 NANOPARTICLES

Nasal administration can be enhanced by using NP-based formulation delivery systems to enhance therapeutic drug absorption. Because NPs can prevent the encapsulated drug from degradation and Pgp efflux pumps, they can improve nasal drug administration and absorption (Prunk et al., 2017).

6.8.1.1.1 Polymeric Nanoparticles (PNPs)

Polymeric nanoparticles (PNPs) have the capability to deliver drugs directly to the target region, and their ability to be loaded with a wide range of drugs makes polymeric NPs (especially those that are biodegradable) widely considered superior to other DDSs utilized in the CNS. Polymeric NPs have various other advantages, including a high surface-to-volume ratio that allows for reduced drug administration frequency and dose. Because of their small size, improved cell absorption, and improved drug accumulation at the target region, such carriers are ideal for intranasal delivery of drugs and improve patient satisfaction (Soppimath et al., 2001; Kreuter et al., 2001; Ahire et al., 2020a).

Biodegradable polymer-based nanoparticles have a wide range of uses in drug delivery and biomedical areas. Biocompatibility, decreased toxicity, biodegradability in various delivery systems, and the ability to encapsulate both hydrophilic and lipophilic drugs are all advantages of biodegradable polymeric NPs (synthetic or natural polymers). Furthermore, NPs that are responsive to environmental stimuli for regulated drug release, in combination with the intranasal route of administration, are innovative therapeutic agent delivery techniques into the brain. These delivery techniques, which use intranasal administration, can bypass the BBB barrier, enhancing drug delivery efficiency and lowering drug adverse effects (George et al., 2019; Rabiee et al., 2021; Ahirrao et al., 2022).

Toxicity and NP aggregation due to nondegradable nature and poor biocompatibility in the brain are two of the most significant disadvantages. Anionic NPs provide a low toxicity potential; whereas, cationic NPs, such as cationic polymeric NPs, can cause platelet aggregation and hemolysis. This involves continuous toxicity monitoring, as the combination of these cationic polymers with biosurfactant causes complications such as dyspnea in immune-compromised patients (de la Harpe et al., 2019; Dhanjal et al., 2021).

6.8.1.1.2 Lipid Nanoparticles

There are two types of lipidic nanoparticles, namely solid lipid nanoparticles (SLNs) and nanostructured lipid carriers (NLC). SLNs, or lipid nanoparticles, are nanoparticles composed of lipids, while NLC is a second generation of lipid nanoparticles and considered an advanced version of SLN. The oils in NLC work by lowering the crystalline degree of SLN's lipid core, preventing the drug from being expelled from the matrix and enhancing drug loading capacity and long-term physical and chemical stability (Muller et al., 2014; Scioli et al., 2020).

They are used for both hydrophilic and hydrophobic drugs trapped in a biocompatible lipid core made up of a lipid or combination of lipids such as compritol 888 ATO, precirol ATO 5, glyceryl monostearate, palmitic acid, stearic acid, and others, and stabilized by surfactant present at the outer shell. SLNs offer a number of benefits, including targeted drug delivery and controlled release, as well as increased bioavailability to reduce dose and undesirable effects. SLNs containing PNPs are less toxic. They are tolerable and biodegradable, with no acute or chronic carrier toxicity, as well as scalability (Yasir et al., 2013; Kumar et al., 2013; Yasir et al., 2008).

Due to an occlusive effect, good application qualities, and SLN adhesion to mucosal membranes, SLNs offer an enhancement to traditional nose-to-brain drug administration since they can protect the encapsulated drug from chemical and biological degradation and may also increase nasal retention time (Kaur et al., 2008).

These nanocarriers have certain drawbacks, such as low capacity to transport hydrophilic drugs; *in vitro* and *in vivo* instability; poor storage stability due to polymorphic lipid transitions that enable the release of encapsulated drug, nanoparticle aggregation, and phase separation. These drawbacks were recognized as the key difficulties to scaling-up and approval of the SLN, NLC, and NE formulations. To avoid these drawbacks, several solutions have been investigated. The addition of anionic or cationic surfactants that enhance the absolute value of ZP, for example, to prevent nanoparticle aggregation during storage, resulted in an increase in the nanoparticles' electrostatic repulsion. The nanoparticles are coated with polyethylene glycol (PEG) or its derivatives to increase the circulation in the bloodstream, which results in improved bioavailability (Cunha et al., 2020; Shobo et al., 2018; Battaglia et al., 2012; Costa et al., 2021).

6.8.1.2 Liposomes

Liposomes are small spherical artificial vesicles made from cholesterol and non-toxic phospholipids. Due to their size and lipophilic and hydrophilic character (besides biocompatibility), liposomes are very effective systems for nose to brain drug delivery (Akbarzadeh et al., 2013). A significant portion of the research has been focused on liposomes in particular. This is due to liposomes' remarkable biocompatibility, ability to carry both hydrophilic and lipophilic drugs, ease of surface modification for targeting or enhancing nasal residence time, and clinical development potential (Gaillard et al., 2012). Liposomes for intranasal delivery have been produced for both hydrophilic and lipophilic drugs as hydrophilic drugs can be entrapped in the aqueous core of liposomes (Keservani et al., 2020), while lipophilic drugs are introduced into the phospholipid bilayer around the core. Lipophilic substances may be indirectly delivered to the brain following intranasal delivery by crossing the BBB after being systemically absorbed, or directly transported to the brain via trigeminal or olfactory nerve pathways. Nasally given hydrophilic and large molecular weight drugs, on the other hand, would mostly employ the direct route because they cannot cross the BBB, even if they are absorbed systemically through the nasal epithelium (Hong et al., 2015, 2019).

Surface positivity of liposomal preparations also helps to increase drug uptake in the brain by increasing/extending drug accumulation over the olfactory region. Furthermore, encapsulating liposomes in a mucoadhesive gel appears to result in a prolonged nasal residence time (Mura et al., 2018).

By transiently promoting tight junction opening between epithelial cells, a permeation enhancer in the drug solution can increase nasal absorption of less permeable drug molecules. However, because of their nasal toxicity, the use of permeation enhancers has been restricted. Also, changes in the nasal environment's tonicity can have a big impact on liposomal formulation stability and drug release rate. Formulations that can limit particle aggregation at nasal pH would be necessary to ensure liposome intranasal stability as cationic liposomes are prone to aggregation in physiological conditions (Davis et al., 2003; Illum, 2002, 2003; Nakanura et al., 2017).

6.8.1.3 MICROSPHERES

Microspheres are biodegradable polymer particles having a matrix-like structure and a diameter of 1–1,000 μm in which the drug is either dissolved or homogeneously disseminated (Schiffter, 2011). Microspheres, despite their small size, show higher drug-carrying capacity but exhibit a smaller residence time. To address these challenges, formulating mucoadhesive microspheres has been considered a significant technique for prolonging the residence duration in the nasal cavity, as it can delay mucociliary clearance of drug molecules from the nasal cavity. Polymers such as chitosan, Carbopol, starch, hyaluronic acid, and polyacrylic acid have all shown promise as mucoadhesive agents for use in nasal administration, either alone or in combination with another excipient (Yaqoob et al., 2021; Sun et al., 2012).

Microsphere technology is one of the specialized systems that is increasingly popular for developing nasal medications because it may provide continuous contact with the nasal mucosa, which can improve absorption and bioavailability. The nasal mucosa gets dried when microspheres are present because the microspheres absorb moisture. This causes reversible cell shrinkage, resulting in a transient physical dissociation of the tight (intercellular) junctions, which improves drug absorption. The rationale for using a microsphere system is that the preparation of bioadhesive microspheres (in powder form) with good bioadhesive qualities would allow these microspheres to swell in contact with nasal mucosa to form a gel, allowing the pace of clearance from the nasal cavity to be controlled, allowing poorly absorbed drugs to be available for a longer time at the absorptive surface (Chaturvedi et al., 2011; Pereswetoff-Morath et al., 1998; Shaikh et al., 2011; Rathanan et al., 2007).

Collectively, these systems provide a prolonged and constant therapeutic effect, produce more reproducible drug absorption, and bioadhesive microspheres offer prolonged contact of drug products with biological tissues, as well as the extension of contact at the target site. Longer residence periods improve drug absorption effects, allowing for greater control of drug release characteristics, reduced dose frequency, and improved patient compliance (Solanki et al., 2018; Rajput et al., 2012).

Nose to brain delivery requires a smaller payload, but controlled release formulations generally contain a higher drug load, and thus any loss of integrity of the release characteristics of the dosage form may lead to dumping of the dose, resulting in the failure of therapy and potential toxicity. Addition of plasticizer and other polymers may cause irritation to the nasal cavity on repetition and may lead to damage of tissue (Poovi et al., 2014).

6.8.1.4 NANOEMULSIONS (NES)

Nanoemulsions (NEs) are defined as the heterogeneous dispersion of two immiscible liquids (water in oil or oil in water) with a mean droplet size of 20–500 nm (Holban et al., 2016). NEs can be made into a variety of dosage forms including creams, gels, sprays, foams, and more, making them an excellent dosage form for nose to brain administration (Sarkar, 2005). NEs have a higher surface area than other formulations and have longer physical stability because of their small droplet size, which prevents destabilizing phenomena including coalescence, creaming, and sedimentation (Bonferoni et al., 2019).

O/W NEs are a potential option for the encapsulation of hydrophobic drugs, shielding them from enzymatic degradation, enhancing their solubility in liquid media, regulating their drug release, and boosting their bioavailability, especially for nose to brain transport (Sessa et al., 2014; Talele et al., 2021).

Coating with bio-adhesive polymer extends the nasal residence time of NEs to overcome nasal clearance, achieve enhanced mucosal absorption, and facilitate high influx of drug from nose to brain. NEs have also been shown to reduce the side effects and toxicity of drugs. The very small globule size of NEs shows greater permeation of the formulation into the brain. Meanwhile, zeta potential plays a significant role in drug retention time. Formulations carrying positive charges depict good attachment to the nasal mucosa, indicating a higher retention time of NEs in the nasal cavity, and NEs can be formulated as controlled release formulations (Ahmad et al., 2017; Viet Nguyen et al., 2019; Samaridou et al., 2018).

Nano-emulsion looks like a promising approach for the nose to brain delivery of drugs, but its stability is still a concern as phase inversion, creaming, and coalescence may arise in NEs, which will restrict the applicability of NEs in nose to brain delivery. Also, the use of surfactants and permeation enhancers may cause ciliotoxicity on repetitive doses.

6.8.1.5 NANOSUSPENSIONS

Nanosuspensions, defined as the dispersion of crystalline particles with diameters less than 1,000 nm in aqueous solutions, can be employed as an alternative formulation approach for intranasal delivery of drugs that are poorly water soluble. Nanosuspensions are commonly referred to as "carrier-free" DDSs as they require a very low concentration of stabilizers (surfactants, polymers, or both) (Patravale et al., 2004; Muller et al., 2011).

Nanosuspension formulations are free of cytotoxic solubilizers and cosolvents, allowing a faster rate of drug dissolution and enhanced apparent solubility, as well as increased adhesiveness to surface/cell membranes, both of which can facilitate particle retention on the nasal mucosa. They also allow for increased drug loading, which is especially beneficial in nasal formulations due to dosage volume restrictions, and show more targeted drug delivery (Muller et al., 1999; Ahire et al., 2018, 2020b).

Nanosuspensions contain very small particle size, thus offering more surface area, which results in a greater dissolution rate. The addition of surfactant also gives higher stability to the formulation. Coating of nanosuspension will lead to the controlled release of the drug and prolonged residence at the site of action. Smaller particle size also tends to permeate the drug into the brain easily and shows enhanced bioavailability (Paila et al., 2019; Patel et al., 2021).

Though nanosuspension technology has grown rapidly in recent decades, the in vivo safety and health implications of nanosuspension should be considered during development and scale-up. Sudden and rapid bursts of drug release may cause toxicity to the nasal membrane, thus coating or tailoring needs to be done for the drug molecule to avoid toxicity (Jacob et al., 2020; Yadollahi et al., 2015).

6.8.1.6 NANOGELS

Nanogels are three-dimensional hydrogel materials with a high capacity to hold water that are created by crosslinked swellable polymer networks that do not dissolve in the aqueous medium. Nanogels can be made from a number of natural or synthetic polymers or a combination of the two. The chemical composition of nanogels can be changed to fine-tune their properties such as size, charge, porosity, amphiphilicity, softness, and degradability. They are generally spherical particles, although recent advances in synthetic techniques have made it possible to make nanogels of various shapes (Rolland et al., 2005; Kersey et al., 2012; Thombre et al., 2022).

Nanogels are highly biocompatible and have a large loading capacity for guest molecules due to their hydrophilic nature, and their distinctive physical features provide them considerable advantages over other forms of nanomaterials for biomedical applications. Nanogels not only preserve the cargo from degradation and elimination, but they also help generate a controlled, triggered response at the target site due to their exceptional qualities such as stimuli-responsive behavior, softness, and swelling (Kabanov et al., 2009; Soni et al., 2016).

Chemically crosslinked polymers are more useful in biomedical applications due to their size stability and reproducibility compared to physically crosslinked nanogels. The network has a large amount of interior space for drug loading in their inflated condition, resulting in a high drug loading capacity. Factors including the chemical nature of the polymer matrix, the degree of crosslinking, external triggers, and charge density influence their swelling ability. Solvent penetrates the open spaces when they swell, causing a change in volume that can be rapid or slow depending on the degree of crosslinking. Their ability to expand and contract is beneficial for drug loading and release. Their soft nature influences their bio-distribution, resulting in increased bodily circulation. Because of their compact size, they may precisely target the appropriate areas, resulting in increased cellular absorption (Vinogradov et al., 2004; Raemdonck et al., 2009; Molina et al., 2015; Aderibigbe et al., 2018).

Nanogels can deliver both hydrophilic and lipophilic drugs, and intranasal delivery can cross the BBB, making them effective for treating brain illnesses. When compared to the administration of a free drug, nanogels have been demonstrated to be more successful in terms of reduced drug toxicity, increased cellular absorption of the drug, high drug loading, and controlled release of the loaded drug at the targeted site (Neamtu et al., 2017; Tahara et al., 2015).

However, the type of material used to prepare nanogels for intranasal administration can result in loss of epithelial cells, the shrinkage of the mucosal layer, and loss of the ciliary layer (Aderibigbe et al., 2018).

6.8.2MMISCELLANEOUS APPROACHES

6.8.2.1 PRODRUG APPROACH

Most drugs that are taken by the IN route are designed as solutions or powders, which require a dissolution process before absorption. Lipophilic drugs permeate easily through biological membranes. The purpose of prodrugs is to increase drug penetration by increasing lipophilicity or water solubility. The prodrug must have appropriate water solubility and stability, adequate hydrophobicity, adequate safety, and a reasonable in vivo transformation to the parent drug. Targeted prodrug strategies are novel directions in the treatment of brain diseases that aim to improve the therapeutic index and reduce cytotoxic therapeutic effects. Prodrugs can selectively concentrate at desired cells and demonstrate activity there. Prodrug-based nano-systems are also of great interest since they can give major benefits such as enhanced chemical stability in vivo, longer drug release times, and reduced toxicity before degradation. Enzymatic activation of prodrugs in hypoxic cells, as well as the usage of nanoparticles, are all important ways for achieving local activation of prodrugs (Zeynep et al., 2021; Xie et al., 2020). The prodrug approach does not increase only the solubility of the drug but also shows site-specific action, longer drug release, and better stability, thus it can be applicable to deliver the drug into the brain via the nasal cavity.

6.8.2.2 *MUCOADHESIVE SYSTEM*

As we discussed earlier in microsphere, the addition of mucoadhesive polymer into the formulation gives better retention of the drug in nasal mucosa with controlled drug release from the formulation at the absorption site. It also delays the nasal clearance. Mucoadhesive DDSs interact with the mucus layer covering the mucosal epithelial surface and mucin molecules, prolonging the residence time of the dosage form at the site of absorption. They promote an intimate contact of the dosage form with the underlying absorption surface and thus improve the therapeutic activity of the drug. The mucoadhesive system shows targeting and localization of the dosage form at a specific site (Boddupalli et al., 2010). Thus, formulating a mucoadhesive system for nose to brain delivery is one of the promising approaches.

Various formulations can be prepared by using bioadhesive polymer as a mucoadhesive system, such as microspheres, nanogels, NEs, liposomes, etc.

The only drawback of these systems is that the bioadhesive polymer should be of natural origin and safe; otherwise, repetitive exposure of this polymer to nasal mucosa may cause damage to the tissues of the nasal cavity.

6.8.2.3 *ENZYME INHIBITORS*

Many enzymes, such as CYP450 isoforms, transferases, carboxylesterases, and others, are found in the nasal cavity and are responsible for drug metabolism. Inhibition of certain enzymes may improve the stability of the compounds in situ, avoiding their biotransformation, and so increasing the quantity of active drug that can be absorbed into the brain to produce an effect (Wong et al., 2010; Martins et al., 2005).

P-glycoproteins (Pgp) are membrane transporters that allow drugs to leave the brain and are located in the BBB, nasal mucosa, olfactory epithelium, and olfactory bulb. Not all medication is Pgp substrates. Pgp substrates may lose their nose-to-brain advantage and be transferred from the brain into the bloodstream with little or no effect (Graff et al., 2005).

Peptidase inhibitors reduce nasal metabolism and increase residence time, allowing for better absorption and bioavailability. Bestatin, amastatin, boroleucine, fusidic acids, and phospholipids are some of the most commonly used enzyme inhibitors (Trevino et al., 2020).

Thus, the enzyme inhibition approach may be useful for the improvement in absorption of drugs from the nasal cavity and also enhanced permeation through the nasal cavity.

6.8.2.4 PERMEATION ENHANCERS

The olfactory and respiratory epithelium's tight connections, as well as their protective mucus lining, operate as selective filters, reducing permeability and diffusion. Drug lipophilicity is important during passive diffusion, whereas a longer nasal residence period is important during active transport. Because of low permeability and poor absorption via the endothelium basement membrane, absorption through the olfactory epithelium is reduced for drugs with molecular weights greater than 1,000 Da. Permeation enhancers have been tested to help medications with big molecular weights absorb better. Increased membrane fluidity and tight junction permeability, hydrophilic pore formation, and reduced viscosity and enzymatic activity are all proposed mechanisms (Ruigrok et al., 2015; Trevino et al., 2020).

Permeation enhancers act by opening tight junctions in addition to improving adhesion of drug to the absorption site and enhancing the absorption of drug through nasal cavity and also prolonging residence time in the nasal mucosa (Bhise et al., 2008).

6.8.2.5 CO-SOLVENT APPROACH

Cosolvents are mixtures of miscible solvents that can drastically affect the solubility of drugs that are poorly aqueous soluble. Water solubility is usually poor in weakly electrolytes and nonpolar compounds. Their solubility is usually improved by adding a water miscible solvent to which the drug is well soluble. Cosolvency is the term for this procedure, and cosolvents are the solvents used to promote solubility (Yalkowsky et al., 1981).

Hydrogen bond donor and/or acceptor groups, as well as small hydrocarbon regions, are found in most cosolvents. Their hydrophilic hydrogen-bonding groups maintain water miscibility, while their lipophilic hydrocarbon regions disrupt water's hydrogen bonding network, lowering water's overall intermolecular attraction. Cosolvents limit water's ability to squeeze out nonpolar, hydrophobic molecules by interrupting self-association and thus, enhancing solubility. The benefits of using cosolvent technology to improve drug solubility in a liquid-based formulation include convenience, which removes the need to mix the solvent before administration; safety, which prevents contamination during the dispensing process; and low cost, which eliminates the need for expensive pharmaceutical technology for dosage form formulation (Millard et al., 2002; Nayak et al., 2012).

Thus, the cosolvent approach can be useful to improve the water solubility of drugs and also provides a higher concentration of the drug at the site of absorption leading to better permeation through the nasal mucosa.

6.9 CONCLUSION

Nose to brain drug delivery is a novel approach for the treatment of diseases and disorders related to the CNS. It is an emerging option for delivering therapeutics to brain tissues for the treatment and prevalence of conditions like migraine, Alzheimer's, and other neurotherapeutic deliveries. For the delivery of protein, peptide, and vaccine, the nasal route is proven to be the optimum route. Now, it is an emerging route for the delivery of antiviral drugs to the CNS to clear off infections from the CNS. CNS drug transport across the nasal route bypasses the blood-brain barrier, which is the biggest advantage associated with it. For drugs with high pre-systemic metabolism, this is an alternative. If we overcome the hurdles of nasal delivery by using optimum formulation ingredients and technology, it is the best solution to deliver potent medicine.

KEYWORDS

- **blood-brain barrier**
- **mucociliary clearance**
- **nasal drug delivery barriers**
- **nose to brain delivery**
- **novel drug deliveries**
- **olfactory epithelium**
- **olfactory nerve pathway**
- **respiratory epithelium**
- **trigeminal nerve pathway**

REFERENCES

Abbas, Z., & Marihal, S., (2014). Gellan gum-based mucoadhesive microspheres of almotriptan for nasal administration: Formulation optimization using factorial design, characterization, and *in vitro* evaluation. *Journal of Pharmacy & Bioallied Sciences, 6*(4), 267.

Abe, K., Irie, T., & Uekama, K., (1995). Enhanced nasal delivery of luteinizing hormone-releasing hormone agonist buserelin by oleic acid solubilized and stabilized in hydroxypropyl-β-cyclodextrin. *Chemical and Pharmaceutical Bulletin, 43*(12), 2232–2237.

Adams, C. P., & Flynn, J., (2006). *Inhibition of Olfactory Neurosensory Function to Treat Eating Disorders and Obesity.* U.S. Patent 7,138,107. Compellis Pharmaceuticals.

Adejare, A., (2020). *Remington: The Science and Practice of Pharmacy.* Academic Press.

Aderibigbe, B. A., & Naki, T., (2018). Design and efficacy of nanogels formulations for intranasal administration. *Molecules, 23*(6), 1241.

Agrawal, M., Saraf, S., Saraf, S., Antimisiaris, S. G., Chougule, M. B., Shoyele, S. A., & Alexander, A., (2018). Nose-to-brain drug delivery: An update on clinical challenges and progress towards approval of anti-Alzheimer drugs. *Journal of Controlled Release, 281*, 139–177.

Ahire, E. D., Talele, S. G., & Shah, H. S., (2020). Nanoparticles as a promising technology in microbial pharmaceutics. In: *Applied Pharmaceutical Science and Microbiology* (pp. 133–158). Apple Academic Press.

Ahire, E., Thakkar, S., Borade, Y., & Misra, M., (2020). Nanocrystal based orally disintegrating tablets as a tool to improve dissolution rate of Vortioxetine. *Bulletin of Faculty of Pharmacy, Cairo University, 58*(1, 2), 11–20.

Ahire, E., Thakkar, S., Darshanwad, M., & Misra, M., (2018). Parenteral nanosuspensions: A brief review from solubility enhancement to more novel and specific applications. *Acta Pharmaceutica Sinica B, 8*(5), 733–755.

Ahirrao, S. P., Sonawane, M. H., Bhambere, D. S., Udavant, P. B., Ahire, E. D., & Kanade, R., (2022). Cocrystal formulation: A novel approach to enhance solubility and dissolution of etodolac. *Biosciences Biotechnology Research Asia, 19*(1), 111.

Ahmad, E., Feng, Y., Qi, J., Fan, W., Ma, Y., He, H., Xia, F., et al., (2017). Evidence of nose-to-brain delivery of nanoemulsions: Cargoes but not vehicles. *Nanoscale, 9*(3), 1174–1183.

Akbarzadeh, A., Rezaei-Sadabady, R., Davaran, S., Joo, S. W., Zarghami, N., Hanifehpour, Y., Samiei, M., et al., (2013). Liposome: Classification, preparation, and applications. *Nanoscale Research Letters, 8*(1), 1–9.

Alex, A. T., Joseph, A., Shavi, G., Rao, J. V., & Udupa, N., (2016). Development and evaluation of carboplatin-loaded PCL nanoparticles for intranasal delivery. *Drug Delivery, 23*(7), 2144–2153.

Al-Ghananeem, A. M., Saeed, H., Florence, R., Yokel, R. A., & Malkawi, A. H., (2010). Intranasal drug delivery of didanosine-loaded chitosan nanoparticles for brain targeting; an attractive route against infections caused by AIDS viruses. *Journal of Drug Targeting, 18*(5), 381–388.

Ali, J., Ali, M., Baboota, S., Kaur, S. J., Ramassamy, C., & Dao, L., (2010). Potential of nanoparticulate drug delivery systems by intranasal administration. *Current Pharmaceutical Design, 16*(14), 1644–1653.

Arnold, M., (2018). Headache classification committee of the international headache society (IHS) the international classification of headache disorders. *Cephalalgia, 38*(1), 1–211.

Battaglia, L., & Gallarate, M., (2012). Lipid nanoparticles: State of the art, new preparation methods and challenges in drug delivery. *Expert Opinion on Drug Delivery, 9*(5), 497–508.

Beukelaar, D., (1999). Pharmacokinetic profile of alniditan nasal spray during and outside migraine attacks. *British Journal of Clinical Pharmacology, 47*(3), 285–290.

Bhattacharya, S., Maelicke, A., & Montag, D., (2015). Nasal application of the galantamine pro-drug Memogain slows down plaque deposition and ameliorates behavior in 5X familial Alzheimer's disease mice. *Journal of Alzheimer's Disease, 46*(1), 123–136.

Bhise, S. B., Yadav, A. V., Avachat, A. M., & Malayandi, R., (2008). Bioavailability of intranasal drug delivery system. *Asian Journal of Pharmaceutics (AJP), 2*(4).

Bi, C., Wang, A., Chu, Y., Liu, S., Mu, H., Liu, W., Wu, Z., et al., (2016). Intranasal delivery of rotigotine to the brain with lactoferrin-modified PEG-PLGA nanoparticles for Parkinson's disease treatment. *International Journal of Nanomedicine, 11,* p.6547.

Boddupalli, B. M., Mohammed, Z. N., Nath, R. A., & Banji, D., (2010). Mucoadhesive drug delivery system: An overview. *Journal of Advanced Pharmaceutical Technology & Research, 1*(4), 381.

Bonferoni, M. C., Rossi, S., Sandri, G., Ferrari, F., Gavini, E., Rassu, G., & Giunchedi, P., (2019). Nanoemulsions for nose-to-brain" drug delivery. *Pharmaceutics, 11*(2), 84.

Bors, L. A., & Erdő, F., (2019). Overcoming the blood–brain barrier. challenges and tricks for CNS drug delivery. *Scientia Pharmaceutica, 87*(1), 6.

Braak, H., & Del, T. K., (2008). Reply to controversies over the staging of α-synuclein pathology in Parkinson's disease. *Acta Neuropathologica, 116*(1), 129–131.

Brambilla, D., Verpillot, R., Le Droumaguet, B., Nicolas, J., Taverna, M., Kóňa, J., Lettiero, B., et al., (2012). PEGylated nanoparticles bind to and alter amyloid-beta peptide conformation: Toward engineering of functional nanomedicines for Alzheimer's disease. *ACS Nano, 6*(7), 5897–5908.

Bruinsmann, F. A., Richter, V. G., De Cristo, S. A. A., Aguirre, T., Raffin, P. A., Stanisçuaski, G. S., & Sonvico, F., (2019). Nasal drug delivery of anticancer drugs for the treatment of glioblastoma: Preclinical and clinical trials. *Molecules (Basel, Switzerland), 24*(23), 4312.

Chattopadhyay, S., Das, S., & Sarma, K. N., (2021). Nose-to-brain drug delivery: An update to the alternative path to successful targeted anti-migraine drugs. *Int. J. Appl. Pharm., 13*(2), 67–75.

Chaturvedi, M., Kumar, M., & Pathak, K., (2011). A review on mucoadhesive polymer used in nasal drug delivery system. *Journal of Advanced Pharmaceutical Technology & Research, 2*(4), 215.

Chen, J., Wang, X., Wang, J., Liu, G., & Tang, X., (2008). Evaluation of brain-targeting for the nasal delivery of ergoloid mesylate by the microdialysis method in rats. *European Journal of Pharmaceutics and Biopharmaceutics, 68*(3), 694–700.

Chen, J., Zhang, C., Liu, Q., Shao, X., Feng, C., Shen, Y., Zhang, Q., & Jiang, X., (2012). Solanum tuberosum lectin-conjugated PLGA nanoparticles for nose-to-brain delivery: *In vivo* and *in vitro* evaluations. *Journal of Drug Targeting, 20*(2), 174–184.

Chen, M. W., Morsy, A. A., Liang, S., & Ng, W. H., (2016). Re-do craniotomy for recurrent grade IV glioblastomas: Impact and outcomes from the National Neuroscience Institute Singapore. *World Neurosurgery, 87*, 439–445.

Chiappetta, D. A., Hocht, C., Opezzo, J. A., & Sosnik, A., (2013). Intranasal administration of antiretroviral-loaded micelles for anatomical targeting to the brain in HIV. *Nanomedicine, 8*(2), 223–237.

Chu, L., Wang, A., Ni, L., Yan, X., Song, Y., Zhao, M., Sun, K., et al., (2018). Nose-to-brain delivery of temozolomide-loaded PLGA nanoparticles functionalized with anti-EPHA3 for glioblastoma targeting. *Drug Delivery, 25*(1), 1634–1641.

Colombo, G., Lorenzini, L., Zironi, E., Galligioni, V., Sonvico, F., Balducci, A. G., Pagliuca, G., et al., (2011). Brain distribution of ribavirin after intranasal administration. *Antiviral Research, 92*(3), 408–414.

Colombo, M., Figueiró, F., De Fraga, D. A., Teixeira, H. F., Battastini, A. M. O., & Koester, L. S., (2018). Kaempferol-loaded mucoadhesive nanoemulsion for intranasal administration reduces glioma growth *in vitro*. *International Journal of Pharmaceutics, 543*(1, 2), 214–223.

Costa, C. P., Moreira, J. N., Lobo, J. M. S., & Silva, A. C., (2021). Intranasal delivery of nanostructured lipid carriers, solid lipid nanoparticles and nanoemulsions: A current overview of *in vivo* studies. *Acta Pharmaceutica Sinica B, 11*(4), 925–940.

Cunha, S., Costa, C. P., Moreira, J. N., Lobo, J. M. S., & Silva, A. C., (2020). Using the quality by design (QbD) approach to optimize formulations of lipid nanoparticles and nanoemulsions: A review. *Nanomedicine: Nanotechnology, Biology and Medicine, 28*, 102206.

Dalpiaz, A., & Pavan, B., (2018). Nose-to-brain delivery of antiviral drugs: A way to overcome their active efflux? *Pharmaceutics, 10*(2), 39.

Dalpiaz, A., Fogagnolo, M., Ferraro, L., Beggiato, S., Hanuskova, M., Maretti, E., Sacchetti, F., et al., (2019). Bile salt-coating modulates the macrophage uptake of nanocores constituted by a zidovudine prodrug and enhances its nose-to-brain delivery. *European Journal of Pharmaceutics and Biopharmaceutics, 144*, 91–100.

Daneman, R., & Prat, A., (2015). The blood–brain barrier. *Cold Spring Harbor Perspectives in Biology, 7*(1), a020412.

Davis, S. S., & Illum, L., (2003). Absorption enhancers for nasal drug delivery. *Clinical Pharmacokinetics, 42*(13), 1107–1128.

De La Harpe, K. M., Kondiah, P. P., Choonara, Y. E., Marimuthu, T., Du Toit, L. C., & Pillay, V., (2019). The hemocompatibility of nanoparticles: A review of cell–nanoparticle interactions and hemostasis. *Cells, 8*(10), 1209.

Dhanjal, D. S., Mehta, M., Chopra, C., Singh, R., Sharma, P., Chellappan, D. K., Tambuwala, M. M., et al., (2021). Novel controlled release pulmonary drug delivery systems: Current updates and challenges. In: *Modeling and Control of Drug Delivery Systems* (pp. 253–272). Academic Press.

Dhuria, S. V., Hanson, L. R., & Frey, II. W. H., (2010). Intranasal delivery to the central nervous system: Mechanisms and experimental considerations. *Journal of Pharmaceutical Sciences, 99*(4), 1654–1673.

Di Stefano, A., Iannitelli, A., Laserra, S., & Sozio, P., (2011). Drug delivery strategies for Alzheimer's disease treatment. *Expert Opinion on Drug Delivery, 8*(5), 581–603.

Djupesland, P. G., (2013). Nasal drug delivery devices: Characteristics and performance in a clinical perspective—A review. *Drug Delivery and Translational Research, 3*(1), 42–62.

Fazil, M., Md, S., Haque, S., Kumar, M., Baboota, S., Kaur, S. J., & Ali, J., (2012). Development and evaluation of rivastigmine-loaded chitosan nanoparticles for brain targeting. *European Journal of Pharmaceutical Sciences, 47*(1), 6–15.

Frey, I. W. H., Panter, S. S., Hanson, L. R. B., & Roeytenberg, A., (2010). *Method of Treating Alzheimer's Disease Comprising Administering Deferoxamine (DFO) to the Upper One-Third of the Nasal Cavity*. U.S. Patent 7,776,312. Health Partners Research Foundation.

Gaillard, P. J., Appeldoorn, C. C., Rip, J., Dorland, R., Van, D. P. S. M., Kooij, G., De Vries, H. E., & Reijerkerk, A., (2012). Enhanced brain delivery of liposomal methylprednisolone improved therapeutic efficacy in a model of neuroinflammation. *Journal of Controlled Release, 164*(3), 364–369.

Gambaryan, P. Y., Kondrasheva, I. G., Severin, E. S., Guseva, A. A., & Kamensky, A. A., (2014). Increasing the efficiency of Parkinson's disease treatment using a poly (lactic-co-glycolic acid)(PLGA) based L-DOPA delivery system. *Experimental Neurobiology, 23*(3), 246.

Ganesan, P., Ramadass, J., Johnson, A. J. D., Sabapathy, L., & Duraikannu, A., (2015). Preparation and characterization of chemical cross-linked oral controlled release of atorvastatin calcium-loaded ethyl cellulose microsphere. *Advanced Science, Engineering and Medicine, 7*(11), 980–989.

George, A., Shah, P. A., & Shrivastav, P. S., (2019). Natural biodegradable polymers-based nano-formulations for drug delivery: A review. *International Journal of Pharmaceutics, 561*, 244–264.

Graff, C. L., Zhao, R., & Pollack, G. M., (2005). Pharmacokinetics of substrate uptake and distribution in murine brain after nasal instillation. *Pharmaceutical Research, 22*(2), 235–244.

Gujarathi, N. A., Bakliwal, A. A., Rane, B., Pathan, V., & Keservani, R. K., (2022). Nanoencapsulated nasal drug delivery system; Chapter 8. In: Nayan, A. G., Juliana, P. A., Raj, K. K., & Anil, K. S., (eds.), *Topical and Transdermal Drug Delivery Systems: Applications and Future Prospects* (pp. 235–257). Academic Press, Taylor & Francis. ISBN: 9781774910702.

Gupta, S., Kesarla, R., Chotai, N., Misra, A., & Omri, A., (2017). Systematic approach for the formulation and optimization of solid lipid nanoparticles of efavirenz by high pressure homogenization using design of experiments for brain targeting and enhanced bioavailability. *Biomed Research International, 2017.*

Hanafy, A. S., Farid, R. M., Helmy, M. W., & El Gamal, S. S., (2016). Pharmacological, toxicological, and neuronal localization assessment of galantamine/chitosan complex nanoparticles in rats: Future potential contribution in Alzheimer's disease management. *Drug Delivery, 23*(8), 3111–3122.

Hirlekar, R. S., & Momin, A. M., (2018). Advances in drug delivery from nose to brain: An overview. *Current Drug Therapy, 13*(1), 4–24.

Holban, A. M., & Grumezescu, A. M., (2016). *Nanoarchitectonics for Smart Delivery and Drug Targeting.* William Andrew.

Hong, S. S., Kim, S. H., & Lim, S. J., (2015). Effects of triglycerides on the hydrophobic drug loading capacity of saturated phosphatidylcholine-based liposomes. *International Journal of Pharmaceutics, 483*(1, 2), 142–150.

Hong, S. S., Oh, K. T., Choi, H. G., & Lim, S. J., (2019). Liposomal formulations for nose-to-brain delivery: Recent advances and future perspectives. *Pharmaceutics, 11*(10), 540.

Illum, L., (1999). Bioadhesive formulations for nasal peptide delivery. *Drugs and the Pharmaceutical Sciences, 98*, 507–539.

Illum, L., (2000). Transport of drugs from the nasal cavity to the central nervous system. *Eur. J. Pharmac. Sci., 11*, 1–18.

Illum, L., (2002). Nasal drug delivery: New developments and strategies. *Drug Discovery Today, 7*(23), 1184–1189.

Illum, L., (2003). Nasal drug delivery—possibilities, problems, and solutions. *Journal Control Release 87*, 187–198.

Illum, L., (2003). Nasal drug delivery—possibilities, problems, and solutions. *Journal of Controlled Release, 87*(1–3), 187–198.

Illum, L., (2006). Nasal clearance in health and disease. *Journal of Aerosol Medicine, 19*(1), 92–99.

Illum, L., Farraj, N. F., & Davis, S. S., (1994). Chitosan is a novel nasal delivery system for peptide drugs. *Pharmaceutical Research, 11*(8), 1186–1189.

Illum, L., Watts, P., Fisher, A. N., Hinchcliffe, M., Norbury, H., Jabbal-Gill, I., Nankervis, R., & Davis, S. S., (2002). Intranasal delivery of morphine. *Journal of Pharmacology and Experimental Therapeutics, 301*(1), 391–400.

Izycka-Swieszewska, E., Bien, E., Stefanowicz, J., Szurowska, E., Szutowicz-Zielinska, E., Koczkowska, M., Sigorski, D., et al., (2018). Malignant gliomas as second neoplasms in pediatric cancer survivors: Neuropathological study. *BioMed Research International, 2018.*

Jacob, S., Nair, A. B., & Shah, J., (2020). Emerging role of nanosuspensions in drug delivery systems. *Biomaterials Research, 24*(1), 1–16.

Jacobs, M. A., Schreuder, R. H., Jap-A-Joe, K., Nauta, J. J., Andersen, P. M., & Heine, R. J., (1993). The pharmacodynamics and activity of intranasally administered insulin in healthy male volunteers. *Diabetes, 42*(11), 1649–1655.

Jadhav, K. R., Gambhire, M. N., Shaikh, I. M., Kadam, V. J., & Pisal, S. S., (2007). Nasal drug delivery system-factors affecting and applications. *Current Drug Therapy, 2*(1), 27–38.

Jafarieh, O., Md, S., Ali, M., Baboota, S., Sahni, J. K., Kumari, B., Bhatnagar, A., & Ali, J., (2015). Design, characterization, and evaluation of intranasal delivery of ropinirole-loaded mucoadhesive nanoparticles for brain targeting. *Drug Development and Industrial Pharmacy, 41*(10), 1674–1681.

Jain, D. S., Bajaj, A. N., Athawale, R. B., Shikhande, S. S., Pandey, A., Goel, P. N., Gude, R. P., Patil, S., & Raut, P., (2016). Thermosensitive PLA based nanodispersion for targeting brain tumor via intranasal route. *Materials Science and Engineering: C, 63*, 411–421.

Janneh, O., Jones, E., Chandler, B., Owen, A., & Khoo, S. H., (2007). Inhibition of P-glycoprotein and multidrug resistance-associated proteins modulates the intracellular concentration of lopinavir in cultured CD4 T cells and primary human lymphocytes. *Journal of Antimicrobial Chemotherapy, 60*(5), 987–993.

Kabanov, A. V., & Vinogradov, S. V., (2009). Nanogels as pharmaceutical carriers: Finite networks of infinite capabilities. *Angewandte Chemie International Edition, 48*(30), 5418–5429.

Kersey, F. R., Merkel, T. J., Perry, J. L., Napier, M. E., & DeSimone, J. M., (2012). Effect of aspect ratio and deformability on nanoparticle extravasation through nanopores. *Langmuir, 28*(23), 8773–8781.

Keservani, R. K., & Gautam, S. P., (2022). Skeletal muscle relaxant activity of different formulation of span 60 niosome. *ARS Pharmaceutica, 63*(1), 32–44.

Keservani, R. K., Sharma, A. K., & Kesharwani, R. K., (2019). *Nanocarriers for Brain Targeting: Principles and Applications.* CRC Press. ISBN:9781771887304.

Krause, K. P., & Müller, R. H., (2001). Production and characterization of highly concentrated nanosuspensions by high-pressure homogenization. *International Journal of Pharmaceutics, 214*(1, 2), 21–24.

Kreuter, J., (2001). Nanoparticulate systems for brain delivery of drugs. *Advanced Drug Delivery Reviews, 47*(1), 65–81.

Krishan, M., Gudelsky, G. A., Desai, P. B., & Genter, M. B., (2014). Manipulation of olfactory tight junctions using papaverine to enhance intranasal delivery of gemcitabine to the brain. *Drug Delivery, 21*(1), 8–16.

Kubek, M. J., (2007). *Pharmacotherapeutic Process and Composition for Central Nervous System Disorders*. United States patent US 7,229,635. Inventor; Indiana University Research, Technology Corp, assignee.

Kublik, H., & Vidgren, M. T., (1998). Nasal delivery systems and their effect on deposition and absorption. *Advanced Drug Delivery Reviews, 29*(1, 2), 157–177.

Kulkarni, A. D., Vanjari, Y. H., Sancheti, K. H., Belgamwar, V. S., Surana, S. J., & Pardeshi, C. V., (2015). Nanotechnology-mediated nose to brain drug delivery for Parkinson's disease: A mini review. *Journal of Drug Targeting, 23*(9), 775–788.

Kumar, M., Misra, A., Babbar, A. K., Mishra, A. K., Mishra, P., & Pathak, K., (2008). Intranasal nanoemulsion based brain targeting drug delivery system of risperidone. *International Journal of Pharmaceutics, 358*(1, 2), 285–291.

Kumar, R., Yasir, M., Saraf, S. A., Gaur, P. K., Kumar, Y., & Singh, A. P., (2013). Glyceryl monostearate based nanoparticles of mefenamic acid: Fabrication and *in vitro* characterization. *Drug Invention Today, 5*(3), 246–250.

Li, J., Zhao, J., Tan, T., Liu, M., Zeng, Z., Zeng, Y., Zhang, L., Fu, C., Chen, D., & Xie, T., (2020). Nanoparticle drug delivery system for glioma and its efficacy improvement strategies: A comprehensive review. *International Journal of Nanomedicine, 15*, 2563.

Lochhead, J. J., & Thorne, R. G., (2012). Intranasal delivery of biologics to the central nervous system. *Advanced Drug Delivery Reviews, 64*(7), 614–628.

Machida, M., Sano, K., Arakawa, M., Hayashi, M., & Awazu, S., (1993). Absorption of recombinant human granulocyte colony-stimulating factor (rhG-CSF) from rat nasal mucosa. *Pharmaceutical Research, 10*(9), 1372–1377.

Madane, R. G., & Mahajan, H. S., (2016). Curcumin-loaded nanostructured lipid carriers (NLCs) for nasal administration: Design, characterization, and *in vivo* study. *Drug Delivery, 23*(4), 1326–1334.

Martins, P. P., Smyth, H. D., & Cui, Z., (2019). Strategies to facilitate or block nose-to-brain drug delivery. *International Journal of Pharmaceutics, 570*, 118635.

Marzolini, C., Mueller, R., Li-Blatter, X., Battegay, M., & Seelig, A., (2013). The brain entry of HIV-1 protease inhibitors is facilitated when used in combination. *Molecular Pharmaceutics, 10*(6), 2340–2349.

McMartin, C., Hutchinson, L. E., Hyde, R., & Peters, G. E., (1987). Analysis of structural requirements for the absorption of drugs and macromolecules from the nasal cavity. *Journal of Pharmaceutical Sciences, 76*(7), 535–540.

Md, S., Khan, R. A., Mustafa, G., Chuttani, K., Baboota, S., Sahni, J. K., & Ali, J., (2013). Bromocriptine loaded chitosan nanoparticles intended for direct nose to brain delivery: Pharmacodynamic, pharmacokinetic and scintigraphy study in mice model. *European Journal of Pharmaceutical Sciences, 48*(3), 393–405.

Millard, J. W., Alvarez-Nunez, F. A., & Yalkowsky, S. H., (2002). Solubilization by cosolvents: Establishing useful constants for the log–linear model. *International Journal of Pharmaceutics, 245*(1, 2), 153–166.

Misra, A., Ganesh, S., Shahiwala, A., & Shah, S. P., (2003). Drug delivery to the central nervous system: A review. *J. Pharm. Pharm. Sci., 6*(2), 252–273.

Mistry, A., Stolnik, S., & Illum, L., (2009). Nanoparticles for direct nose-to-brain delivery of drugs. *International Journal of Pharmaceutics, 379*(1), 146–157.

Mohammadi, M., & Yazdanparast, R., (2011). Modulation of H_2O_2-induced mitogen-activated protein kinases activation and cell death in SK-N-MC cells by EUK134, a Salen derivative. *Basic & Clinical Pharmacology & Toxicology, 108*(6), 378–384.

Molina, M., Asadian-Birjand, M., Balach, J., Bergueiro, J., Miceli, E., & Calderón, M., (2015). Stimuli-responsive nanogel composites and their application in nanomedicine. *Chemical Society Reviews, 44*(17), 6161–6186.

Morimoto, K., Miyazaki, M., & Kakemi, M., (1995). Effects of proteolytic enzyme inhibitors on nasal absorption of salmon calcitonin in rats. *International Journal of Pharmaceutics, 113*(1), 1–8.

Mulam, T. R., Kshirsagar, D. J., & Kakad, S. P., (2021). *Formulation and Optimization of Ritonavir Nasal Nanosuspension for Brain Targeting.* Indian Drugs.

Muller, R. H., Becker, R., Kruss, B., & Peters, K., (1999). *Pharmaceutical Nanosuspensions for Medicament Administration as Systems with Increased Saturation Solubility and Rate of Solution.* U.S. Patent 5,858,410. Medac Gesellschaft fuer Klinische Spezialpraeparate mbH.

Müller, R. H., Gohla, S., & Keck, C. M., (2011). State of the art of nanocrystals–special features, production, nanotoxicology aspects and intracellular delivery. *European Journal of Pharmaceutics and Biopharmaceutics, 78*(1), 1–9.

Müller, R. H., Staufenbiel, S., & Keck, C. M., (2014). Lipid nanoparticles (SLN, NLC) for innovative consumer care & household products. *Household and Personal Care Today, 9*, 18–24.

Munjal, S., Brand-Schieber, E., Allenby, K. et al. (2017). A multicenter, open-label, long-term safety and tolerability study of DFN-02, an intranasal spray of sumatriptan 10 mg plus permeation enhancer DDM, for the acute treatment of episodic migraine. *J. Headache Pain., 18*(31), https://doi.org/10.1186/s10194-017-0740-3.

Mura, P., Mennini, N., Nativi, C., & Richichi, B., (2018). *In situ* mucoadhesive-thermosensitive liposomal gel as a novel vehicle for nasal extended delivery of opiorphin. *European Journal of Pharmaceutics and Biopharmaceutics, 122*, 54–61.

Nakamura, T., Noma, Y., Sakurai, Y., & Harashima, H., (2017). Modifying cationic liposomes with cholesteryl-PEG prevents their aggregation in human urine and enhances cellular uptake by bladder cancer cells. *Biological and Pharmaceutical Bulletin, 40*(2), 234–237.

Nayak, M. K., & Mohanty, S., (2012). Schedule risk analysis of ICT infrastructure projects. *Int. J. Comput. Appl., 38*, 1–5.

Neamtu, I., Rusu, A. G., Diaconu, A., Nita, L. E., & Chiriac, A. P., (2017). Basic concepts and recent advances in nanogels as carriers for medical applications. *Drug Delivery, 24*(1), 539–557.

Oakley, C. B., & Kossoff, E. H., (2014). Migraine and epilepsy in the pediatric population. *Current Pain and Headache Reports, 18*(3), 1–12.

Pailla, S. R., Talluri, S., Rangaraj, N., Ramavath, R., Challa, V. S., Doijad, N., & Sampathi, S., (2019). Intranasal zotepine nanosuspension: Intended for improved brain distribution in rats. *DARU Journal of Pharmaceutical Sciences, 27*(2), 541–556.

Pandey, V., Gadeval, A., Asati, S., Jain, P., Jain, N., Roy, R. K., Tekade, M., et al., (2020). Formulation strategies for nose-to-brain delivery of therapeutic molecules. In: *Drug Delivery Systems* (pp. 291–332). Academic press.

Pardeshi, C. V., & Belgamwar, V. S. (2013). Direct nose to brain drug delivery via integrated nerve pathways bypassing the blood-brain barrier: an excellent platform for brain targeting.

Expert Opin Drug Deliv. 10(7), 957–972. doi: 10.1517/17425247.2013.790887. Epub 2013 Apr 16. PMID: 23586809.

Pardeshi, C. V., Rajput, P. V., Belgamwar, V. S., & Tekade, A. R., (2012). Formulation, optimization, and evaluation of spray-dried mucoadhesive microspheres as intranasal carriers for valsartan. *Journal of Microencapsulation, 29*(2), 103–114.

Patel, H. P., Chaudhari, P. S., Gandhi, P. A., Desai, B. V., Desai, D. T., Dedhiya, P. P., Vyas, B. A., & Maulvi, F. A., (2021). Nose to brain delivery of tailored clozapine nanosuspension stabilized using (+)-Alpha-tocopherol polyethylene glycol 1000 succinate: Optimization and *in vivo* pharmacokinetic studies. *International Journal of Pharmaceutics, 600*, 120474.

Patravale, V. B., Date, A. A., & Kulkarni, R. M., (2004). Nanosuspensions: A promising drug delivery strategy. *Journal of Pharmacy and Pharmacology, 56*(7), 827–840.

Paul, W., & Sharma, C. P., (2000). Chitosan, a drug carrier for the 21st century: A review. *STP Pharma Science, 10*(1), 5–22.

Pereswetoff-Morath, L., (1998). Microspheres as nasal drug delivery systems. A*dvanced Drug Delivery Reviews, 29*(1, 2), 185–194.

Picone, P., Ditta, L. A., Sabatino, M. A., Militello, V., San, B. P. L., Di Giacinto, M. L., Cristaldi, L., et al., (2016). Ionizing radiation-engineered nanogels as insulin nanocarriers for the development of a new strategy for the treatment of Alzheimer's disease. *Biomaterials, 80*, 179–194.

Prunk, M., & Kos, J., (2017). Nanoparticle-based delivery of protease inhibitors to cancer cells. *Current Medicinal Chemistry, 24*(42), 4816–4837.

Rabiee, N., Ahmadi, S., Afshari, R., Khalaji, S., Rabiee, M., Bagherzadeh, M., Fatahi, Y., et al., (2021). Polymeric nanoparticles for nasal drug delivery to the brain: Relevance to Alzheimer's disease. *Advanced Therapeutics, 4*(3), 2000076.

Raemdonck, K., Demeester, J., & De Smedt, S., (2009). Advanced nanogel engineering for drug delivery. *Soft Matter, 5*(4), 707–715.

Rajput, S., Agrawal, P., Pathak, A., Shrivasatava, N., Baghel, S. S., & Baghel, R. S., (2012). A review on microspheres: Methods of preparation and evaluation. *World Journal of Pharmacy and Pharmaceutical Sciences, 1*, 422–438.

Rao, S. B., & Sharma, C. P., (1997). Use of chitosan as a biomaterial: Studies on its safety and hemostatic potential. *Journal of Biomedical Materials Research: An Official Journal of the Society for Biomaterials and the Japanese Society for Biomaterials, 34*(1), 21–28.

Rathanan, M., Kumar, D. S., Shirwaikar, A., Kumar, R., Kumar, D. S., & Prasad, R. S., (2007). Preparation of mucoadhesive microspheres for nasal delivery by spray drying. *Indian Journal of Pharmaceutical Sciences, 69*(5), 651.

Robbins, M. S., (2021). Diagnosis and management of headache: A review. *JAMA, 325*(18), 1874–1885.

Rolland, J. P., Maynor, B. W., Euliss, L. E., Exner, A. E., Denison, G. M., & DeSimone, J. M., (2005). Direct fabrication and harvesting of monodisperse, shape-specific nanobiomaterials. *Journal of the American Chemical Society, 127*(28), 10096–10100.

Ross, T. M., Zuckermann, R. N., Reinhard, C., & Frey, II. W. H., (2008). Intranasal administration delivers peptoids to the rat central nervous system. *Neuroscience Letters, 439*(1), 30–33.

Ruigrok, M. J., & De Lange, E., (2015). Emerging insights for translational pharmacokinetic and pharmacokinetic-pharmacodynamic studies: Towards prediction of nose-to-brain transport in humans. *The AAPS Journal, 17*(3), 493–505.

Sagar, V., Pilakka-Kanthikeel, S., Pottathil, R., Saxena, S. K., & Nair, M., (2014). Towards nanomedicines for neuro AIDS. *Reviews in Medical Virology, 24*(2), 103–124.

Samaridou, E., & Alonso, M. J., (2018). Nose-to-brain peptide delivery–The potential of nanotechnology. *Bioorganic & Medicinal Chemistry, 26*(10), 2888–2905.

Sarker, D. K., (2005). Engineering of nanoemulsions for drug delivery. *Current Drug Delivery, 2*(4), 297–310.

Schiffter, H. A., (2011). *The Delivery of Drugs–Peptides and Proteins*, 587–604.

Scioli, M. S., Muraca, G., & Ruiz, M. E., (2020). Solid lipid nanoparticles for drug delivery: Pharmacological and biopharmaceutical aspects. *Frontiers in Molecular Biosciences*, 319.

Sekerdag, E., Lüle, S., Pehlivan, S. B., Öztürk, N., Kara, A., Kaffashi, A., Vural, I., et al., (2017). A potential non-invasive glioblastoma treatment: Nose-to-brain delivery of farnesyl thiosalicylic acid incorporated hybrid nanoparticles. *Journal of Controlled Release, 261*, 187–198.

Semwal, R., Upadhyaya, K., Semwal, R. B., & Semwal, D. K., (2018). Acceptability of nose-to-brain drug targeting in context to its advances and challenges. *Drug Delivery Letters, 8*(1), 20–28.

Seremeta, K. P., Chiappetta, D. A., & Sosnik, A., (2013). Poly (ε-caprolactone), Eudragit® RS 100 and poly (ε-caprolactone)/Eudragit® RS 100 blend submicron particles for the sustained release of the antiretroviral efavirenz. *Colloids and Surfaces B: Biointerfaces, 102*, 441–449.

Sessa, M., Balestrieri, M. L., Ferrari, G., Servillo, L., Castaldo, D., D'Onofrio, N., Donsì, F., & Tsao, R., (2014). Bioavailability of encapsulated resveratrol into nanoemulsion-based delivery systems. *Food Chemistry, 147*, 42–50.

Shaikh, R., Raj, S. T. R., & James, G. M., (2011). *Mucoadhesive Drug Delivery Systems, 3*(1), 89–100.

Sharma, S., Lohan, S., & Murthy, R. S. R., (2014). Formulation and characterization of intranasal mucoadhesive nanoparticulates and thermo-reversible gel of levodopa for brain delivery. *Drug Development and Industrial Pharmacy, 40*(7), 869–878.

Shobo, A., Pamreddy, A., Kruger, H. G., Makatini, M. M., Naicker, T., Govender, T., & Baijnath, S., (2018). Enhanced brain penetration of pretomanid by intranasal administration of an oil-in-water nanoemulsion. *Nanomedicine, 13*(9), 997–1008.

Silberstein, S. D., Shrewsbury, S. B., & Hoekman, J., (2020). Dihydroergotamine (DHE)–then and now: A narrative review. *Headache: The Journal of Head and Face Pain, 60*(1), 40–57.

Solanki, N., (2018). Microspheres an innovative approach in drug delivery system. *MOJ Bioequivalence Bioavailab, 5*, 56–58.

Soni, K. S., Desale, S. S., & Bronich, T. K., (2016). Nanogels: An overview of properties, biomedical applications, and obstacles to clinical translation. *Journal of Controlled Release, 240*, 109–126.

Soppimath, K. S., Aminabhavi, T. M., Kulkarni, A. R., & Rudzinski, W. E., (2001). Biodegradable polymeric nanoparticles as drug delivery devices. *Journal of Controlled Release, 70*(1, 2), 1–20.

Sousa, F., Dhaliwal, H. K., Gattacceca, F., Sarmento, B., & Amiji, M. M., (2019). Enhanced anti-angiogenic effects of bevacizumab in glioblastoma treatment upon intranasal administration in polymeric nanoparticles. *Journal of Controlled Release, 309*, 37–47.

Sun, Y., Shi, K., & Wan, F., (2012). Methotrexate-loaded microspheres for nose to brain delivery: *In vitro/in vivo* evaluation. *Journal of Drug Delivery Science and Technology, 22*(2), 167–174.

Swami, R., Singh, I., Jeengar, M. K., Naidu, V. G. M., Khan, W., & Sistla, R., (2015). Adenosine conjugated lipidic nanoparticles for enhanced tumor targeting. *International Journal of Pharmaceutics, 486*(1, 2), 287–296.

Tahara, Y., & Akiyoshi, K., (2015). Current advances in self-assembled nanogel delivery systems for immunotherapy. *Advanced Drug Delivery Reviews, 95*, 65–76.

Taki, H., Kanazawa, T., Akiyama, F., Takashima, Y., & Okada, H., (2012). Intranasal delivery of camptothecin-loaded tat-modified nanomicells for treatment of intracranial brain tumors. *Pharmaceuticals, 5*(10), 1092–1102.

Talegaonkar, S., & Mishra, P. R., (2004). Intranasal delivery: An approach to bypass the blood-brain barrier. *Indian Journal of Pharmacology, 36*(3), 140.

Talele, S. G., Ahire, E. D., Talele, G. S., & Derle, D. V., (2021). An innovative approach as self-emulsifying drug delivery system for phytoconstituents. In: *Enhancing the Therapeutic Efficacy of Herbal Formulations* (pp. 69–84). IGI Global.

Tang, S., Wang, A., Yan, X., Chu, L., Yang, X., Song, Y., Sun, K., et al., (2019). Brain-targeted intranasal delivery of dopamine with borneol and lactoferrin co-modified nanoparticles for treating Parkinson's disease. *Drug Delivery, 26*(1), 700–707.

Thombre, N. A., Niphade, P. S., Ahire, E. D., & Kshirsagar, S. J., (2022). Formulation development and evaluation of microemulsion based lornoxicam gel. *Biosciences Biotechnology Research Asia, 19*(1), 69–80.

Trevino, J. T., Quispe, R. C., Khan, F., & Novak, V., (2020). Non-invasive strategies for nose-to-brain drug delivery. *Journal of Clinical Trials, 10*(7).

Van, W. M., Wauthoz, N., Rosière, R., Amighi, K., Mathieu, V., Lefranc, F., Van, G. S. W., & De Vleeschouwer, S., (2013). Formulations for intranasal delivery of pharmacological agents to combat brain disease: A new opportunity to tackle GBM? *Cancers, 5*(3), 1020–1048.

Vaz, G. R., Hädrich, G., Bidone, J., Rodrigues, J. L., Falkembach, M. C., Putaux, J. L., Hort, M. A., et al., (2017). Development of nasal lipid nanocarriers containing curcumin for brain targeting. *Journal of Alzheimer's Disease, 59*(3), 961–974.

Veronesi, M. C., Alhamami, M., Miedema, S. B., Yun, Y., Ruiz-Cardozo, M., & Vannier, M. W., (2020). Imaging of intranasal drug delivery to the brain. *American Journal of Nuclear Medicine and Molecular Imaging, 10*(1), 1.

Viet, N. K., Laidmäe, I., Kogermann, K., Lust, A., Meos, A., Viet Ho, D., Raal, A., et al., (2019). Preformulation study of electrospun haemanthamine-loaded amphiphilic nanofibers intended for a solid template for self-assembled liposomes. *Pharmaceutics, 11*(10), 499.

Vinogradov, S. V., & Kabanov, A. V., (2004). Synthesis of nanogel carriers for delivery of active phosphorylated nucleoside analogues. In: Papers presented at the meeting. *American Chemical Society. Division of Polymer Chemistry* (Vol. 228, No. Pt 2, p. 296). NIH Public Access.

Wong, Y. C., & Zuo, Z., (2010). Intranasal delivery—Modification of drug metabolism and brain disposition. *Pharmaceutical Research, 27*(7), 1208–1223.

Xie, A., Hanif, S., Ouyang, J., Tang, Z., Kong, N., Kim, N. Y., Qi, B., et al., (2020). Stimuli-responsive prodrug-based cancer nanomedicine. *EBioMedicine, 56*, 102821.

Yadollahi, R., Vasilev, K., & Simovic, S., (2015). Nanosuspension technologies for delivery of poorly soluble drugs. *Journal of Nanomaterials, 2015*.

Yalkowsky, S. H., & Roseman, T. J., (1981). Solubilization of drugs by cosolvents. *Techniques of Solubilization of Drugs, 12*, 91–134.

Yaqoob, M., Jalil, A., & Bernkop-Schnürch, A., (2021). Mucoadhesive polymers: Gateway to innovative drug delivery. In: *Modeling and Control of Drug Delivery Systems* (pp. 351–383). Academic Press.

Yasir, M., & Sara, U. V. S., (2013). Preparation and optimization of haloperidol loaded solid lipid nanoparticles by Box–Behnken design. *Journal of Pharmacy Research, 7*(6), 551–558.

Yasir, M., & Sara, U. V. S., (2014). Solid lipid nanoparticles for nose to brain delivery of haloperidol: *In vitro* drug release and pharmacokinetics evaluation. *Acta Pharmaceutica Sinica B, 4*(6), 454–463.

Yates, R., Sörensen, J., Bergström, M., Antoni, G., Nairn, K., Kemp, J., Långström, B., & Dane, A., (2005). Distribution of intranasal 11C-zolmitriptan assessed by positron emission tomography. *Cephalalgia, 25*(12), 1103–1109.

Yee, K. K., & Rawson, N. E., (2003). *Methods of Treating Anosmia and Repopulating Olfactory Nerves with Retinoids.* U.S. Patent 6,506,801. Monell Chemical Senses Center.

Zarcone, D., & Corbetta, S., (2017). Shared mechanisms of epilepsy, migraine, and affective disorders. *Neurological Sciences, 38*(1), 73–76.

Zhang, L., McClements, D. J., Wei, Z., Wang, G., Liu, X., & Liu, F., (2020). Delivery of synergistic polyphenol combinations using biopolymer-based systems: Advances in physicochemical properties, stability, and bioavailability. *Critical Reviews in Food Science and Nutrition, 60*(12), 2083–2097.

PART III

Novel Drug Delivery System in Fatal Disease

CHAPTER 7

Novel Drug Delivery Systems in the Treatment of Cancer: A Better Future Hope

NITIN R. SHIRSATH,[1] MAHESH P. MORE,[2] PRASHANT K. DESHMUKH,[2] VISHAL S. BAGUL,[3] JAYVADAN K. PATEL,[4] BHAVESH V. BHAVSAR,[5] and PRASHANT B. PATIL[4,6]

[1]*Department of Pharmaceutics, H. R. Patel Institute of Pharmaceutical Education and Research, Karwand Naka, Shirpur, Maharashtra, India*

[2]*Department of Pharmaceutics, Dr. Rajendra Gode College of Pharmacy, Malkapur, Maharashtra, India*

[3]*Department of Pharmacognosy, H. R. Patel Institute of Pharmaceutical Education and Research, Karwand Naka, Shirpur, Maharashtra, India*

[4]*Nootan Pharmacy College, Faculty of Pharmacy, Sankalchand Patel University, Visnagar, Gujarat, India*

[5]*Vice President, KriGen Pharmaceuticals LLC, 800 Edwards Brothers Dr, Lillington, North Carolina, USA*

[6]*Department of Pharmaceutical Chemistry, H. R. Patel Institute of Pharmaceutical Education and Research, Karwand Naka, Shirpur, Maharashtra, India*

ABSTRACT

Cancer is the world's second-largest cause of death. Cancer is a broad term that describes a variety of disorders defined by the uncontrollable proliferation of aberrant cells that can infiltrate and destroy normal human tissue. Cancer

Novel Drug Delivery Systems in the Management of Chronic Diseases.
Raj K. Keservani, Umesh D. Laddha, Eknath D. Ahire, and Sanjay J. Kshirsagar (Eds.)

has the potential to spread throughout your entire body. Surgery, radiation, and chemotherapy are being used to treat malignancies. Each of these methods has its own set of benefits and drawbacks. Other cancer treatment methods must be developed to improve therapeutic potency in terms of affinity, efficiency, and drug targeting. The use of nanotechnology to deliver medications is a novel and more effective cancer therapy technique. Nanoparticles, microparticles (MPs), hydrogels (HGs), phytosomes are the examples of novel drug delivery systems (NDDS). In this chapter, we have discussed the various inorganic molecules based as well as natural and synthetic polymers-based system that are used in development of novel anticancer formulations. Other Novel drug delivery-based strategies like Ligand/receptor-based targeting, penetrating peptide, Chemical conjugation cancer chemotherapeutics also discussed in this chapter.

7.1 CANCER

Cancer is a wide-ranging term that encompasses more than 100 different diseases that manifest in diverse regions of the body. The unregulated creation of new damaged or dysfunctional cells within the body is a feature of all malignancies. We have billions of cells since we are made up of them, cancer can begin and spread practically everywhere in the body (Way, 2017; Park et al., 2015). Cancerous tumors can infect surrounding tissues and spread to other regions of the body, causing new tumors to grow (a process called metastasis). Cancerous tumors are also known as malignant tumors. Many cancers, including leukemias, produce solid tumors, whereas blood cancers do not (Liotta, 1992). Benign tumors do not penetrate or disseminate into the tissues around them. Malignant tumors, on the other hand, frequently recur after being removed. Benign tumors, on the other hand, might grow to be gigantic. Benign brain tumors, for example, might cause major symptoms or even death (Weinberg & Hanahan, 2000). Cancer is a genetic disease, which means that changes in genes that control how our cells behave, notably how they grow and divide, cause it. The body usually eliminates cells with damaged DNA before they become cancerous. Our bodies' ability to do so, however, diminishes as we age. One of the reasons people are more prone to acquire cancer later in life is because of this.

7.2 GLOBAL AND INDIAN BURDEN OF CANCER

Cancer is the largest cause of death globally. In 2020, an estimated 19.3 million new cases of cancer will be reported. According to the GLOBOCAN

2020, worldwide estimated cancers in both sexes with all ages are breast cancer (BC) (11.7%), lung (11.4%), colorectum (10%), prostate (7.3%), stomach (5.6%), liver (4.7%), cervix uteri (3.1%), and other cancers (46%) (Figure 7.1). The most dominant malignancies like lung cancer (18%), colorectum cancer (9.4%), liver cancer (8.3%), BC (6.9%), stomach cancer (7.7%), esophagus (5.5%), pancreas (4.7%), and 39.5% of other cancers have the greatest death rates. The most prevalent cases in the last 5 years are lung cancer (LC) at about 5.2%, colorectum cancer (10.4%), prostate (9.8%), BC (15.4%), stomach cancer (3.6%), esophagus (5.5%), bladder (3.4%), thyroid (3.9%), and other cancers (48.3%) (Sung et al., 2021; Pantwalawalkar et al., 2022).

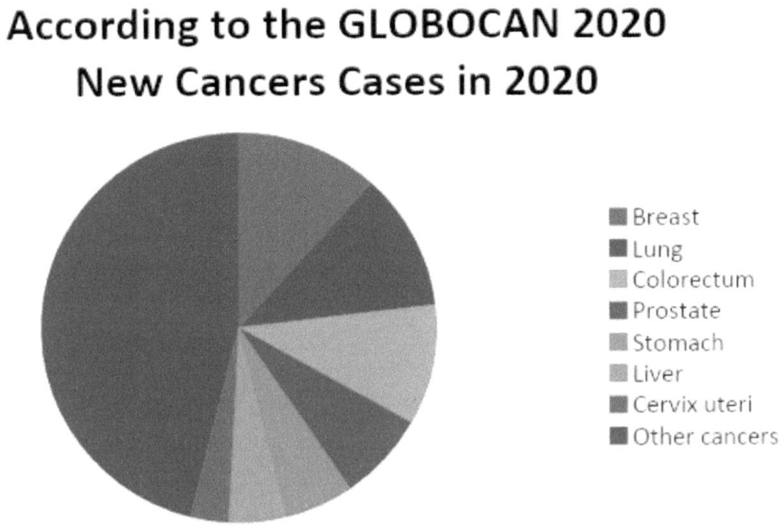

FIGURE 7.1 Worldwide numbers of new cancer cases in 2020.

According to GLOBOCAN 2020, the estimated numbers of incidence cases in India for BC is 178361, Lip, oral cavity 135929, Cervix uteri 123907, Lung 72510, Colorectum 65358, Oesophagus 63180, Stomach 60222, Leukemia 48419, Ovary 45701, and Non-Hodgkin lymphoma (NHL) 35828 (Figure 7.2). Due to cancers, the estimated numbers of mortality in India are for BC is 90408, Lip, oral cavity 75290, Cervix uteri 77348, Lung 66279, Colorectum 38161, Oesophagus 58342, Stomach 53253, Leukemia 35392, Ovary 32077, and NHL 20390 (Cao et al., 2021; Bray et al., 2018).

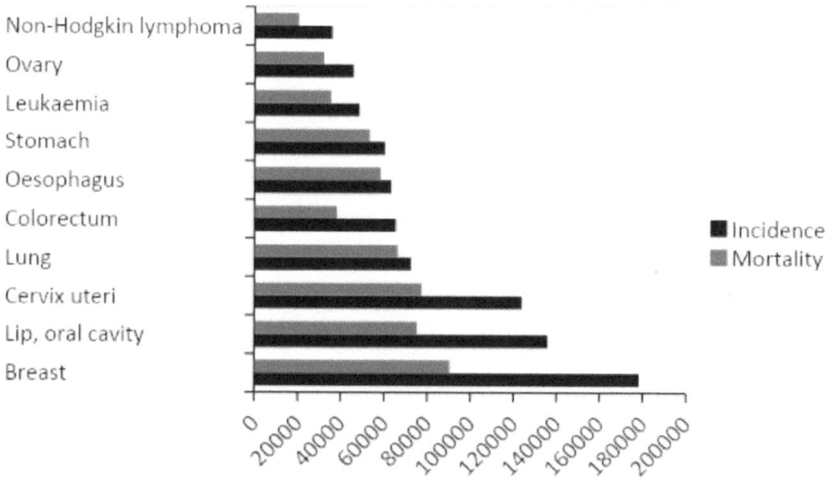

FIGURE 7.2 Estimated incidence and mortality of cancer in India.

In India, novel cancer incidence trends have been reported, including an increase in LC among women and colon cancer among men. Although the general prevalence of cervical cancer is declining, women in urban areas are acquiring malignancies of the uterus, ovary, and lung. The incidence of colon, rectum, and prostate cancer in men has been shown to be much higher in the cities of Delhi, Chennai, and Bengaluru (de Martel et al., 2020; Chandramohan & Thomas, 2018). The overall figure of cancer cases in India is predicted to climb from 979,786 in 2010 to 1,148,757 in 2020. In 2010, 190,244 men were diagnosed with tobacco-related cancers, with that figure anticipated to climb to 225,241 by 2020. Female cases are expected to rise from 75,289 in 2010 to 93,563 in 2020. In 2010, it was estimated that 107,030 men and 86,606 women will be diagnosed with gastrointestinal (GI) cancer. Head and neck cancer had 122,643 and 53,148 cases in 2010, respectively, whereas lymphoid and hematopoietic cancer had 62,648 and 41,591 cases (Sharma, 2020). According to predictions, the figure of diagnosed with reproductive malignancies in women would rise from 153,850 in 2010 to 182,602 in 2020. By 2020, the figure of diagnosed with BC in women's is predicted to increase by more than 100,000 (Arbyn et al., 2020).

7.3 GAP IN CURRENT TREATMENT

Cancer is one of the leading causes of death worldwide. In 2022, 19.3 million new cancer diagnoses were made, and 14.9% of Indian people died. In the next two decades, the figures of new cases of cancer will increase to 22 million or more. To cure cancer, a variety of treatments are used; however, in most cases, the response to therapy is poor, resulting in considerable side effects as well as drug resistance over time. It is vital to find new strategies to control cancer because of the growing cancer mortality rate and the absence of sophisticated forms of chemical treatment and radiation. On its own, traditional cancer treatment delivery is inadequate, and certain drugs have low solubility, a short half-life, and restricted absorption. When pharmaceuticals are lowered to nanoscale, they can deliver various benefits such as better blood circulation, increased drug affinity, and improved therapeutic efficacy. Nanoparticulate drug delivery research is necessary; this is an alternative technique for improving efficacy, patient compliance, and effective medication targeting in cancer treatment by altering/eliminating the issue of standard dosing. An effective medication delivery technology for a range of diseases and disorders is nanoparticulate-based drug delivery. It also enhances patient compliance, efficiency, and bioavailability (Shirsath & Goswami, 2019; Nangare et al., 2022).

7.4MAPPLICATIONS OF NOVEL DRUG DELIVERY IN DIFFERENT TYPES OF CANCER

7.4.1 SYNTHETIC POLYMER-BASED DRUG NOVEL DRUG DELIVERY

7.4.1.1 POLYLACTIC-CO-GLYCOLIC ACID (PLGA)

Polylactic-co-glycolic acid (PLGA) is a biocompatible and biodegradable co-polymer comprising lactic and glycolic acids linked by ester bonds. The FDA and European Medicine have authorized PLGA as a medication delivery vehicle for parenteral administration. Different ways can be recycled to make PLGA nanoparticles. To create sphere-shaped nanoparticles, one of the most commonly used ways is emulsion-solvent evaporation. Therapeutic substances, medicines, or genes are enclosed within the core or adsorbed onto the nanoparticle surface. To aid effective transfection, PLGA-based nanoparticles shield siRNA from degradation and bypass cellular barriers (Avgoustakis, 2005). PLGA nanocapsules can also be used to encapsulate hydrophilic and/or hydrophobic medicines for co-distribution of nucleic acid and chemotherapeutic therapies. Through fluid phase pinocytosis and clathrin-mediated endocytosis, PLGA nanoparticles may be ingested into cells. Because PLGA nanoparticles have a negative surface charge, they may be modified with hydrophilic and cationic polymers for electrostatic adsorption of gene molecules, such as PEI and chitosan (Prakobvaitayakit & Nimmannit, 2003). One of the most efficacious biodegradable PNPs is poly(lactic-co-glycolic acid) (PLGA) (NPs). Because of its controlled and sustained release (SR) qualities, minimal toxicity, and biocompatibility with tissue and cells, the US FDA has authorized it for use in drug delivery systems (DDS). With increasing expertise in the application of PLGA nanoparticles in cancer treatment, these nanoparticles have a bright future ahead of them, with great effectiveness and low adverse effects (Tabatabaei et al., 2014). Jain et al. (2011) manufactured nanocarriers using the biocompatible and biodegradable polymer poly(lactic-co-glycolic acid) (PLGA), which has been authorized by the US Food and Drug Administration (FDA), are being investigated for the controlled delivery of medicinal medicines, proteins, peptides, oligonucleotides, and genes. Surface functionalization of PLGA nanoparticles has paved the way for a variety of engineered PLGA-based nanocarriers that can demonstrate a wide range of combined properties and functions, depending on reticular requirements, such as increased oral bioavailability, site-specific drug delivery, and tailored release characteristics. His study focused on recent advancements in PLGA-based nanotechnology, with a special emphasis on cancer therapies. They begin with a quick overview of cancer nanotechnology and then go through developmental features as well as the *in vitro* and *in vivo* effectiveness of PLGA-based nanocarriers for targeted medication or gene delivery (Jain et al., 2011).

Targeted nanomedicine for cancer therapy has garnered a lot of traction and is being studied intensively. To explore selectivity towards cancer cells and cellular internalization, PLGA nanoparticles were produced by Amin et al. (2016) and hematin, an iron-containing blood porphyrin, was adhered to the surface of the nanoparticles. Hematin has previously been proven to aid cancer cell growth and proliferation. FTIR was used to confirm hematin conjugation. HeLa cells were used to investigate tumor selectivity and uptake. Hematin conjugated particles had a stronger affinity for cancer cells than control particles (zeta potential: 15.19 mV). The particles were absorbed by heme carrier protein 1, according to the results. These findings show that hematin might be a useful cancer-targeting material for nanotherapeutics. Yang et al. (2007) created antibody conjugated magnetic PLGA nanoparticles for BC diagnostics and treatment. For the BC finding and treatment, DOX-magnetic PLGA nanoparticles coupled with well-tailored antibodies were produced. The nanoparticle formulation or drug release kinetics was not hampered by magnetic nanocrystals incorporated in PNPs. Using magnetic resonance imaging, the multimodal nanoparticles displayed extraordinary cancer cell affinity and ultrasensitivity. Furthermore, the anticancer medications that had been loaded were released and maintained for three weeks (Amin et al., 2016).

7.4.1.2 POLY(L-LYSINE) (PLL)

As a highlighted cell-penetrating peptide (CPP), poly-L-lysine (PLL) produces high intestinal membrane penetration and increases medicinal moieties' systemic availability. PLL is generally thought to be safe, and the FDA has sanctioned it for use as a commercial food preservative in a number of states (Mirtič & Grdadolnik, 2013). PLL has become a great candidate for use in a variety of arenas, including antimicrobial agents, drug delivery carriers, and adjuvants in biomedical fields, due to its excellent polycationic property linked to its abundant amino terminals in biological conditions and the tendency to self-assemble when altered as an amphipathic molecule (Orive et al., 2006). Yang et al. (2018) prepared PLL-based nanoparticles for targeted therapy to liver cancer. The theranostic efficacy of a PLL nanoparticle (NP) system for the regulated release of curcumin (CUR) by pH stimuli was compared to that of CUR solution in vitro and in vivo. Through a condensation procedure, deoxycholic acid (DOCA), methoxy polyethylene glycol (MPEG), and cyanine 5.5 (cy5.5) remained conjugated to the amine cluster of PLL (PLL-DOCA-MPEG-cy5.5), followed by dialysis encapsulation of CUR (PLL-DOCA-MPEG-cy5.5/CUR NPs). Proton nuclear magnetic resonance (1H NMR), transmission electron microscopy (TEM), and dynamic light scattering (DLS) were used to determine the composition, morphology, and size distribution of PLL-DOCA-MPEG-cy5.5 NPs. Changes in the charge and size of the NPs at low pH resulted in increased cellular absorption of CUR into the Hep3B – Human hepatoma cell lines via electrostatically absorptive endocytosis, according to in vitro testing. PEGylation with MPEG was shown to be particularly efficient in increasing the EPR effect by prolonging blood circulation time. In addition, the inclusion of Cy5.5 in the NPs resulted in effective in vivo and ex vivo biodistribution pictures. According to their research (Yang et al., 2018), CUR NPs/PLL-DOCA-MPEG-cy5.5 may have interesting uses in cancer theranosis.

Biocompatible amino-functionalized magnetic nanoparticles as carriers of particular antibodies capable of detecting and/or targeting cancer cells were developed by Khmara et al. (2018). To determine the ideal PLL/Fe3O4 weight ratio, magnetic nanoparticles with varied PLL/Fe3O4 content were synthesized and analyzed. The UV/VIS results agreed with the ideal PLL/Fe3O4 weight ratio of 1.0 based on both DLS studies and zeta potential. Carbonic anhydrase IX (CA IX) is a cancer biomarker that is induced by hypoxia, a physiological stress present in solid tumors and linked to aggressive tumor behavior. It was coupled with magnetic nanoparticles with an appropriate PLL concentration. CA IX is found on the cell surface, with the epitope of antibody-binding facing out into the extracellular space, making it ideal for antibody-based tumor cell targeting. They found that when PLL/Fe3O4 magnetic nanoparticles are coupled with a CA IX-specific antibody, they have cytotoxic properties that vary by cell type and attach to cells expressing CA IX (Khmara et al., 2017).

7.4.1.3 POLY(ETHYLENE GLYCOL)-B-POLY(E-CAPROLACTONE) (PEG-PCL)

PEG-PCL is a kind of amphiphilic polyether-polyester copolymer that's commonly used to encapsulate chemotherapeutics. It's easy to manufacture by commencing e-caprolactone ring-opening polymerization (ROP) with PEG and stannous octoate (Sn(Oct)2) as a catalyst. PEG-PCL is a strong match for DDSs of injectable because of its great biocompatibility, biodegradability, and low toxicity. Nanotechnology gives cancer researchers new tools to work with. As an improved drug delivery technology, biodegradable PNPs show promise in cancer therapy (Bogdanov et al., 1998). Biodegradable and amphiphilic poly(e-caprolactone)/poly(ethylene glycol) (PCL/PEG) copolymers show promise as drug delivery vehicles. PCL/PEG copolymeric nanoparticles have gotten a lot of interest recently as a promising nanoplatform for anticancer drug delivery (Wei et al., 2009). Gou et al. (2011) developed PCL/PEG copolymeric nanoparticles for cancer treatment. Authors were the use of nanoparticles made of a PCL/PEG copolymer for anticancer medication delivery. PCL/PEG copolymeric nanoparticles overcoming problems of hydrophobic drug insolubility in water, leads to targeting chemotherapeutic drugs to tumors, and diagnostic agents, vaccines, and delivering genes (Gou et al., 2011). Eatemadi et al. (2016) also prepared PCL-PEG-PCL nanoparticle for BC treatment. Nanotechnology has the potential to revolutionize cancer treatment. This field introduces nanovectors/nanocarriers for anticancer targeted delivery of medications, as well as imaging applications. Chrysin, a flavonoid found in nature, has recently been shown to have essential roles in chemical defenses, nitrogen fixation, anti-inflammatory, and antioxidant activities. The prepared PCL-PEG-PCL nanoparticle not only will combine nano-chrysin therapy cancer cell cytotoxicity, but it will also serve as a supplemental and potentially complicated treatment for BC (Eatemadi et al., 2016).

Behl et al. (2020) prepared biodegradable PEG-PCL nanoparticles for anticancer activity. Because of their amphiphilic nature, biodegradability, and biocompatibility, as well as their semi-crystalline structure with a low glass transition temperature, poly(ethylene glycol) methyl ether-block-poly(β-caprolactone) copolymers are excellent biomedical materials. PEG-PCL copolymers are ideal candidates for continuous delivery applications over a period of more than a year due to their gradual drug release characteristic. This distinct characteristic has prompted their use in drug delivery of anticancer applications in the form of microspheres (Gujarathi et al., 2022), nanogels, nanospheres, polymersomes, micelles, and other nanoaggregates. PEG-PCL copolymers have been used to encapsulate a wide range of anticancer medicines and bioactives in order to generate effective DDSs of anticancer as well as provide regulated drug release characteristics. PEG-PCL copolymer-based nanoparticles have made significant progress as drug delivery vehicles for anticancer in recent years, hydrophobic medicines have a higher drug loading capacity due to their increased drug loading capacity, improved bioavailability, and resistance to phagocyte overpowering, reduced blasted discharge and increased drug scattering time within the bloodstream during systemic inoculation. Anticancer medications can be delivered to a specific place using these nano-formulated drug delivery devices in a targeted manner. *In vivo*, the created nano-formulations show promise in the cancer-specific treatment, with better anticancer activity and lower toxicity. The creation and synthesis of several types of PEG-PCL copolymers (both traditional and greener ways) as well as physicochemical characterization techniques utilized largely for PEG-PCL co-polymeric systems are highlighted in this study. In this work, they have also summarized the usage of PEG-PCL co-polymer nano-formulations with several anticancer medicines as drug delivery platforms (Behl et al., 2020). Zhang et al. (2016) prepared PCL–PEG–PCL PNPs for doxorubicin (DOX) conveyance alongside BC. Thin-film hydration and an ultrasonic dispersion approach were used to successfully integrate DOX, an anthracycline anticancer medicine, into PNPs based on poly(-caprolactone)–poly(ethylene glycol)–poly(-caprolactone) (PCL–PEG–PCL) amphiphilic triblock copolymers. The polymeric NPs with loading DOX that were produced had spherical forms with a core-shell topology. When the concentration and incubation period of DOX-loaded

NPs were increased in an in vitro cytotoxicity experiment against EMT6 cells, the cytotoxic impact eventually approximated that of free DOX. Ex vivo DOX fluorescence imaging further demonstrated that DOX-loaded NPs had very effective targeting and accumulation at the implanted location of an EMT6 xenograft tumor in vivo, owing to an improved penetration and retention mechanism. These findings point to DOX-loaded polymeric NPs made from the triblock copolymers of PCL–PEG–PCL as a viable nanosized drug delivery method for cancer treatment (Zhang et al., 2016).

7.4.1.4 POLY(ETHYLENE GLYCOL)-B-POLY(LACTIC ACID) (PEG-PLA)

Another synthetic block copolymer that has been extensively studied for drug administration is PEG-PLA. PLA comes in three different stereochemical forms: poly(L-lactide), poly(D-lactide), and poly(D, L-lactide) (PDLLA) (Khodaverdi et al., 2019). Tuning L- and D-isomer racemization can affect PLA's mechanical strength and breakdown rate. The semicrystalline PLLA polymer from L-lactide is hard and transparent with a Tg of 53.8°C, whereas the amorphous PDLA polymer from D-lactide has a Tg of 55.8°C, poor mechanical strength, and a rapid degradation rate due to the difference in crystallinity. Due to numerous compensations such as high drug loading efficiency, low toxicity, unique size factor, and ability to sustain more drug payload with tunable surfaces, polyethylene glycol-polylactic acid (PEG-PLA) polymeric micelles have received a lot of kindness in contemporary years in cancer treatment. Furthermore, the prolonged drug release performance of such unique micelles with minimized size effects promotes increased chemotherapeutic agent bioavailability (Mishra et al., 2016).

One of the most significant roadblocks to the use of RNA interference (RNAi) in cancer treatments has been the delivery of slight interfering RNA (siRNA). Yang et al. (2011) presents a stealthy cationic lipid aided polymeric nanoparticle system for effective siRNA encapsulation and administration that was made with poly(ethylene glycol)-b-poly(d,l-lactide), siRNA, and a cationic lipid utilizing a double emulsion-solvent evaporation approach. The cationic lipid increased the encapsulation effectiveness of siRNA into nanoparticles to over 90%, and the loading weight ratio of siRNA increased to 4.47%, while the nanoparticle diameter was approximately 170 to 200 nm. The siRNA was successfully absorbed by cancer cells and escaped from the endosome, resulting in considerable gene knockdown. Following delivery of particular siRNAs by nanoparticles, considerable down regulation of luciferase countenance in HepG2-luciferase cells that stably produce luciferase and reduction of polo-comparable kinase 1 (Plk1) expression in HepG2 cells were observed. Furthermore, in cooperation cancer cells as HepG2 and MDA-MB-435s, nanoparticles containing siRNA targeting the Plk1 gene were observed to trigger impressive death. In an ortho-topic murine liver cancer model, systemic administration of particular siRNA via nanoparticles effectively decreased luciferase expression and lowered tumor development in an MDA-MB-435s murine xenograft model, showing therapeutic promise in disease therapy (Yang et al., 2011). β-Sitosterol (β-Sit) is a dietary phytosterol that has been demonstrated n to have anticancer action against a variety of malignancies. However, due to its low water solubility, it has limited bioavailability and therapeutic effectiveness. Andima et al. (2018) were utilized poly(lactide-co-glycolic acid) (PLGA) and block copolymers of poly(ethylene glycol)-block-poly(lactic acid) (PEG-PLA) to encapsulate β-Sit into nanoparticles in order to improve its anticancer activity *in vitro*. A simple emulsion-solvent evaporation approach was used to make β-sitosterol-loaded PLGA and PEG-PLA nanoparticles (β-Sit-PLGA and β-Sit-PEG-PLA). MTT and Flow cytometry tests were used to assess their cellular absorption and antiproliferative activities in contradiction of human BC cells as MCF-7 and MDA-MB-231, respectively. In contrast to β-Sit-PEG-PLA nanoparticles, flow cytometry analysis (FACS) revealed that β-Sit-PLGA nanoparticles were readily taken up by the cells. As a result, β-Sit-PLGA nanoparticles were chosen to test antiproliferative efficacy. In a concentration range of 6.64–53.08 g/mL, cell viability was condensed by up to 80% compared to untreated cells. They presented that encapsulating β-sitosterol in PLGA nanoparticles is a potential method for enhancing its anticancer action in contradiction of BC cells (Andima et al., 2018). Afsharzadeh et al. (2020) present a new PEG-PLA nanoparticle-based targeted self-assembled system (PEG-PLA NPs) combining docetaxel and galbanic acid (GBA), which was targeted in prostate cancer cell lines utilizing ((S)-2-(3-((S)-5-amino-1-carboxypentyl) ureido) pentanedioic acid (ACUPA). Drug-loaded PEG-PLA NPs and ACUPA-PEG-PLA anti-proliferation against LNCaP (PSMA+) and PC3 (PSMA) cells was compared using the MTT test. When PEG-PLA NPs with a normal size of 130–140 nm stayed exposed to pH 5.5, they produced more GBA and docetaxel than when they were exposed to pH 7.5. ACUPA-modified PEG-PLA may efficiently improve drug absorption in PSMA+ prostate cancer cells, according to spectrofluorometric analyzes. In LNCaP cells, cytotoxicity experiments revealed that targeted NPs loaded with various doses of GBA, and a constant quantity of docetaxel (4 nM) had higher toxicity than free GBA and nontargeted NPs. These findings imply that an ACUPA-conjugated PEG-PLA nanosystem encapsulating GBA and docetaxel might be an auspicious delivery vehicle for cancer-targeting PSMA with a petite half-life circulation and low therapeutic effectiveness (Afsharzadeh et al., 2020).

7.4.2 NATURAL POLYMER-BASED NOVEL DRUG DELIVERY

7.4.2.1 CHITOSAN

Chitosan is an N-deacetylated chitin derivative; however, N-deacetylation is nearly never complete. Between chitin and chitosan, no clear nomenclature for the degree of N-deacetylation has been established. Cancer is one of the world's most lethal diseases, and the efficiency of contemporary chemotherapeutic techniques is restricted due to a slew of connected issues. Most anticancer medicines have poor water solubility, which leads to multidrug resistance and low bioavailability, short plasma half-life, and serious side effects to healthy tissues due to unspecific distribution of chemotherapeutic drugs have a number of significant drawbacks in cancer chemotherapy (Dodane & Vilivalam, 1998). Nanoscience has advanced to the point that cancer cells may be targeted specifically and selectively. Chitosan is a naturally occurring biocompatible and mucoadhesive biopolymer that has been widely employed for the successful delivery of anti-cancer medications at the target location in recent years. In their polymeric matrix, chitosan nanoparticles may contain both hydrophilic and lipophilic chemotherapeutic medicines, proteins/peptides, and various genetic elements such as DNA, miRNA, siRNA, and others. These nanoparticles upsurge the solubility and stability of encapsulated compounds in biological systems because of their unique polymeric matrix. By managing the release rates and targeting the delivery process, chitosan nanoparticles have the potential to enhance the pharmacokinetic profile as well as the therapeutic efficacy of encapsulated medicines. These nanoparticles use passive or active targeting mechanisms to selectively distribute encapsulated medications to desired areas, reducing unwanted side effects (Ahmad et al., 2022a).

Effective medication distribution in the pancreatic cancer (PAC) treatment is now a serious challenge. Nanomedicine plays an important part in the treatment of cancer by delivering anticancer medications to malignant tumor cells in a targeted way, resulting in greater efficacy by lowering anticancer drug toxicity to normal, sensitive locations. Arya et al. (2011) developed and characterized a targeted system for the therapy of PAC represented by

Herceptin conjugated gemcitabine loaded chitosan nanoparticles (HER2-Gem-CS-NPs). As a result of greater cellular binding, the likelihood with eventual uptake and extended intracellular retention, the targeted NPs demonstrated better antiproliferative activity as well as an accelerated S-phase arrest, resulting in apoptosis, when compared to free gemcitabine and unconjugated gemcitabine-loaded nanoparticles. As a result, HER2-Gem-CS-NPs can administer gemcitabine to PAC patients in an efficient and targeted manner (Arya et al., 2011).

Tan et al. (2017) prepared Chitosan nanoparticles for colon cancer treatment. Exogenous DNA can be put into nanoparticles to allow for the generation of immune-stimulatory cytokines. NKG2D recognizes MHC class I chain-related protein in humans and retinoic acid early induced transcript-1 in mice, both of which operate as tumor-associated antigens. Through lymphocyte activation, biologic medicines based on interleukin 21 (IL-21) have shown anticancer activity. The NKG2D-IL-21 fusion protein recognizes tumor cells via the NKG2D moiety and activates T lymphocytes via the IL-21 moiety, according to theory. Natural killer (NK) and CD8+ T-lymphocytes can be activated in tumor cells that have been pretransfected with dsNKG2D-IL-21 gene nanoparticles. Mice intramuscularly treated with gene nanoparticles had higher serum IL-21 levels. After being given intravenously for 4–24 hours, dsNKG2D-IL-21 gene nanoparticles accumulated in tumor tissues. By activating NK and T cells in vivo, dsNKG2D-IL-21 gene nanoparticles inhibited tumor development and extended the lifespan of tumor-bearing animals. As a result, the dsNKG2D-IL-21 gene nanoparticles had effective anticancer properties and might be employed for tumor treatment (Tan et al., 2017). According to the study by Alizadeh et al. (2019), Chitosan is a low-toxic and natural polysaccharide derived from the deacetylation of chitin. Chitosan-based nanoparticles have excellent biodegradation and biodistribution in the biological environment, making them one of the most promising nanocarriers for delivering various therapeutic drugs to tumor cells, particularly ovarian cancer cell lines. Chitosan's free amine groups may be covalently and non-covalently modified and linked to various polymers and targeting moieties, allowing it to reach the tumor location via passive and active targeting techniques. Ovarian cancer is one of the most frequent and resistant diseases in women, with a terrible prognosis, and scientists have been working on innovative approaches to improve treatment outcomes in recent decades. They overviewed the current breakthroughs in various modifications of chitosan-based nanoformulations used for therapy on ovarian cancer (Alizadeh et al., 2019).

Folic acid's carboxyl moiety may be covalently attached to chitosan molecules, retaining a strong affinity for colorectal cancer (CRC) cells with overexpression of the folate receptor. One of the most frequent types of cancers is CRC, yet it often remains unnoticed due to current colonoscopy procedures. Improved colonoscopy procedures for identifying dysplasia and malignancies will reduce mortality. Yang et al. (2010) created a folic acid-attached chitosan nanoparticle to deliver 5-aminolaevulinic acid (5-ALA) to improve CRC cell detection *in vivo*. Chitosan can be conjugated with folic acid to form a folic acid chitosan conjugate, which can then be loaded with 5-ALA to form nanoparticles (fCNA). Chitosan can be efficiently conjugated with folic acid to make a folic acid chitosan conjugate, which may then be loaded through 5-ALA to produce nanoparticles (fCNA). The data show that fCNA is more easily taken up by HT29 and Caco-2 cell lines after a brief period of absorption, most likely via receptor-mediated endocytosis, and that PpIX accumulates in cancer cells as a function of folate receptor expression and folic acid change. As a result, the chitosan-folic acid combination appears to be an excellent vector for delivering 5-ALA to the intestinal mucosa for fluorescent endoscopic detection (Yang et al., 2010).

7.4.2.2 GELATIN

Gelatin Collagen, the primary protein found in animal skin, bones, hides, and white connective tissues, is partially hydrolyzed to produce gelatin. Acid processing of collagenous raw material produces type A gelatin; alkaline or lime processing produces type B gelatin. Gelatin is classed as a derived protein since it is made from collagen through a controlled partial hydrolysis process and does not occur naturally (Poppe, 1992). Reversible gel-to-sol transition, viscosity of heated aqueous solutions, ability to act as a protective colloid, water permeability, and insolubility in cold water yet perfect solubility in hot water are characteristics of gelatin. It is also beneficial to your health. The pharmaceutical and photography industries both utilize these qualities. Gelatin also produces strong, homogeneous, transparent, and fairly flexible coverings that swell and absorb water rapidly, making it excellent for the production of photographic films and medicinal capsules (Ledward, 2000).

Because of its anti-cancer properties, a grape phytochemical, resveratrol (RSV), has received a lot of attention. RSV has a number of drawbacks, however, including instability, low water solubility, and a short biological half-life, all of which limit its use in the medical, food, and pharmaceutical sectors. In their research, Karthikeyan et al. (2013) encapsulated RSV in gelatin nanoparticles (GNPs) and examined its anticancer potential in NCI-H460 LC cells. According to spectrophotometric measurements, the highest amount of RSV entrapped in GNPs was 93.6%. Kinetics of in vitro drug release revealed that RSV was released from GNPs in a burst followed by a steady and persistent release. The synthesized RSV-GNPs had a much faster and more efficient absorption rate than free RSV. Furthermore, in NCI-H460 cells, RSV-GNPs therapy had a greater antiproliferative effect than free RSV treatment. RSV-GNPs treated cells produced more ROS, caused more DNA damage, and had a higher apoptotic rate than free RSV treated cells. The developed RSV-GNPs mixture elicited no harmful reaction in an erythrocyte aggregation experiment. According to HPLC analyses, RSV-GNPs were more bioavailable and had a longer half-life than free RSV. Therefore, the system of GNPs carriers might be a potential strategy for controlling distribution and improving the therapeutic index of RSV, which is poorly water soluble (Karthikeyan et al., 2013).

Nanoscale drug carriers serve a crucial role in modulating drug distribution, permeability, and retention. Despite the fact that natural biomaterials are ideal for drug delivery and controlled release for anticancer medication, which have been encapsulated in a variety of carriers. Gulfam et al. (2012) created gliadin and gliadin with gelatin composite nanoparticles using an electrospray deposition technique for drug administration and controlled release (e.g., cyclophosphamide). For 48 hours, cyclophosphamide was slowly released from gliadin nanoparticles. The gliadin-gelatin composite nanoparticles, on the other hand, released cyclophosphamide rapidly. Furthermore, Western blotting research indicated that cells of BC cultivated with cyclophosphamide-loaded 7% nanoparticles of gliadin for 24 hours turned apoptotic. As a result, the gliadin-based nanoparticle might be a strong instrument for anticancer medications-controlled release and delivery (Gulfam et al., 2012). Meghani et al. (2018) designed gelatin-oleic nanoparticles for cancer therapy. Targeted anti-cancer medication delivery using fattigation-platform-based gelatin-oleic nanoparticles was achieved by bio-orthogonal click of chemistry and principles of metabolic glycoengineering. The cell surface was treated with a precursor of sialic acid (Ac4ManNAz). 1-(3-dimethylaminopropyl)-3-ethyl carbodiimide hydrochloride/N-hydroxy succinimide (EDC/NHS) chemistry was used to conjugate gelatin and oleic acid, followed by the covalent attachment of di-benzo cyclooctyne (DBCO) to the cell surface in a click reaction. The intracellular absorption of

GON-DBCO-DOX was about two times greater than that of GON-DOX and Caelyx, according to flow cytometry results. As a result, the newly developed GON-DBCO-DOX produced for delivering drugs that is both safe and effective that actively targeted anticancer drugs (Meghani et al., 2018). Lu et al. (2011) found that insufficient drug delivery to tumor cells is a key cause of failure intravesical therapy for nonmuscle invasive bladder cancer, mainly due to drug concentration dilution during treatment due to urine output. To solve this issue, they created paclitaxel GNPs with controlled drug concentrations. In dogs, the notion that a continuous, therapeutic concentration may be achieved in urine, bladder tissue, and tumors was tested. Drug dilution by freshly generated urine may be overcome by constant drug release from paclitaxel GNPs, and maintained drug levels in tumors may reduce treatment frequency (Lu et al., 2011).

Hamarat et al. (2016) developed gemcitabine – magnetic GNPs for targeted drug delivery in LC. The two-step desolvation process was used to coat with nanoparticles of magnetic iron oxide (IONPs) with gelatin type B. Drug loading through adsorption was investigated under a variety of circumstances, including temperature, contact duration, pH, and the starting concentration of gemcitabine. Constraints and constants for Langmuir isotherm curves were also computed. According to the Langmuir isotherm, the Gibbs free energy of the adsorption process at 25°C was 4.74 kJ/mol. This result was 7.86 kJ/mol at 37°C, on the other hand. At pH values of 5 and 7.4, *in vitro* drug release was tested with gemcitabine-loaded magnetic GNPs and free gemcitabine, and the results were compared (Hamarat et al., 2016).

7.4.2.3 GELLAN GUM

Despite the development of cancer treatments, off-target cell toxicity remains one of the supreme difficult problems to overcome in cancer treatment. Drug delivery method research is a burgeoning topic of study that covers a wide range of materials and processing processes (Sworn & Stouby, 2021). Gellan gum is a natural polymer with physicochemical features that empower drug loading for long-term release in a variety of environments and anatomical sites. Gellan gum is an anionic exopolysaccharide generated by *Sphingomonas elodea* by fermentation and gels in the presence of cations. In addition, it is harmless and biocompatible. For the use of gellan gum in drug delivery applications, a range of physical and chemical gelation processes have been reported to yield a variety of form factors with changeable mechanical and physicochemical properties, such as hydrogels (HGs), nanohydrogels, beads, films, or patches. The resultant formulations have promising in terms of drug delivery, such as increased drug bioavailability, solubility, and release over time without jeopardizing biocompatibility or introducing side effects (Villarreal-Otalvaro & Coburn, 2021).

Methotrexate (MTX)-loaded gellan gum microparticles (MPs) were created by Dhanka et al. (2018). To create bare gellan gum microparticles (GG MPs) and MTX loaded gellan gum microparticles, they used basic water-in-oil (W/O) emulsion solvent diffusion technique (MTX-GG MPs). Thermogravimetric analysis (TGA) characterization revealed that MTX-GG MPs had marginally better thermal stability than GG MPs. In a 24-hour *in vitro* release testing, MTX-GG MPs released 84% of the medication. GG MPs and MTX-GG MPs caused 1.0% hemolysis in red blood cells of humans (RBCs) in a hemolysis investigation. GG MPs and MTX-GG MPs were shown to be biocompatible in cell viability experiments on L929 cells (Dhanka et al., 2018).

Because gold nanoparticles (AuNPs) are now being investigated for medication distribution and other biomedical uses, it is vital to investigate their destiny inside the body. Dhar et al. (2011) looked at the cellular absorption and AuNPs toxicity made using gellan gum, a microbial polysaccharide used as a reducing and capping agent. The cellular absorption of AuNPs was studied in mouse embryonic fibroblast cells (NIH3T3) and a human glioma cell line (LN-229). According to the cellular uptake research, the gellan gum-reduced AuNPs were found in cancer cells (LN-229) but not in normal mouse embryonic fibroblast cells (NIH3T3). Subacute 28-day oral toxicity tests in rats were used to assess the toxicity of AuNPs. The rats did not display any hematological or biochemical abnormalities after receiving subacute dose of gum-reduced AuNPs. In comparison to the control group, the weight and normal architecture of several organs were unchanged. The findings show that, in addition to establishing the targeted absorption of nanoparticles into malignant cells, the gellan gum-abridged AuNPs are toxicity-free in animals after oral administration (Dhar et al., 2011). Kathle et al. (2018) looked at how chitosan-gellan nanocapsules (CGNCs) can be used to encapsulate and release Tamoxifen citrate (TMC) for the treatment of BC. For the manufacture of CGNCs, a polyelectrolyte complex coacervation approach was adopted. Michigan Cancer Foundation-7 (MCF-7) BC cells were used to test in vitro cell viability. A smooth surface, a wide range of nanosizes, a positive surface charge, and long-term drug release were all features of CGNCs. In MCF-7 cells, TMC-loaded CGNCs were similarly found to be more cytotoxic than the free drug. As a result, CGNCs may be ideal for treating BC since they transport the medicine to the site of action for a long time (Kathle et al., 2018).

7.4.3 INORGANIC-BASED NOVEL DRUG DELIVERY

ZnONPs (zinc oxide nanoparticles) have been widely used as an effective anticancer tool. Boskabadi et al. (2020) green-synthesized ZnONPs in order to explore their anticancer effects on MCF7 and MDA-MB231 (human breast) and HT-29 (colon cancer) cell lines. The effects of these chemicals were then studied in cancer (HT-29, MDA-MB231, and MCF7) and normal (Huvec) cell lines. The nanoparticles' apoptotic and antioxidant properties were also found and confirmed. The apoptotic and cytotoxic effects of 40-nm ZnONPs are cell-selective (BAX upregulation, sub-G1 peaks increase, and fluorescent labeled apoptotic cells). The findings show that the unique biocompatible ZnONPs might be used in human colon and BC cells as a safe natural apoptosis inducer. To confirm its cell-selective anticancer efficacy, further genes' countenance and cell lines are needed (Boskabadi et al., 2020). NK cell-based immunotherapy has been exposed to have promising results in the treatment of LC with non-small cell, one of the world's most frequent cancers. It has been established that the greater the NK cells infiltration into the tumor, the better the therapeutic impact. Wu et al. (2018) created a magnetic nanoparticle (NP)-labeled NK cell-based immune-cell delivery method for clear LC cells that can be accumulated at the tumor site utilizing a small external magnetic device. For magnetic targeted treatment, they produced NPs of super-para-magnetic iron oxide with a core Fe_3O_4 magnetic and a polydopamine (PDA) shell. Fe_3O_4@PDA NPs have good physiological stability and biocompatibility, making them easy to take up by the NK cells. The occurrence of NPs had no effect on the NK cells biology. NK cells with Fe_3O_4@PDA NP decreased tumor development, reduced Ki-67 expression, and boosted apoptosis in A549 cancer cells, according to *in vitro* and *in vivo* investigations. Under a magnetic field, NK cells with Fe_3O_4@PDA NP displayed greater iron density of intra-tumoral and enhanced concentration of $CD56^+$ NK cells, according to H&E staining. They observed that Fe_3O_4@PDA NPs are a potential magnetic nanomaterial with the capability to influence immune cells and thereby limit tumor progression (Wu et al., 2018).

Ahmadi et al. (2018) were produced mesoporous silica nanocarriers and capped with chitosan natural polymer to accept therapeutic molecule capsules. This nanocarrier functions as the pH responsive shield to boost curcumin solubility and anticancer activities in the cell lines of U87MG glioblastoma cancer. The drug-loading content and encapsulation efficiency were assessed at 88.1 ± 4.76% and 8.81 ± 0.47%, respectively. In 96 hours, the CS-MCM-41 released curcumin slowly and consistently at a low pH (42.72 ± 2.29%) compared to the ambient pH (19.54 ± 1.36%). After 72 hours of treatment with unrestricted curcumin and curcumin-loaded CS-MCM-41, the IC_{50} was 15.20 g/mL ($p<0.05$) and 5.21 g/mL ($p<0.05$), respectively (Ahmadi et al., 2018). In cancer nanotechnology, Shakeri-Zadeh et al. (2021) created gold nanoparticle-mediated bubbles. For decades, microbubbles (MBs) have been investigated extensively in the field of biomedicine. The most common springs of energy utilized to create MBs are ultrasound and laser. Traditional acoustic approaches produce MBs that are poorly localized. To create MBs with a focused continuous laser, a high level of energy is necessary, which might be damaging to healthy tissues. To minimize the energy verge and the bubbles' region of action, nanoparticles with plasmonic light-responsive, such as AuNPs, are preferentially employed with continuous laser. It is also widely known that using pulsed lasers rather than continuous lasers reduces the required AuNPs dosages and lowers the laser power threshold. When well-confined bubbles form in biological contexts, they have their own mechanical and optical characteristics. The mechanical effects of a bubble's collapse can be felt in the surrounding region. Cargo delivery towards cancer cells, in addition to surgery, death, and transfection, can all benefit from such a capacity. Furthermore, the bubbles' exceptional light scattering capabilities makes them ideal for cancer imaging. His review starts with a prolog of the fundamental characteristics of AuNPs-mediated bubbles, followed by a discussion of their growing applications in the realm of cancer nanotechnology (Shakeri-Zadeh et al., 2021).

Folate (folic acid) is a critical component of the DNA synthesis process and is required for cancer cell replication. The fact that is sometimes folate receptors are overexpressed in cancer cells at higher levels than normal cells is the foundation of the folate targeting technique. Samadian et al. (2016) developed Folate-conjugated gold nanoparticle for cancer therapy target. Traditional cancer therapy procedures have numerals of downsides, including non-specificity and limited efficiency in distinguishing between healthy and malignant cells. Targeted nanomaterials attach to cancer cells selectively and influence them with just negligible side properties on healthy cells. Several folates conjugated nanomaterials have been reported in numerous papers, with folate conjugated AuNPs holding considerable potential in targeted cancer treatment. AuNPs have been discovered as potential prospects for innovative cancer therapy techniques due to their biocompatibility, ease of production and functionalization, chemo-physical stability, and optical tunable features. In the preceding decade, there has been a great detonation in AuNPs research, with a rapid increase in publications in biomedicine field. Even though there are several studies on "gold nanoparticles" and "folate targeting" in the biomedical literature, there are just a scarce rumor on "folate conjugated AuNPs." They examined and showed recent biomedical improvements that have been made possible by folate conjugated AuNPs (Samadian et al., 2016).

Liver cancer is the second greatest cause origin of cancer-related death worldwide, and it is the sixth most commonly diagnosed cancer. Because of its antioxidant and anticancer qualities, selenium nanoparticles (Se-NPs) have gained popularity as nano-carriers for gene therapy. Naidoo et al. (2022) created, functionalized, and characterized SeNPs that could efficiently bind, protect, and transport pCMV–Luc DNA to HepG2 cells. The SeNPs were made by reducing ascorbic acid and then functionalizing them by PLL to stabilize and give them positive charges. The SeNPs were then coated by lacto-bionic acid (LA) to target the abundantly expressed asialoglycoprotein receptors (ASGPRs) on surface of the hepatocytes. SeNPs were all in spherical in rang of nanoscale between (130 nm), and capable of binding, compacting, and shielding pDNA against nuclease destruction. In the cell lines studied, the functionalized SeNP nanocomplexes showed limited cytotoxicity (30%) and improved transfection efficiency. Inclusive, these functionalized SeNPs have properties that make them good gene nanocarriers for treating liver cancer (Naidoo et al., 2022).

The screening of natural-origin medications with fewer systemic side effects than synthesized molecules is the current objective of anticancer drug research. Curcumin is an essential polyphenolic pigment that has conventionally been used as a spice in Indian cuisine. The chemical possesses a wide range of pleiotropic effects, including potent antioxidant and anti-inflammatory capabilities. However, due to its poor solubility and bioavailability, its therapeutic potential is restricted. By casing biodegradable polyelectrolyte multi-layers such as dextran (DXS) and poly(L-lysine) with Fe_3O_4 nanoparticles (nano-Fe_3O_4), Mancarella et al. (2015) created a layer-by-layer functionalization of Fe_3O_4 nanoparticles (nano-Fe_3O_4) (PLL). To produce a tall upload of curcumin in Fe_3O_4 nanoparticles, physicochemical tests were carried out. The therapeutic effectiveness of Nano-Fe_3O_4 was later demonstrated against the SKOV-3-ovarian cancer cell line (Mancarella et al., 2015).

7.5 LIGAND/RECEPTOR-BASED TARGETING

The particular chemical communication between the ligands and receptors are a key driver of cell-to-cell communication, and their dysregulation has been linked to illnesses including cancer, autoimmunity, and neurodegeneration (Villar-Cheda et al., 2014). With work dating back to the initial twentieth century, the ability to directly alter protein–protein interactions have motivated the creation of targeted treatments against ligands and receptors (Villar-Cheda et al., 2014; Kim & Cochran, 2017). Drug-target interactions must be studied in order to get a better knowledge of biological processes and to design novel therapeutic compounds. Improved ligand-binding assays have paralleled advancements in reagents, detection, and instrumentation technologies, as well as the expansion of beneficial therapeutic goals of interest and the growing recognition of the importance of biochemical aspects of drug-target interactions in determining drug molecules' clinical performance (Fang, 2012).

7.5.1 ANTIBODY-DRUG CONJUGATES (ADC)

Antibody-drug conjugates (ADCs), which are antibody-targeted drug payloads with the goal of increasing therapeutic index, have seen a recent rise in approvals, with three of the seven United States authorizations in the 2019. For relapsed, metastatic solid tumors, trastuzumab emtansine, fam-trastuzumab deruxtecan, and enfortumab vedotin are currently accessible, while more than 150 ADCs are at various phases of research. Anticancer medicines such as mitomycin C, idarubicin, anthracycline, Nacetyl melphalan, DOX, vinca alkaloids, and MTX were largely attached to murine mAbs by noncleavable linkers (using amide or a succinimide spacer) in the first generation of ADCs (Dosio et al., 2011; Endo & Tsurugi, 1987; Pietersz et al., 1987).

The failure of the first-generation ADCs has been attributed to the following factors:

1. **Drug Potency:** Patients' serum concentrations were outside of the therapeutic range.
2. **Target-Antigen Expression:** Because tumor cells only express a modest number of antigen molecules (explained below), the amount of drug molecules supplied by the ADCs inside the cell seemed to be less than the threshold concentration necessary for cell death.

3. **Internalization of ADC Molecules:** The quantity of ADC molecules transported inside the cell is frequently less than the amount attached to the cell surface.
4. **Tumor Localization:** The rate of radiolabeled mAbs being localized at the tumor in patients was found to be modest (0.003–0.08% of administered doses/g tumor). Mouse tumor xenografts, on the other hand, accumulate at a significantly greater rate (15–20% of the administered doses/g tumor).
5. **Linker Stability:** The linkers utilized were either too stable (which resulted in low potency and effectiveness) or too labile (which resulted in poor target selectivity and systemic toxicity). The hydrazone linker's short half-life (43 hours) raised concerns about unintended toxicity from premature off-target drug release.
6. **Immunological Responses:** The early ADCs employed murine or chimeric mAbs, which elicited an immune response and the formation of antibodies of human antimouse (HAMAs), which prevented therapy cycles from repeating (Phillips et al., 2008; Abdollahpour-Alitappeh et al., 2019).

7.5.2 APTAMERS (APTS)

Aptamers (Apts) are single-stranded nucleic acid oligonucleotides with high specificity and affinity for their targets, often known as chemical antibodies. They're usually chosen by a time-consuming *in vitro* technique known as methodical evolution by exponential enrichment (SELEX) of ligands. Due to their improved features over conventional antibodies, such as their smaller physical size, lower immunogenicity, and lower toxicity, Apts have recently emerged as a new class of agents for delivering therapeutic medicines to cancer cells by targeting certain cancer-associated hallmarks. Apts may be also structurally changed to make them more flexible, allowing them to conjugate other agents like nanomaterials and therapeutic RNA agents, expanding their cancer therapy possibilities (Zhou et al., 2016). Apts are responsible to explored for cancer therapy theirs Apts (DNA/RNA), functions, and anti-tumor effect listed in Table 7.1.

TABLE 7.1 Aptamers Explored for Cancer Therapy

Aptamers (DNA/RNA)	Functions	Anti-Tumor Effect	References
AFP-apt (RNA)	Inhibits AFP-involved pathways.	Inhibits *in vitro* HCC proliferation.	Lee & Lee (2012)
Pegaptanib (RNA)	VEGF-associated tumor inhibition and vessel formation.	Potential therapeutic agent for solid cancers characterized by extensive angiogenesis.	Ng et al. (2006)
AS1411 (DNA)	Inhibits nucleolin-associated cell processes and NF-κB or Bcl-2 signaling.	Inhibits *in vitro* tumor growth of a variety of cancer cells. Inhibits in vivo tumor growth of AML, LC, RC, BC, and PAC.	Bates et al. (2009); Ireson & Kelland (2006)
A30 (RNA)	Inhibits HRG-dependent tyrosine phosphorylation of HER2 and MAPK signaling.	Inhibits *in vitro* tumor growth of BC cells and serves as a potential inhibitor for HER-3 over-expressed cancers such as LC, BC, GC, and PAC.	Chi-Hong et al. (2003)
PNDA-3 (DNA)	Blocks interaction between periostin and its cell surface receptors.	Inhibits tumor cell growth and metastasis *in vitro* and *in vivo* of breast cancer (BC).	Lee et al. (2013)
PPAR-apt (RNA)	Inhibits PPAR-dependent VEGF signals.	Inhibits *in vitro* tumor growth of CRC.	Kwak et al. (2009)

Abbreviations: AFP: alpha fetoprotein; AGE: advanced glycation end; AML: acute myelocytic leukemia; Apt: aptamer; BC: breast cancer; Bcl-2: B-cell lymphoma 2; CRC: colorectal cancer; GC: gastric cancer; HCC: hepatocellular carcinoma; HMCCs: hematological cancer cells; HRG: heregulin; LC: lung cancer; NF-κB: nuclear factor-κB; PAC: pancreatic cancer; PAUF: pancreatic adenocarcinoma up-regulated factor; PC: prostate cancer; PIC: pancreatic islet cancer; and RC: renal cancer.

Apts obligates a numeral of benefits over traditional antibodies, including reduced size, increased stability, ease of alteration, and absence of immunogenicity. Apts are attractive weapons for cancer treatment because of these features. The positive results of several Apts currently in clinical trials (such as AS1411 and NOX-A12) have paved the way for aptamer-based molecular cancer treatment (Zhou et al., 2016).

Although Apt technology offers a lot of potential in the biomedical area, it still has a lot of technical obstacles to overcome. Targeted therapies based on antibodies have a high degree of target specificity and affinity. However, their immunogenicity is a major worry, as is their high manufacturing cost, both of which have limited their clinical use. When compared to the protein antibodies, oligonucleotide Apts mainly have a numerous advantage, including simple chemical synthesis, virtual nonimmunogenicity, smaller size, and faster tissue penetration, ease of modification with different functional moieties, low production cost, and high biological stability, as discussed in this review. As a result, Apts have appeared as an auspicious new class of molecular ligands that might be used instead of or in addition to protein antibodies. In conclusion, Apt technology has a high commercial value and may be recycled in a variability of biomedical sectors, such as *in vitro* cancer cell detection, *in vivo* tumor imaging, and targeted therapy for cancer (Sun et al., 2014).

7.5.3 SIRNA

Small interfering RNAs (siRNAs) have emerged as new nucleic acid medications specifically for the treatment of incurable illnesses such as malignancies as our understanding of the molecular processes of endogenous RNAi has grown. Although various candidates of siRNA for the treatment of ophthalmic and respiratory illnesses are in clinical trials, there are hurdles in developing siRNAs for anti-cancer treatments since systemic delivery will be necessary in the vast majority of instances (Oh & Park, 2009). Recent research has given some insight on the molecular processes of RNAi, in which dsRNA causes homologous mRNA to be silenced. Dicer, an enzyme found in the human cells cytoplasm, begins RNA silencing by breaking down long dsRNA to produce short interfering RNA (siRNA) with a length of 21–23 nucleotides. The siRNAs are then integrated into an RNA-induced silencing complex (RISC) and unraveled into single-stranded RNA (ssRNA), with the sense strand ssRNA then degraded (Doi et al., 2003). SiRNA, a double-stranded RNA-based molecule, shows a lot of promise as a biopharmaceutical therapy. SiRNA may not interact with chromosomal DNA because RNAi interferes with translation rather than transcription. Because there is no DNA contact, there are few worries about undesirable gene changes that may occur as a outcomes of DNA-based gene therapy. Because siRNA interacts with mRNA rather than protein, it can limit the creation of dangerous proteins before they are produced. Another advantage of siRNA as a therapeutic medication is that it may be used to silence an extensive range of target proteins to cure disorders (de Fougerolles et al., 2007).

7.5.3.1 CHALLENGES OF SIRNA THERAPY

The possibility of an ‘off-target’ impact, which is the suppression of a gene whose expression should not be targeted because the gene shares partial homology with the siRNA, is one such difficulty. Inadvertently silencing non-target genes might cause challenges with data interpretation and even harm. To avoid this problem, great consideration should be given to the design and assortment of effective siRNAs. Internal repeating sequences, secondary structure, GC content, base preference at certain places in the sense strand, and optimal siRNA length (19-22 bps) are the essential characteristics for selecting siRNAs. Several siRNA firms provide online design methods that take into account mRNA target sequence, secondary structure, and siRNA duplex end-stabilities (Kim & Rossi, 2007).

Immune stimulation, or the innate immune system's detection of an siRNA duplex, is another challenge for siRNA treatment (Bumcrot et al., 2006; de Fougerolles et al., 2007).

7.5.4 PEPTIDES

Peptides are extremely adaptable tools with enormous promise for cancer diagnostics and therapeutics development. They can be used in a variety of ways to treat cancers, including mimicking natural proteins to boost or inhibit signaling, targeting medicines particularly to cancer cells, and employing them as a tool for transbarrier delivery. The following summary focuses on the most significant and far-reaching breakthroughs in peptide cancer treatments. Several peptides are now being tested in clinical trials as cancer treatments or diagnostic tools. Gonadotropin-releasing hormone (GnRH) agonists, for example, are the first-line therapy for advanced prostate cancer in order to induce androgen suppression (Kurrikoff et al., 2019). Anticancer peptides are a phrase used to describe a collection of peptides that are generated from antimicrobial peptides but have cancer-selective toxicity due to their membranolytic action (Table 7.2).

TABLE 7.2 Peptides Used in Cancer

Class	Name	Target
GnRH agonists	Buserelin, gonadorelin, leuprolide, nafarelin, goserelin, triptorelin, and histrelin.	Treatment of prostate and breast cancer.
GnRH antagonists	Cetrorelix, abarelix, degarelix	Treatment of prostate and breast cancer.
Somatostatin agonists	Octreotide, depreotide, lanreotide, edotreotide	Treatment and diagnostics of growth hormone producing tumors.

Because of their benefits in manufacturing, such as automated chemical synthesis capabilities and the ease with which alterations may be added, peptides are desired components in cancer therapy (Kurrikoff et al., 2019). In the next 5–10 years, we predict the area to expand dramatically, with very efficient peptide or nonpeptide cancer targeting drugs with low side effects produced as a consequence of tight collaboration between chemists, biologists, radiochemists, radiation physicists, and clinicians (Aina et al., 2002).

7.6 CELL-PENETRATING PEPTIDE (CPP)

Peptide that penetrates cells (CPP) CPPs, formerly known as protein transduction domains, are a vast class of short amino acid sequences (5–30 residues) capable of penetrating biological membranes and delivering a variety of chemicals within cells, such as tiny molecules, nucleic acids, proteins, viruses, imaging agents, and drugs (Heitz et al., 2009).

Electrostatic binding to glycosaminoglycans (GAG) such as heparin sulfate (HS), heparin, and chondroitin sulfate B is the initial point of interaction between CPPs (particularly cationic CPPs) and the cell surface. These glycoproteins, which having a large number of negative charges, are found all across cell membranes (Simon Davis & Parish, 2013; Raucher & Ryu, 2015).

7.6.1 CPPS FOR DELIVERY OF CHEMOTHERAPEUTIC AGENTS

Some anticancer medications, including as Taxol (Aroui et al., 2009), MTX (Raucher & Ryu, 2015), and DOX, are conjugated to CPPs, which enhances membrane permeability, drug delivery, drug half-time circulation, and tumor cell accumulation. Numerous CPPs can be conjugate Dox, which might increase the drug's intracellular delivery. TAT and Pen are the most often utilized, as they have been shown to trigger apoptosis in hamster (CHO) and human cancer cells – HUVEC, MDAMB231, MCF-7 cells at low dosages (Aroui et al., 2009). This peptide with tryptophan and arginine residues was shown to have low cytotoxicity and enhanced medication absorption and delivery in cells (Nasrolahi Shirazi et al., 2013).

7.6.2 CPP AND NUCLEAR ACIDS DELIVERY FOR ANTI-CANCER THERAPY

Gene therapy might stay used for treating variety of ailments, including cancer and other genetic and monogenetic problems. Because the molecules used in this therapy are big and hydrophilic, they cannot pass cell membranes and must be combined with a delivery vector. Initially, viral vehicles were utilized with great success, but at a tremendous cost in terms of toxicity and immunogenicity. The majority of nonviral vectors (liposomes, cationic polymers, and so on) are cationic and low in toxicity. They can form electrostatic connections with DNA's negative charges, although their transfection effectiveness is lower than that of viral carriers. CPPs combine oligonucleotides plus plasmid DNA into nanoparticles with positive charges that interact with biological membranes, allowing CPP-cargo complexes to be internalized (Borrelli et al., 2018).

7.6.3 CPP AND PROTEIN DELIVERY FOR ANTI-CANCER THERAPY

The utilization of therapeutic proteins and vaccine peptides in anticancer therapy has lately been a major focus of translational research. A humoral antibody response is the initial reaction to a peptide-based vaccination. Antigen dispensation via the MHC class I alleyway is required for the initiation of cytotoxic T-cell lymphocytes (CTLs), and CPPs can promote the cytosolic absorption of vaccine-peptides via MHC I presentation. In animal models, LAH4 can non-covalently bound to the cell-penetrating peptide, improved cellular conveyance of a tyrosinase related protein 2 (TRP2) peptide vaccinations, resulting in an anticancer effect in tumor tissues expressing TRP2 (Dinca et al., 2016; Borrelli et al., 2018). Polyethylene-glycol (PEG) manacles linked to the surface by a linker that may be broken through stimuli from tumor tissues, altering local environmental circumstances and inactivating PEG protection have been used to sterically shield CPPs (Kuai et al., 2011).

The lack of cell type specificity, which is related to the fact that most CPPs are swallowed by all cell types, as well as their poor circulatory stability and low bioavailability in target tissues due to poor tissue penetration and/or cellular uptake, has been a key barrier to CPP-based therapeutics. To boost CPP selectivity to sick tissues, several techniques have been explored, including conjugating CPPs with homing peptides or other targeting ligands including RGD peptides, hyaluronic acid and folic acid. Because these active ligands are frequently substantially expressed in a variability of tumor types but not in normal cells, medication delivery has no effect on healthy tissues. It's wealth noting that conjugating a CPP with cargo might occasionally result in novel epitopes that elicit immunological responses, despite the fact that a single CPP did not (Borrelli et al., 2018).

7.7 CHEMICAL CONJUGATION CANCER CHEMOTHERAPEUTICS

The chemical conjugation process is an important consideration in pharmaceutical chemistry to enhance the therapeutic potential of drug candidate. The chemical conjugation is process deals with attachment of inert chemical moiety to drug candidate via covalent or non-covalent interaction. Chemical conjugation converts the active moiety into an inert form, which may have pharmacodynamic characteristics. Through chemical, ionic, or enzymatic means, the drug is converted into a pharmacologically active medication. In the treatment of cancer like heterogenous disease, in severe condition anticancer drugs are delivered through intravenous infusion. Most of the anticancer drugs are unstable in solution or dispersed forms and start degradation. Prodrug improves the cancer therapy by improving the stability of drug in solution or dispersion condition. Prodrug provides suitable route for administration with improved pharmacokinetic properties. Prodrug alters the rate of metabolism and provide long circulation time with improved selective targeting towards cancer cell (Figure 7.3) (Connors, 1986).

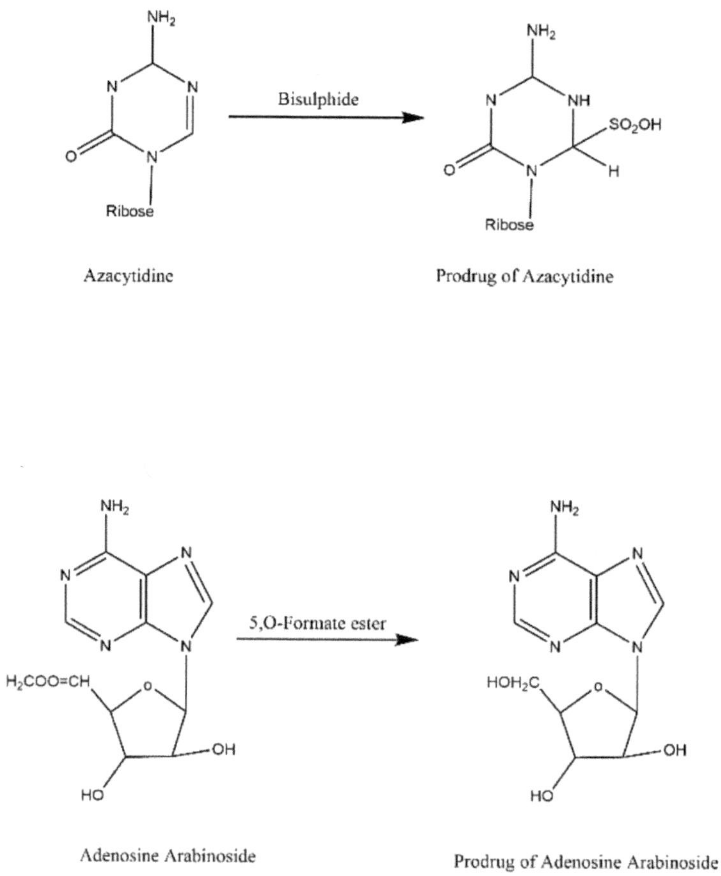

FIGURE 7.3 Transformation of prodrugs of azacytidine and adenosine arabinoside.

Few chemical modifications are elaborated here to understand the prodrug designing approaches are easy and important for improving therapeutic efficacy of chemotherapeutic agent in cancer therapy. For example, azacytidine (AZT) is use for the treatment of myeloid leukemia and preferably delivered through intravenous infusion instead of oral administration (oral dose limiting toxicity). The AZT has reversion and later irreversible conversion into formyl and guanyl urea derivatives, respectively. Both formyl and urea derivatives did not possess any anticancer activity. The concentration of AZT steadily declines after a longer infusion. The addition of bisulfide across the double bonds in AZT at pH 2.5 stabilizes the solution states. Because of its improved stability in acidic conditions, the prodrug of AZT after treatment with bisulfide encourages oral administration. Adenosine arabinoside (ASAB) has antiviral and anticancer activities and efficiently employed as intravenous infusion. The limited solubility (0.5 mg/mL) and high dose (2–5 g) requirement may require several volumes of solution needs to be infused. The several liters of solution require longer time and dose dumping may be observed. As systemic concentration may vary with physiological limits. The conversion of ASAB into stable and soluble prodrug increases solubility by 60 folds after transmission into 5'O-formate ester. The prodrug of ASAB did not require enzymatic deamination it is directly converted to active form without any substrate requirement for activation (Repta et al., 1975).

Synthetic conjugation of prodrug is inactive in nature and itself has low cytotoxicity to cells. But after conversion into active form via enzymatic action, promotes its potent inhibition activity. During the treatment of prostate cancer using diethylstilbestrol diphosphate (prodrug) converted to active form in presence of enzyme phosphatase into diethylstilbestrol (Bagshawe et al., 1991).

7.7.1 PRODRUGS OF ALKYLATING AGENTS

Cyclophosphamide derivatives are generally use as prodrug form of alkylating agent. The prodrug cyclophosphamide activated at early stages in the liver from 4-hydrozycyclophsphamide with intermediate aldophosphamide. Aldehyde dehydrogenase in liver inactivates the aldophosphamide and avoids release into blood. Another alkylating agent was Dacarbazine metabolized into liver and forms N-demethylation derivative. The Dacarbazine was used in very less due to toxicity. The formation of methyl carbonium induces toxic effect in cell and generated with release of diazomethane in acidic environment (Waller & George, 1989).

Alkylating nitrogen mustards are best suitable for targeting to tumor cell using prodrug therapy approach. Nitrogen mustards are bind to cellular macromolecules and form crossing between DNA and inhibit the growth. The nitrogen mustards have non-specific binding and toxic to normal as well as tumor cells. Based on high reactive and poor reactivity of nitrogen mustard has shown higher or lesser toxicity to tumor cell, respectively. The variable toxicity of alkylating agent may perceive the variable pharmacokinetic effect. For improving target specificity, the prodrugs of alkylating agents are designed in such way that, prodrugs are poor candidate and non-toxic in uncleaved stage. The presence of enzyme can immediately convert prodrug into active form. The concentration such enzymes are more in tumor cell compared to normal cell and chances of accumulation and cleavage is again more in tumor cell. The target specific activity is more compared to bare drug candidate. The prodrug in this case is activated in presence of enzyme azoreductase (Connors, 1986). In the case of hepatocellular carcinoma (HCC), the concentration of azoreductase was more and azobenzenes type prodrugs were designed for targeting. The azobenzenes are converted to active form like alkylating agent and reactive in tumor cell. Initially, the azobenzenes are inert in nature and did not show any pharmacological activity. The alkylating agent may also diffuse from the tumor tissue and available in systemic circulation. After entering the alkylating agent into blood circulation, it hydrolyzes before reaching to bone marrow (Bukhari et al., 1973). Ascites cells are efficient in conversion of tetrazolium salts to farmazane. The presence of conjugated ring structure in nitrogen mustard containing tetrazolium salts, they are inactive in nature. After reduction to farmazane by ascites cells reactivity of alkylating moiety was increased with reduction of amine (Tsou & Su, 1963).

7.7.2 PRODRUGS OF ANTIMETABOLITES

Antimetabolite is a class of antiviral and anticancer drugs typically an analog of natural metabolites. Natural analog of purines and pyrimidines bases, folate analog, nucleoside or nucleotide, etc. Inhibition of RNA or DNA synthesis considered as important strategy in killing cancer cell (Peters, 2014). The enzyme responsible for altering the activation of prodrug and also interfere in activity of active form. The uridine phosphorylase is enzyme responsible for formation of active form of 5-fluorouracil (5FU) from 5-deoxy-5-fluorouridine (5FUd). The 5FU is cytotoxic metabolite inhibit the growth of leukemia. The L1210 leukemia cells did not able to convert 5FUd to 5FU due to presence of unrecognized activity of 5-O-tosyl towards uridine phosphorylase. The 5FU conversion was poorly addressed in the case of L1210 leukemia cells (Ajmera & Danenberg, 1982; Au et al., 1983). Valproic acid-based prodrug of Gemcitabine (VAG) is under clinical development with lead compound was designed by Eli Lilly & Company. The attached valproic acid at N4 position of gemcitabine cleaved in presence of enzyme carboxylesterase 2 (CES2) convert into active form. The presence of higher concentration of CES2 in liver and GIT leads to the early activation and toxicity profile of VAG and gemcitabine was similar and considered as promising prodrug (Veltkamp et al., 2008).

7.7.3 PRODRUGS OF ANTHRACYCLINES/ANTI-TUMOR ANTIBIOTICS

Easy way to reduce the toxicity of anthracylin derivates is too encapsulated along with polymeric substrate. The polymer bound anthracycline is available in conjugated form and avoid direct release of drug. DOX and daunoribicine (DAN) are aggressively use in the management of cancer. The DOX and DAN are used for treatment of multiple tumors along with blood cancer treatment. The DOX and DAN have dose limiting toxicity may limit it use in the treatment. On later stages of treatment patients may develop cardiac toxicity or myeloma suppression. Anthracylin derivatives are studied from several decades and researcher working to resolve the problems associated with toxicity. The anthracyclin has shown its action on cell surface, intercalation, and inhibition of DNA synthesis, intercalation, and inhibition of RNA synthesis (Tritton, 1991; Takakura & Hashida, 1995). In most cases, lysosomotropic agents are used for conjugation with DAN. At C14 position specifically use for conjugation with additional sugar moiety was also utilized for macromolecular interaction (Zunino et al., 1982).

Covalent functionalization can be possible at amino sugar residue of anthracycline molecule using free spacer group (*cis*-acotinyl spacer). The lysosomal enzymes act on the spacer group and release the drug for their activity. The conjugation did not cleave in extracellular space due to non-hydrolytic nature of bond and only release in the presence of lysosomal enzyme (Bouma et al., 1986). Polymeric surfactant micelle is also governing interest in delivering drug to the site of action. The amino group DOX and carboxylic group of polyethylene glycol (PEG)–polyaspartic acid mainly produce the amide bond through covalent functionalization. The presence hydrophilic and hydropobic regions convert the structure into amphiphilic character. The micelle has least size around 50 nm and acquires long circulation time. The conjugated structure did not release drug into extracellular environment and only release at the site of action (Yokoyama et al., 1990).

Platinum based drugs like cisplatin develops drug resistance in cancer patients. As alternative therapy, cisplatin, and DOX were delivered as combination therapy. The combination therapy may not provide much effectiveness due to abundant glutathione with microenvironment pH may reduce effectiveness of drug. The cisplatin prodrug Pt(IV) reduces drug resistance and kills the cancer cell effectively. The long lipid chain with carboxylic group present at axial position with cis-aconitic anhydride-DOX (CAD) improves therapeutic outcome. To the generates self-assembled structure, the polyethylene glycol–block–poly(L-lysine) (PEPL) crosslinked to Pt (IV) and CAD. In acidic tumor microenvironment Pt(IV) is separated from CAD and converted into active Pt (II) and DOX and kill tumor cell. The pH/redox sensitive drug release only shown in cancer cell while in normal cell limited release was observed and non-specific cell killing was alleviated (Zhang et al., 2019).

7.7.4 PRODRUGS OF TOPOISOMERASE INHIBITORS

Camptothecin was modified to irinotecan prodrug form and it mainly water-soluble moiety modified via carbamate linkage between dipiperidine promoiety and camptothecin. The bioconvertible camptothecin occurs in presence of enzymes principally in liver and partly at the site of tumor cell. Irinotecan is topoisomerase I inhibitor cleaved in presence of carboxylesterases. The lactone and carboxylate forms are available while lactone form is more pharmacologically active (Rautio et al., 2008).

7.7.5 ENZYME PRODRUG THERAPY

Most prodrug therapies are effective in presence of enzyme and provided success scientifically and clinically. Based on the activity of known enzyme the enzymologist shares the interlinking with chemist and design suitable prodrug. The selection of class of prodrug may be easy in understanding the catalytic interaction between enzyme and prodrug. The selection of enzyme was very critical step and one or more of the following ideal characteristics can be considered for evaluation of prodrug. The enzymes possess non-reversible reaction, must have high specificity and act at neutral pH or biological pH. The enzyme may be stable at body temperature and did not lose its activity. The enzyme must be stable at conjugated conditions and did not have any co-factor. The monomeric enzymes are preferred over dimeric or polymeric domains with absence of common inhibitors. Alternatively, mammalian or microbial enzymes also used for analysis but differ the specificity compared to human enzyme. Microbial enzymes are not preferable for prodrug therapies due to immunogenic responses (Zelikin, 2020; Zawilska et al., 2013).

7.7.6 MACROMOLECULAR PRODRUG

For delivery the drug in an inactive form, drug is bind to macromolecules as protecting domain. The interaction preferably a covalent attachment to chemotherapeutic drug but in recent times stearic, hydrogen bonding, supramolecular attachments were highlighted. The use of macromolecules is to preferable avoiding direct contact of drug toward biological media and somatic cells. Additionally, macromolecules avoid preexposure and burst release into normal cells. The macromolecule conjugated chemotherapeutic agents are inactive in nature and by virtue of chemical or enzymatic reaction converted to active form. Sources of macromolecules are numerous, scientists have opportunity to select from natural resource, go for semisynthetic or synthetic polymer may also use for conjugation process. The macromolecules must be inert in nature, nonimmunogenic, did not have ability of accumulation in organs or tissue, embedded functional groups for interaction with drug and biological components, target specificity and unable to divert the parent activity of drug. The ideal macromolecules possess two or more characteristics defined above. Different classes of macromolecules with examples are highlighted in Table 7.4.

TABLE 7.4 Sources of Macromolecules for Interaction

Sl. No.	Parent Class	Subclass	Examples
1.	Natural	Proteins	Albumin
2.		Polypeptides	Pullulan, chitin, chitosan
3.		Nucleic acids	DNA, cDNA, pDNA
4.		Antibodies	Monoclonal antibodies
5.		Lectins	Mannose, concanavalin A
6.		Glycoproteins	Aisalfetuin
7.		Hormones	Thyrotropin
8.		Bacterial source	Hyaluronic acid
9.		Animal	Lactoferrin
10.	Semisynthetic	Biopolymers	Chitosan oligosaccharide, carboxymethyl chitosan
11.	Synthetic	polyamino acids	Polyglutamic acid, polylysine

Synergistic inhibition potential was reported in conjugation with certain components. The activity enhancing macromolecules are advantageous in future therapies (Takakura & Hashida, 1995).

7.7.7MANTIBODY ENZYME PRODRUG THERAPY COMBINATION

Novel active targeting approaches are emerging therapeutic use by researcher to notify and improve the therapeutic efficiency of conventional drug candidates. The antibodies have high specificity and targetability towards cancer cell combined with low molecular weight drug (Senter et al., 1993). Complexes consist of two different modules containing Antibody bind to enzymes and conjugated with low molecular weight drug. When the antibody recognizes the cell surface antigen, it releases the enzyme – drug complex (prodrug) into extracellular compartment. Later, the enzyme converts prodrug to drug and enters inside the cellular spaces. The entered drug shows therapeutic activity. The advantage of this type of therapy was drug is also inhibiting the nearby cancer cells having low antigen presentation (Takakura & Hashida, 1995). Etoposide phosphate is a prodrug of etoposide conjugated with monoclonal antibodies and alkaline phosphates enzyme. The *in vivo* animal study suggests that intraperitoneal administration shows profound inhibition potential with presence of prodrug at active site within 18–24 h (Senter et al., 1988). Palpable human tumor xenograft was transplanted in mice model with drug resistant characteristics shows potential inhibition in the presence of monoclonal antibody (MAB) – carboxypeptidase G2 enzyme complex with prodrug conjugate (Bagshawe et al., 1991; Senter et al., 1993).

7.8 THERAPIES

7.8.1 RADIOTHERAPY

Radiotherapy is a cancer treatment in which cancer cells are exposed to high doses of radiation in order to kill them and shrink tumors. X-rays employ low-level radiation to see within your body, such as in X-rays of your teeth or damaged bones. The radiation therapy kills or delays cancer cells' development by destroying their DNA at high dosages. Cancer cells that have had their DNA damaged beyond repair will either cease proliferating or die. When damaged cells die, the body breaks them depressed and eliminates them (Figure 7.4).

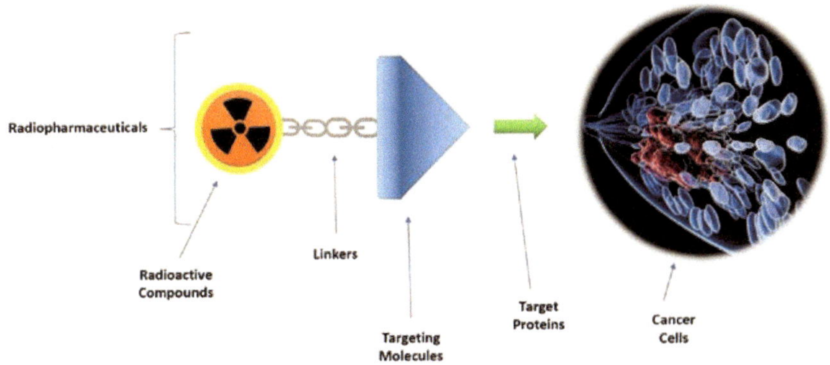

FIGURE 7.4 Radiopharmaceuticals consist of a radioactive molecule, a targeting molecule, and a linker.

Source: Adapted from National Cancer Institute. https://www.cancer.gov/news-events/cancer-currents-blog/2020/radiopharmaceuticals-cancer-radiation-therapy

Roxana et al. records quantifiable ultra-sound characterization of radiotherapy responses on cancer mouse models, the study concludes that changes in cell structure following radiation have a considerable impact on ultrasound spectrum characteristics. This lays the underpinning for future research into the use of ultrasonography in cancer patients to noninvasively personalize therapies based on their reactions to certain interventions (Vlad et al., 2009).

Peiyu Xue et al. give a review on the synergy of the ellagic acid in cancer treatment when combined with radiotherapy and chemotherapy. Finally, due to its biological properties such as antioxidant and anti-inflammatory properties, ellagic acid has synergistic effects in radiation/chemotherapy for conduction of treatment of malignancies and has a tremendous potential in decreasing the adverse effects of the different radiotherapies and chemotherapies. Ellagic acid could be act as an auspicious cancer therapeutic adjuvant. Novel ellagic acid delivery strategies that overcome the low ellagic acid solubility and bioavailability should be researched in the near future to fully exploit the medicinal potential of ellagic acid (Xue et al., 2022). Radiotherapy-Activated Cancer-Associated Fibroblasts are defined by Joke Tommelein et al. for promotion to tumor progression by activating the insulin-like growth factor-1 receptor in cancer-associated fibroblasts and investigating changes in CRC cell proliferation, transcriptome, metabolome, and kinome in response to paracrine signals released by irradiation cancer-associated fibroblasts. DNA damage as well as p53 activation and cell-cycle arrest with release of the mediators of paracrine, together with insulin like growth factor-1, are all linked to cell-cycle arrest., were all observed in cancer-associated fibroblasts after preoperative irradiation. Preoperative radiotherapy-activated cancer-associated fibroblasts subsequently improved CRC cell survival and a metabolic transition favoring glutamine intake via insulin-like growth factor-1 receptor activation. According to his findings, paracrine insulin-like growth factor-1/insulin-like growth factor-1 receptor signaling enhances CRC progression, indicating a preclinical rationale for targeting this activation loop (Tommelein et al., 2018).

To overcome tumor hypoxia, Rui Zhang et al. created catalase-loaded cisplatin pro-drug liposomes (Keservani & Gautam, 2020) to overcome tumor hypoxia for the improvement of chemoradiotherapy cancer treatment. In judgment to the control groups, the catalase-loaded cisplatin pro-drug produced liposomes cause the most DNA damage after X-ray radiation in between cancer treatments. *In vivo* tumor treatment with catalase-loaded cisplatin pro-drug produced liposomes nanoparticles also shows a significantly enhanced therapeutic effect in chemo-radiotherapy. Later, by combining cisplatin grounded chemotherapy and catalase persuaded tumor hypoxia relief for combination chemo-radiotherapy with high synergistic efficacy, an exquisite form of liposome-based nanoparticles (Keservani et al., 2017) is produced in this work, promising for clinical translation in cancer treatment (Zhang et al., 2017). According to Sandeep Chavan et al. Gandush Churna's gargle treatment formulation, which contains *Terminalia chebula* Retz. *Terminalia bellirica* (Gaertn) Roxb., *Phyllanthus emblica* L., and *Curcuma longa* L., and its decoction Gandush Kwath, it could be an effective supplemental treatment for improving oral hygiene and reducing radiation side effects in individuals with oral cancer (Chavan et al., 2021).

7.8.2 GENE THERAPY

The transportation of genetic elements with adenovector and corrected gene into cancer cells (Jarouliya et al., 2018) is known as gene therapy (Figure 7.5). To assure the transport of genetic materials, a vector is used as a carrier. Afterward, therapeutic genes are delivered to cells; they work by a variety of processes, including silencing, up- or down-regulation, repair, or alteration of the target genes. Cell death and tumor necrosis may be caused by suicide genes. Silencing genes stops cells from growing and tumors' from shrinking. While gene therapy may improve the responsiveness to other combination medicines (e.g., chemotherapy, immunotherapy, or radiation) (Mitra et al., 2015).

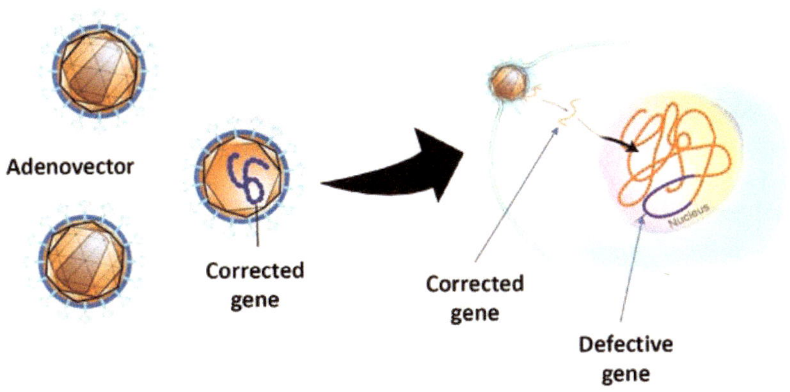

FIGURE 7.5 Gene therapy consists of an adenovector and corrected gene.

Source: Adapted from Gate Scientific. https://gatescientific.com/technique-geeks-blog/f/gene-therapy-pt-1-the-bad-beginning

Liang et al. published systemic p53 gene therapy of cancer with immune-lipoplexes, and it targeted by anti-transferrin receptor scFv. According to his primary research, Compared to the transferrin (Tf) molecule, the anti-transferrin receptor single-chain antibody (TfRscFv) has a variety of advantages: (i) scFv is much smaller than Tf, resulting in a smaller immunolipoplex that penetrates solid tumors better; (ii) unlike Tf, scFv is a recombinant protein, not a blood product; and (iii) large-scale production and strict quality control of recombinant scFv, as well as scFv-immunolipoplex, are both feasible. The ability of this tumor-targeted and efficient p53 gene delivery technology to sensitize tumors to chemotherapy could lower the effective dose of the drug, resulting in less severe side effects and a lower risk of recurrence. Furthermore, this method can be used to treat both primary and recurring cancers, as well as metastatic disease (Xu et al., 2001).

Interferon-gene therapy using cationic liposomes to deliver the interferon gene into tumor cells varied from standard interferon-protein therapy, according to a review by Jun Yoshida et al. Interferon gene transfer has stood revealed to trigger apoptosis in tumor cells resistant to interferon protein, as a glioma, melanoma, and renal cell carcinoma (Yoshida et al., 2004). Gene therapy techniques in mice have produced ultimate circulating endostatin levels of 8–33 ng/ml, according to Andrew et al. literature search. After administering a recombinant adenovirus, they found endostatin levels in plasma as 1,770 ng/ml. Adenoviral infection decreased the progress of MC-38 adenocarcinoma, which is comparatively resilient to infection. These findings support gene delivery strategies that employ the host as a "factory" to manufacture high amounts of antiangiogenic drugs in the bloodstream (Feldman et al., 2000). Collins and colleagues present a review of the clinical development of the Toca 511 vector of retroviral replication for cancer gene therapy. Toca-511 has demonstrated highly auspicious results in initial stage clinical studies, according to his review study. This vector advanced to the registrational study of Phase III; however, the results, which were released in late 2019, were largely unfavorable. However, the median prodrug dose schedule was not ideal, and several prespecified subgroups showed promising possible efficacy. More clinical research is needed, as well as the development of RRV with additional transgene payloads (Collins et al., 2021). According to Lin Zhang's research, DNA vector-based RNAi, in which RNAi sequences targeting murine VEGF isoforms are introduced downstream of an RNA polymerase III promoter, has potential applications in isoform-specific "knock-down" of VEGF. Large molecular weight VEGF isoforms were especially reduced *in vitro* in the presence of isoform-specific RNAi constructs. Furthermore, the H1 promoter may be preferable to the U6 promoter when used for vector-based RNAi of VEGF isoforms. This method offers a fresh way to investigate the role of multiple VEGF isoforms, and it could lead to VEGF isoform-specific cancer treatment (Zhang et al., 2003). Stephen et al. describe a single-arm, open-label, repeat-dose clinical trial using intravenous nadofaragene firadenovec gene therapy for BCG-unresponsive non-muscle-invasive bladder cancer. He looked at 198 patients to see if they were eligible. A total of 157 participants in the study were enlisted and received at least one dosage of the study drug. Six patients were omitted from the efficacy analyses because they did not fit the criterion of BCG insensitive and unresponsive non-muscle invasive bladder cancer; in the per-protocol efficacy analysis, the remaining 151 patients were included. Within 3 months of the first dose, 55 (534%) of 103 patients with carcinoma *in situ* (with or without a high-grade Ta or T1 tumor) achieved a full response, which was maintained in 25 (455%) of 55 patients at 12 months (Boorjian et al., 2021).

7.9 OTHER DELIVERY SYSTEMS

In other delivery system for the cancer treatment, micro-particles, HGs, micro-bubbles, and phytosomes ordinarily stated (Figure 7.6). They are frequently used to target medications, improve bioavailability and stability, and provide long-lasting drug release (Ahire et al., 2020a, b).

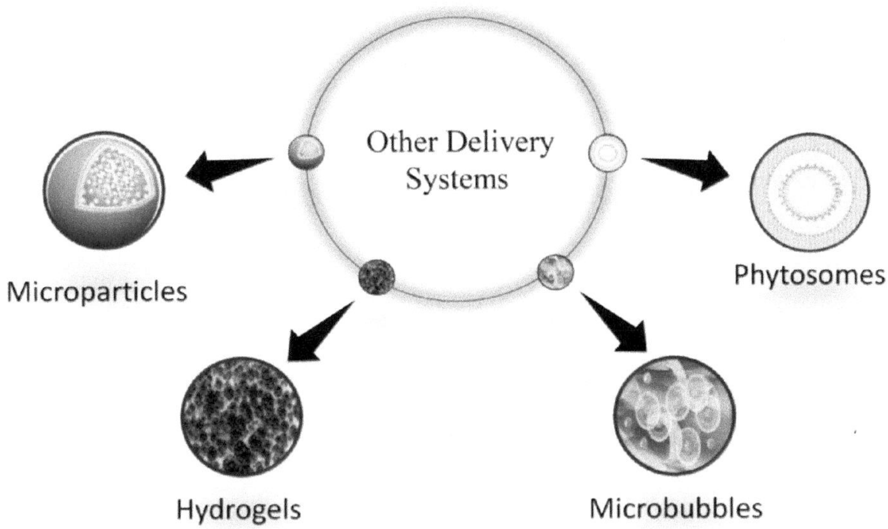

FIGURE 7.6 Other delivery systems.

7.9.1 MICRO-PARTICLES

Microencapsulation is a commonly used technology to target drugs, increase bioavailability and stability, and provide sustained drug release. Micro-particles are beneficial for breast cancer (BC) treatment because they lower medication toxicity to non-targeted cells and tissues while also increasing effectiveness (Hani et al., 2011; Rajput & Agrawal, 2010). Bosio et al. developed DOX-loaded calcium carbonate hybrid nanoporous micro-particles that may effectively target cancer cells and deliver regulated medication release (Bosio et al., 2014). Patino et al. investigated the impact of micro-particle surface modification on interactions with normal and malignant breast epithelial cell lines of human being. A considerable difference in cellular absorption efficiency, endocytic route, and cytotoxicity was reported between two cell lines, indicating the need of surface modification of micro-particles for distinct cell 11 targeting's (Patiño et al., 2015). By using a single emulsion solvent evaporation process, PLGA-based micro-particles loaded with bacterially produced prodigiosin were created. The produced micro-particles result in the regulated distribution of medicine, which can be employed to prevent regrowth following surgical excision of triple negative BC. Prodigiosin promotes localization and apoptosis, as well as a reduction in cell viability (Obayemi et al., 2016). Davoodi et al. used atomization technique for coaxial electrohydrodynamic to create double-walled micro-particles encapsulated with cisplatin and PTX, which were then embedded into an alginate-branched polyethyleneimine (PEI) hydrogel injectable to achieve localized delivery of dual-drug and anticancer drugs-controlled release with greater anticancer response (Davoodi et al., 2016).

Tissue factor of tumor derived expressing micro-particles serve as biomarkers that aid in the prevention and treatment of cancer-related venous thromboembolism problems (Nomura et al., 2015). In the clinic, PLGA micro-particles encapsulated Tumor lysates vaccine therapy designed to prevent unprompted lung metastases in between the mice 4T1 adenocarcinoma breast tumors has been found to be efficacious and well-tolerated (Gross et al., 2014) The emulsion/extraction approach was used to formulate PLGA-based biodegradable anastrozole micro-particles, and this work demonstrated that micro-particle drug carriers may be a better method for long-term anti-cancer treatment (Zidan et al., 2006). Injectable sustained-release curcumin micro-particles were explored as a unique technique to fight cancer by Shahanj et al. who discovered that therapy with curcumin micro-particles had a substantial effect on tumor angiogenesis (Shahani et al., 2010). The chemotherapeutic activity of the medication was also improved by magnetic iron-carbon composite micro-particle formulations (Rudge et al., 2000). Tamoxifen, a no steroidal antiestrogen cross linked by alginate/chitosan micro-particles, was developed to target the lymphatic scheme for the specific treatment of metastatic estrogen receptor-positive BC, and its efficacy was verified *in vitro* (Coppi & Iannuccelli, 2009). Micro-particles with nucleic acids have been shown to overcome multidrug resistance in both hematological and nonhematological malignancies (Jaiswal et al., 2012).

7.9.2 HYDROGELS (HGS)

Higher hydrophobicity, high drug loading capacity, size flexibility, biocompatibility, and simplicity of handling and production are all the compensational rewards of HGs. HGs are made up of hydrophilic polymers dispersed in water that form a polymeric mesh that conceals drug molecules inside, and they have a tendency to inflate, allowing the drug to be released outside of the polymeric meshes for disintegration and dissolution (Larrañeta et al., 2018; Ahmad et al., 2022b). The measurement of angiogenic indicators in serum using shape coded hydrogel MPs indicated the monitoring of numerous clinical conditions. By removing the proteins, bio-inert PEG HGs improved the sensitivity of the approach for the detection of VEGF, FGF, and PDGF in single and multiplexed experiments (Al-Ameen et al., 2015). The creation of

DNA adducts exhibited double strand in fatal DNA, and AhR ligands of 5F203 and GW 610 benzothiazole HGs (Thombre et al., 2022) showed anticancer action. Designing prodrugs with high nanomolar potency was proposed by *in vitro* research and clinical data (Stone et al., 2015). To explore *in vitro* cancer biology investigations, a PEG loaded chitosan thermoreversible hydrogel was designed as a tumor microenvironment. The findings revealed increased cancer cell resistance to treatment, increased tumor cell aggregation, and increased release of growth agents such as cytokines (Alexander et al., 2014). Weight gain experiments were used to develop poly[N-isopropylamide] integrated HGs employing co-polymers of P(NIPA) conjugated Acrylamide and Butyl Methacrylate and homo-polymer of P(NIPA). The properties of edema caused by prodigiosin consumption were compared to swelling in a controlled setting (Danyuo et al., 2016). As a thermosensitive, injectable, and multi-compartment hydrogel, PEGylated fluorocarbon nanoparticles were coupled with PEGylated hydrocarbon nanoparticles. Peri-tumoral injection of remedy cocktail loaded MCH formulation demonstrated increased attention for sustained delivery of bioactive substances in an *in vitro* and *in vivo* model (Kai et al., 2015). Anti-tumor activity was studied using *in vitro* three-dimensional culture biomimetic models (3D) of glycosaminoglycan-based hydrogel that mimicked the *in vivo* microenvironment. These models were less sensitive to chemotherapy than the 2D cultures, according to the findings (Bray et al., 2015). A standard proliferation experiment was used to assess three-dimensional cultures of alginate core-shell hydrogel beads produced to inhibit MCF-7 BC cell proliferation. It has fixated its exertions on medicines with a high multicellular resistance index (Yu et al., 2015). Ionic cross-linked HGs were designed to speed up the release of biomolecules and offer reversible self-healing using ultrasound. The findings suggested that this approach, which is used to distribute medications in chemotherapy in a time-controlled manner, is effective (Segovia et al., 2015). Poly(methacrylic acid) integrated pH-responsive natural HGs capsules can behave as shape-responsive carriers when chemical stimulation is applied. It resembles red blood cells in size and shape, and it offers a novel medication delivery method (Kozlovskaya et al., 2014).

7.9.3 MICRO-BUBBLE

Micro-bubbles are bubbles with a diameter of less than 1 mm but greater than 1 meter. Micro-bubbles are becoming more used as a contrast agent for oncological molecular imaging with ultrasound impact. Micro-bubbles are generally low intensity pulsed ultrasound were used to augment curcumin cytotoxicity in human cancer cells, including apoptosis induction and cell motility inhibition (Li et al., 2016). The effects of ultrasound-mediated micro-bubble stimulation and radiation treatment on breast tumors and blood vessel endothelium were explored, and the results revealed that combined ultrasound and radiation therapy can yield stronger anticancer and antivascular effects (Lai et al., 2016; Yan et al., 2013). The researchers developed PTX-loaded liposomes coupled to micro-bubble complexes for ultrasound-triggered targeted chemotherapy against BC cells, and they found that higher release was accomplished with ultrasound exposure, resulting in a considerable increase in PTX antitumor efficacy (Yan et al., 2013). Ultrasound stimulated micro-bubbles containing anticancer agents are combined with antiangiogenic metronomic chemotherapy to produce antivascular effects, and this study found that the combination of Ultrasound stimulated micro-bubble treatments and metronomic cyclophosphamide inhibited tumor growth significantly (Todorova et al., 2013).

Escoffre et al. investigated the application and efficiency of combined DOX with micro-bubble-assisted ultrasound in BC cells and found that improved BC cell death and glioblastoma cells with induced apoptosis can be achieved (Escoffre et al., 2011). Park et al. used an ultrasound gene therapy approach with micro-bubbles to test the adenine nucleotide translocase-2 short hairpin RNA for anticancer effects using PEI intraperitoneal injection and found stronger antitumor effects *in vivo* (Park et al., 2015). Sorace et al. elucidate study as micro-bubbles loaded with DOX and PTX for enhanced drug delivery to tumor cells when triggered with ultrasound, found that PTX loaded ultrasound contrast agent has 20 times the potential of DOX to achieve a sustained and targeted drug delivery (Cochran et al., 2011). The anticancer activity of LyP-1 peptide coated PTX loaded lipid micro-bubbles used in ultrasound-assisted chemotherapy to target breast carcinoma cells was significantly increased (Yan et al., 2011). Angiogenesis of tumor with molecular imaging consuming micro-bubbles ultrasonic contrast agent was created for sensitive detection of an intravascular targets and tested in a BC animal model *in vivo*, with results revealing the feasibility of low mechanical index molecular ultrasound imaging (Anderson et al., 2011). On the cultivated cell membrane, a comparison of the formation of small pores of two phospholipids founded vectors liposomes and phospholipids micro-bubbles revealed that micro-bubbles appeared to be a unique technique for drug and gene delivery and might boost formation of small pores in cell membranes (Zhao et al., 2006). The researchers created biotinylated bleomycin conjugated with streptavidin-derivatized micro-bubbles for carbohydrate-dependent cancer cell targeting and discovered that the bleomycin micro-bubble conjugation only targeted cancer cells without harming or binding to normal breast cell lines (Chapuis et al., 2009). Ultrasonic micro-bubbles bind to LyP-1 and can be employed as bio-acoustic indicators or drug carrier vehicles in a microfluidic system to target cancer cells (Li et al., 2009). Micro-bubbles coupled low intensity pulsed ultrasound with curcumin cytotoxicity and drug delivery to the BC cells of human and discovered that it improved antitumor impact significantly; suggesting that this combined treatment is a promising strategy for antineoplastic treatment (Li et al., 2016).

7.9.4 PHYTOSOMES

Because of their varied pharmacological properties, phytoconstituents have been employed in medicine since ancient times. Despite the fact that many phytoconstituents and plant extracts have great bioactivity *in vitro*, they have no *in vivo* activities due to inadequate lipid solubility, incorrect molecular size, or both, resulting in poor absorption and bioavailability. Phytosomes are new formulation techniques that can assist solve these issues. Phytosomes are made by binding a standardized the plant extract or its constituents to phospholipids, primarily phosphatidylcholine, to form a lipid suitable complex. Many prominent herbal medications, such as Ginko biloba, hawthorn, olive oil, grape seed, green tea, and ginseng, have been treated with phytosomes technology. Phytosomes are a new medication delivery technology that contains hydrophilic bio-activated phytoconstituents of the herbs that are encased in phospholipids and bonded together (Dhir et al., 2016; Sharma & Roy, 2010). Indena S.p.A. in Italy was the first to develop phytosome technology. In both water and lipids, the active pharmacological components in polyphenols are sparingly soluble. This characteristic makes it challenging to synthesize polyphenolic APIs into commercial pharmaceuticals. Bombardelli & Spelta developed a new medication delivery mechanism called phytosomes in 1991 (Azeez et al., 2018). Phytosomes are micelles generated when phospholipids and water interact. With the addition of polyphenolic plant extracts, the interaction is amplified. The charged phosphate head of phospholipids conjugates the polar functional groups of lipophilic substances by hydrogen bonding and polar contact, generating a micelle configuration (Lu et al., 2019). To increase luteolin bioavailability and passive targeting in BC cells, they developed nano-phytosomes containing luteolin, and co-treating MDA-MB 231 cells with nanoparticles containing luteolin and DOX resulted in the highest percentage of cell death, according to the findings (p0.05) (Sabzichi et al., 2014). Sabzichi and his coauthors used phytosome-based medication delivery of luteolin to block the Nrf2-mediated DOX signals in cancer cells (Ahire et al., 2020c, Surana et al., 2021). They discovered that phytosomal-curcumin suppresses cell proliferation and migration in BC cells stimulated by thrombin via AMP-Kinase, implying that more investigation into this novel anticancer drug's therapeutic potential in the treatment of BC is required (Hashemzehi et al., 2018).

Usama et al. created, developed, and characterized scorpion venom conjugated curcumin phytosomes. The produced phytosomes had a smaller vesicle size and a higher positive surface charge, both of which are important for anticancer action. Cell cycle arrest with apoptosis stimulation, mitochondrial membrane permeability, and inflammatory mediator were all investigated in fabricated phytosomes. When PC3 cells – human prostatic cancer cells were treated with SV conjugated CUR phytosomes, the cell cycle was arrested at the pre-G1 phase, apoptosis, and necrosis were increased, mitochondrial membrane permeability was reduced, Bcl-2, NF-kB, TNF-levels were reduced, and Bax, caspase, and p53 levels were increased (Al-Rabia et al., 2022). Alhakamy et al. combined refluxing and anti-solvent precipitation, a novel phytosomal delivery method for thymoquinone was created. A three specified factors and levels Box-Behnken design was used to optimize this thymoquinone delivery system. The profile for SR with much better anticancer potential could be obtained with thymoquinone using this phytosomal nanocarrier platform, as thymoquinone phytosomes greatly increased reactive oxygen species formation in A549 cells (Alhakamy et al., 2020).

7.10 CONCLUSION

Cancer is a disease caused by uncontrolled cell growth and division caused by cellular alterations. According to recent studies, the number of cancer diagnoses and deaths is increasing every day. The conventional method of treatment has some limitations such as less affinity with cancerous cells, less elimination half-life, adverse effect, low solubility, and oral bioavailability. Novel drug delivery system (NDDS) will eliminate the side effects of conventional therapy; will increase the drug's solubility/dissolution rate with better drug targeting and oral bioavailability. After executing their effect, the novel system was degraded and expelled from the body. As a consequence, the human body system has no unfavorable or adverse effects. For the preparations of novel formulations, various natural, synthetic, and inorganic substances are available that alter the drug's behaviors. Finally, we conclude that NDDS will improve drug efficiency and drug targeting and seem to be suitable approaches for effective cancer treatment.

It is hoped that the novel drug delivery-based therapy will be a new opportunity for effective treatment of cancer.

ACKNOWLEDGMENT

The authors are thankful to the institute for providing facilities such as a library, journals, and internet sources for writing the chapter.

CONFLICT OF INTEREST

The authors declare no conflict of interest.

KEYWORDS

- **cancer**
- **chemical conjugation**
- **drug delivery**
- **hydrogels**
- **microparticles**
- **nanoparticles**
- **phytosomes**
- **targeting**
- **therapy**

REFERENCES

Abdollahpour-Alitappeh, M., Lotfinia, M., Gharibi, T., Mardaneh, J., Farhadihosseinabadi, B., Larki, P., Faghfourian, B., Sepehr, K. S., Abbaszadeh-Goudarzi, K., & Abbaszadeh-Goudarzi, G., (2019). Antibody–drug conjugates (ADCs) for cancer therapy: Strategies, challenges, and successes. *Journal of Cellular Physiology, 234*, 5628–5642.

Afsharzadeh, M., Hashemi, M., Babaei, M., Abnous, K., & Ramezani, M., (2020). PEG-PLA nanoparticles decorated with small-molecule PSMA ligand for targeted delivery of galbanic acid and docetaxel to prostate cancer cells. *Journal of Cellular Physiology, 235*, 4618–4630.

Ahire, E. D., Talele, S. G., & Shah, H. S., (2020). Nanoparticles as a promising technology in microbial pharmaceutics. In: *Applied Pharmaceutical Science and Microbiology* (pp. 133–158). Apple Academic Press.

Ahire, E., Talele, S. G., & Talele, G. S., (2020). Microbial pigments: A green microbial technology. In: *Biologically Active Natural Products* (pp. 225–244). Apple Academic Press.

Ahire, E., Thakkar, S., Borade, Y., & Misra, M., (2020). Nanocrystal based orally disintegrating tablets as a tool to improve dissolution rate of vortioxetine. *Bulletin of Faculty of Pharmacy, Cairo University, 58*(1, 2), 11–20.

Ahmad, M. Z., Rizwanullah, M., Ahmad, J., Alasmary, M. Y., Akhter, M. H., Abdel-Wahab, B. A., Warsi, M. H., & Haque, A., (2022a). Progress in nanomedicine-based drug delivery in designing of chitosan nanoparticles for cancer therapy. *International Journal of Polymeric Materials and Polymeric Biomaterials, 71*, 602–623.

Ahmad, Z., Salman, S., Khan, S. A., Amin, A., Rahman, Z. U., Al-Ghamdi, Y. O., Akhtar, K., et al., (2022b). Versatility of hydrogels: From synthetic strategies, classification, and properties to biomedical applications. *Gels, 8*, 167.

Ahmadi, N. N., Hassani, K. H., Beygzadeh, M., Teimourian, S., & Kazemzad, M., (2018). Delivery of curcumin by a pH-responsive chitosan mesoporous silica nanoparticles for cancer treatment. *Artificial Cells, Nanomedicine, and Biotechnology, 46*, 75–81.

Aina, O. H., Sroka, T. C., Chen, M. L., & Lam, K. S., (2002). Therapeutic cancer targeting peptides. *Peptide Science: Original Research on Biomolecules, 66*, 184–199.

Ajmera, S., & Danenberg, P. V., (1982). Synthesis and biological activity of 5'-substituted 5-fluoropyrimidine nucleosides. *Journal of Medicinal Chemistry, 25*, 999–1002.

Al-Ameen, M. A., Li, J., Beer, D. G., & Ghosh, G., (2015). Sensitive, quantitative, and high-throughput detection of angiogenic markers using shape-coded hydrogel microparticles. *Analyst, 140*, 4530–4539.

Alexander, A., Khan, J., Saraf, S., & Saraf, S., (2014). Formulation and evaluation of chitosan-based long-acting injectable hydrogel for PEGylated melphalan conjugate. *Journal of Pharmacy and Pharmacology, 66*, 1240–1250.

Alhakamy, N. A., Badr-Eldin, S. M., Fahmy, U. F., Alruwaili, N. K., Awan, Z. A., Caruso, G., Alfaleh, M. A., et al., (2020). Thymoquinone-loaded soy-phospholipid-based phytosomes exhibit anticancer potential against human lung cancer cells. *Pharmaceutics, 12*, 761.

Alizadeh, L., Zarebkohan, A., Salehi, R., Ajjoolabady, A., & Rahmati-Yamchi, M., (2019). Chitosan-based nanotherapeutics for ovarian cancer treatment. *Journal of Drug Targeting, 27*, 839–852.

Al-Rabia, M. W., Alhakamy, N. A., Rizg, W. Y., Alghaith, A. F., Ahmed, O. A., & Fahmy, U. A., (2022). Boosting curcumin activity against human prostatic cancer PC3 cells by utilizing scorpion venom conjugated phytosomes as promising functionalized nanovesicles. *Drug Delivery, 29*, 807–820.

Amin, M. L., Kim, D., & Kim, S., (2016). Development of hematin conjugated PLGA nanoparticle for selective cancer targeting. *European Journal of Pharmaceutical Sciences, 91*, 138–143.

Anderson, C. R., Hu, X., Tlaxca, J., decleves, A. E., Houghtaling, R., Sharma, K., Lawrence, M., et al., (2011). Ultrasound molecular imaging of tumor angiogenesis with an integrin targeted microbubble contrast agent. *Investigative Radiology, 46*, 215.

Andima, M., Costabile, G., Isert, L., Ndakala, A. J., Derese, S., & Merkel, O. M., (2018). Evaluation of β-Sitosterol loaded PLGA and PEG-PLA nanoparticles for effective treatment of breast cancer: Preparation, physicochemical characterization, and antitumor activity. *Pharmaceutics, 10*, 232.

Arbyn, M., Weiderpass, E., Bruni, L., De Sanjosé, S., Saraiya, M., Ferlay, J., & Bray, F., (2020). Estimates of incidence and mortality of cervical cancer in 2018: A worldwide analysis. *The Lancet Global Health, 8*, e191–e203.

Aroui, S., Brahim, S., De Waard, M., Bréard, J., & Kenani, A., (2009). Efficient induction of apoptosis by doxorubicin coupled to cell-penetrating peptides compared to unconjugated doxorubicin in the human breast cancer cell line MDA-MB 231. *Cancer Letters, 285*, 28–38.

Arya, G., Vandana, M., Acharya, S., & Sahoo, S. K., (2011). Enhanced antiproliferative activity of Herceptin (HER2)-conjugated gemcitabine-loaded chitosan nanoparticle in pancreatic cancer therapy. *Nanomedicine: Nanotechnology, Biology and Medicine, 7*, 859–870.

Au, J. L. S., Rustum, Y. M., Minowada, J., & Srivastava, B. S., (1983). Differential selectivity of 5-fluorouracil and 5′-deoxy-5-fluorouridine in cultured human B lymphocytes and mouse L1210 leukemia. *Biochemical Pharmacology, 32*, 541–546.

Available: https://www.cancer.gov/about-cancer/understanding/what-is-cancer (accessed on 6 December 2023).

Avgoustakis, K., (2005). Polylactic-co-glycolic acid (PLGA). *Encyclopedia of Biomaterials and Biomedical Engineering, 1*, 1–11.

Azeez, N. A., Deepa, V. S., & Sivapriya, V., (2018). Phytosomes: Emergent promising nano vesicular drug delivery system for targeted tumor therapy. *Advances in Natural Sciences: Nanoscience and Nanotechnology, 9*, 033001.

Bagshawe, K., Sharma, S., Springer, C., Antoniw, P., Boden, J., Rogers, G., Burke, P., et al., (1991). Antibody directed enzyme prodrug therapy (ADEPT): Clinical report. *Disease Markers, 9*, 233–238.

Bates, P. J., Laber, D. A., Miller, D. M., Thomas, S. D., & Trent, J. O., (2009). Discovery and development of the G-rich oligonucleotide AS1411 as a novel treatment for cancer. *Experimental and Molecular Pathology, 86*, 151–164.

Behl, A., Parmar, V. S., Malhotra, S., & Chhillar, A. K., (2020). Biodegradable diblock copolymeric PEG-PCL nanoparticles: Synthesis, characterization, and applications as anticancer drug delivery agents. *Polymer, 207*, 122901.

Bogdanov, B., Vidts, A., Van, D. B. A., Verbeeck, R., & Schacht, E., (1998). Synthesis and thermal properties of poly (ethylene glycol)-poly (ϵ-caprolactone) copolymers. *Polymer, 39*, 1631–1636.

Boorjian, S. A., Alemozaffar, M., Konety, B. R., Shore, N. D., Gomella, L. G., Kamat, A. M., Bivalacqua, T. J., et al., (2021). Intravesical nadofaragene firadenovec gene therapy for BCG-unresponsive non-muscle-invasive bladder cancer: A single-arm, open-label, repeat-dose clinical trial. *The Lancet Oncology, 22*, 107–117.

Borrelli, A., Tornesello, A. L., Tornesello, M. L., & Buonaguro, F. M., (2018). Cell penetrating peptides as molecular carriers for anti-cancer agents. *Molecules, 23*, 295.

Bosio, V. E., Cacicedo, M. L., Calvignac, B., León, I., Beuvier, T., Boury, F., & Castro, G. R., (2014). Synthesis and characterization of $CaCO_3$–biopolymer hybrid nanoporous microparticles for controlled release of doxorubicin. *Colloids and Surfaces B: Biointerfaces, 123*, 158–169.

Boskabadi, S. H., Balanezhad, S. Z., Neamati, A., & Tabrizi, M. H., (2020). The green-synthesized zinc oxide nanoparticle as a novel natural apoptosis inducer in human breast (MCF7 and MDA-MB231) and colon (HT-29) cancer cells. *Inorganic and Nano-Metal Chemistry, 51*, 733–743.

Bouma, J., Beijnen, J., Bult, A., & Underberg, W., (1986). Anthracycline antitumor agents. *Pharmaceutisch Weekblad, 8*, 109–133.

Bray, F., Ferlay, J., Soerjomataram, I., Siegel, R. L., Torre, L. A., & Jemal, A., (2018). Global cancer statistics 2018: GLOBOCAN estimates of incidence and mortality worldwide for 36 cancers in 185 countries. *CA: A Cancer Journal for Clinicians, 68*, 394–424.

Bray, L. J., Binner, M., Holzheu, A., Friedrichs, J., Freudenberg, U., Hutmacher, D. W., & Werner, C., (2015). Multi-parametric hydrogels support 3D *in vitro* bioengineered microenvironment models of tumor angiogenesis. *Biomaterials, 53*, 609–620.

Bukhari, A., Connors, T., Gilsenan, A., Ross, W., Tisdale, M., Warwick, G., & Wilman, D., (1973). Cytotoxic agents designed to be selective for liver cancer. *Journal of the National Cancer Institute, 50*, 243–247.

Bumcrot, D., Manoharan, M., Koteliansky, V., & Sah, D. W., (2006). RNAi therapeutics: A potential new class of pharmaceutical drugs. *Nature Chemical Biology, 2*, 711–719.

Cao, W., Chen, H. D., Yu, Y. W., Li, N., & Chen, W. Q., (2021). Changing profiles of cancer burden worldwide and in China: A secondary analysis of the global cancer statistics 2020. *Chinese Medical Journal, 134*, 783–791.

Chandramohan, K., & Thomas, B., (2018). Cancer trends and burden in India. *The Lancet Oncology, 19*, e663.

Chapuis, J. C., Schmaltz, R. M., Tsosie, K. S., Belohlavek, M., & Hecht, S. M., (2009). Carbohydrate dependent targeting of cancer cells by bleomycin – microbubble conjugates. *Journal of the American Chemical Society, 131*, 2438–2439.

Chavan, S., Bhuvad, S., Kumbhlakar, B., Auti, J., Walunj, T., Pathak, S., Tanpure, R., et al., (2021). Antimicrobial and antioxidant potential of a standardized Ayurvedic formulation explains its clinical efficacy as gargles in post-radiotherapy oral cancer patients. *Journal of Herbal Medicine, 30*, 100510.

Chi-Hong, B. C., Chernis, G. A., Hoang, V. Q., & Landgraf, R., (2003). Inhibition of heregulin signaling by an aptamer that preferentially binds to the oligomeric form of human epidermal growth factor receptor-3. *Proceedings of the National Academy of Sciences, 100*, 9226–9231.

Cochran, M. C., Eisenbrey, J., Ouma, R. O., Soulen, M., & Wheatley, M. A., (2011). Doxorubicin and paclitaxel loaded microbubbles for ultrasound triggered drug delivery. *International Journal of Pharmaceutics, 414*, 161–170.

Collins, S. A., shah, A. H., Ostertag, D., Kasahara, N., & Jolly, D. J., (2021). Clinical development of retroviral replicating vector Toca 511 for gene therapy of cancer. *Expert Opinion on Biological Therapy, 21*, 1199–1214.

Connors, T., (1986). Prodrugs in cancer chemotherapy. *Xenobiotica, 16*, 975–988.

Coppi, G., & Iannuccelli, V., (2009). Alginate/chitosan microparticles for tamoxifen delivery to the lymphatic system. *International Journal of Pharmaceutics, 367*, 127–132.

Danyuo, Y., Dozie-Nwachukwu, S., Obayemi, J., Ani, C., Odusanya, O., Oni, Y., Anuku, N., et al., (2016). Swelling of poly (N-isopropylacrylamide) P (NIPA)-based hydrogels with bacterial-synthesized prodigiosin for localized cancer drug delivery. *Materials Science and Engineering: C, 59*, 19–29.

Davoodi, P., Ng, W. C., Yan, W. C., Srinivasan, M. P., & Wang, C. H., (2016). Double-walled microparticles-embedded self-cross-linked, injectable, and antibacterial hydrogel for controlled and sustained release of chemotherapeutic agents. *ACS Applied Materials & Interfaces, 8*, 22785–22800.

De Fougerolles, A., Vornlocher, H. P., maraganore, J., & Lieberman, J., (2007). Interfering with disease: A progress report on siRNA-based therapeutics. *Nature Reviews Drug Discovery, 6*, 443–453.

De Martel, C., Georges, D., Bray, F., Ferlay, J., & Clifford, G. M., (2020). Global burden of cancer attributable to infections in 2018: A worldwide incidence analysis. *The Lancet Global Health, 8*, e180-e190.

Dhanka, M., Shetty, C., & Srivastava, R., (2018). Methotrexate loaded gellan gum microparticles for drug delivery. *International Journal of Biological Macromolecules, 110*, 346–356.

Dhar, S., Mali, V., Bodhankar, S., Shiras, A., Prasad, B., & Pokharkar, V., (2011). Biocompatible gellan gum-reduced gold nanoparticles: Cellular uptake and subacute oral toxicity studies. *Journal of Applied Toxicology, 31*, 411–420.

Dhir, S., Khar, R., Chakraborthy, G., & Saurabh, M., (2016). Phytosomes: A brief overview. *Journal of Pharmaceutical Research, 15*, 56–62.

Dinca, A., Chien, W. M., & Chin, M. T., (2016). Intracellular delivery of proteins with cell-penetrating peptides for therapeutic uses in human disease. *International Journal of Molecular Sciences, 17*, 263.

Dodane, V., & Vilivalam, V. D., (1998). Pharmaceutical applications of chitosan. *Pharmaceutical Science & Technology Today, 1*, 246–253.

Doi, N., Zenno, S., Ueda, R., Ohki-Hamazaki, H., Ui-Tei, K., & Saigo, K., (2003). Short-interfering-RNA-mediated gene silencing in mammalian cells requires dicer and eIF2C translation initiation factors. *Current Biology, 13*, 41–46.

Dosio, F., Brusa, P., & Cattel, L., (2011). Immunotoxins and anticancer drug conjugate assemblies: The role of the linkage between components. *Toxins, 3*, 848–883.

Eatemadi, A., Daraee, H., Aiyelabegan, H. T., Negahdari, B., Rajeian, B., & Zarghami, N., (2016). Synthesis and characterization of chrysin-loaded PCL-PEG-PCL nanoparticle and its effect on breast cancer cell line. *Biomedicine & Pharmacotherapy, 84*, 1915–1922.

Eknath Ahire, Shreya Thakkar, Mahesh Darshanwad, & Manju Misra, (2018). Parenteral nanosuspensions: a brief review from solubility enhancement to more novel and specific applications, *Acta Pharmaceutica Sinica B, 8*(5), 733–755 https://doi.org/10.1016/j.apsb.2018.07.011.

Endo, Y., & Tsurugi, K., (1987). RNA N-glycosidase activity of ricin A-chain. Mechanism of action of the toxic lectin ricin on eukaryotic ribosomes. *Journal of Biological Chemistry, 262*, 8128–8130.

Escoffre, J. M., Piron, J., Novell, A., & Bouakaz, A., (2011). Doxorubicin delivery into tumor cells with ultrasound and microbubbles. *Molecular Pharmaceutics, 8*, 799–806.

Fang, Y., (2012). Ligand–receptor interaction platforms and their applications for drug discovery. *Expert Opinion on Drug Discovery, 7*, 969–988.

Feldman, A. L., Restifo, N. P., Alexander, H. R., Bartlett, D. L., Hwu, P., Seth, P., & Libutti, S. K., (2000). Antiangiogenic gene therapy of cancer utilizing a recombinant adenovirus to elevate systemic endostatin levels in mice. *Cancer Research, 60*, 1503–1506.

Gou, M., Wei, X., Men, K., Wang, B., Luo, F., Zhao, X., Wei, Y., & Qian, Z., (2011). PCL/PEG copolymeric nanoparticles: Potential nanoplatforms for anticancer agent delivery. *Current Drug Targets, 12*, 1131–1150.

Gross, B. P., Wongrakpanich, A., Francis, M. B., Salem, A. K., & Norian, L. A., (2014). A therapeutic microparticle-based tumor lysate vaccine reduces spontaneous metastases in murine breast cancer. *The AAPS Journal, 16*, 1194–1203.

Gujarathi, N. A., Patil, T., Rane, B., Babu, A., & Keservani, R. K., (2022). Micro- and nano-systems in transdermal drug delivery system; Chapter 2. In: Nayan, A. G., Juliana, P. A., Raj, K. K., & Anil, K. S., (eds.), *Topical and Transdermal Drug Delivery Systems: Applications and Future Prospects* (pp. 41–78). Apple Academic Press, Taylor & Francis. ISBN: 9781774910702.

Gulfam, M., Kim, J. E., Lee, J. M., Ku, B., Chung, B. H., & Chung, B. G., (2012). Anticancer drug-loaded gliadin nanoparticles induce apoptosis in breast cancer cells. *Langmuir, 28*, 8216–8223.

Hamarat, S. S., Yasa, M., Cihnioglu, A. O., Abdulhayoglu, M., Yilmaz, H., & Ak, G., (2016). Development of gemcitabine-adsorbed magnetic gelatin nanoparticles for targeted drug delivery in lung cancer. *Artificial Cells, Nanomedicine, and Biotechnology, 44*, 943–949.

Hani, U., Bhat, R. S., & Shivakumar, H. G., (2011). Formulation design and evaluation of metronidazole microspheres in a bio-adhesive gel for local therapy of vaginal candidiasis. *Lat. Am. J. Pharm, 30*, 161–167.

Hashemzehi, M., Behnam-Rassouli, R., Hassanian, S. M., Moradi-Binabaj, M., Moradi-Marjaneh, R., Rahmani, F., Fiuji, H., et al., (2018). Phytosomal-curcumin antagonizes cell growth and migration induced by thrombin through AMP-Kinase in breast cancer. *Journal of Cellular Biochemistry, 119*, 5996–6007.

Heitz, F., Morris, M. C., & Divita, G., (2009). Twenty years of cell-penetrating peptides: From molecular mechanisms to therapeutics. *British Journal of Pharmacology, 157*, 195–206.

Ireson, C. R., & Kelland, L. R., (2006). Discovery and development of anticancer aptamers. *Molecular Cancer Therapeutics, 5*, 2957–2962.

Jain, A. K., Das, M., Swarnakar, N. K., & Jain, S., (2011). Engineered PLGA nanoparticles: An emerging delivery tool in cancer therapeutics. *Critical Reviews™ in Therapeutic Drug Carrier Systems, 28.*

Jaiswal, R., Gong, J., Sambasivam, S., Combes, V., Mathys, J. M., Davey, R., Grau, G. E., & Bebawy, M., (2012). Microparticle-associated nucleic acids mediate trait dominance in cancer. *The FASEB Journal, 26*, 420–429.

Jarouliya, U., Kesharwani, R. K., & Kesharwani, R. K., (2018). Advances of nanotechnology in cancer therapy. In: *Nanobiomaterials* (pp. 291–328). Apple Academic Press. ISBN: 9781771885911.

Kai, M. P., Keeler, A. W., Perry, J. L., Reuter, K. G., Luft, J. C., O'Neal, S. K., Zamboni, W. C., & Desimone, J. M., (2015). Evaluation of drug loading, pharmacokinetic behavior, and toxicity of a cisplatin-containing hydrogel nanoparticle. *Journal of Controlled Release, 204*, 70–77.

Karthikeyan, S., Prasad, N. R., Ganamani, A., & Balamurugan, E., (2013). Anticancer activity of resveratrol-loaded gelatin nanoparticles on NCI-H460 non-small cell lung cancer cells. *Biomedicine & Preventive Nutrition, 3*, 64–73.

Kathle, P. K., Gautam, N., & Kesavan, K., (2018). Tamoxifen citrate loaded chitosan-gellan nanocapsules for breast cancer therapy: Development, characterization and *in-vitro* cell viability study. *Journal of Microencapsulation, 35*, 292–300.

Keservani, R. K., & Gautam, S. P., (2020). Formulation and evaluation of baclofen liposome vesicles using lecithin. *ARS Pharmaceutica, 61*(3), 175–180.

Keservani, R. K., Sharma, A. K., & Kesharwani, R. K., (2017). *Drug Delivery Approaches and Nanosystems*, Two-Volume Set. CRC Press. ISBN:9781771885850.

Khmara, I., Koneracka, M., Kubovcikova, M., Zavisova, V., Antal, I., Csach, K., Kopcansky, P., et al., (2017). Preparation of poly-L-lysine functionalized magnetic nanoparticles and their influence on viability of cancer cells. *Journal of Magnetism and Magnetic Materials, 427*, 114–121.

Khodaverdi, E., Tayarani-Najaran, Z., Minbashi, E., Alibolandi, M., HOSSEINI, J., Sepahi, S., Kamali, H., & Hadizadeh, F., (2019). Docetaxel-loaded mixed micelles and polymersomes composed of poly (caprolactone)-poly (ethylene glycol)(PEG-PCL) and poly (lactic acid)-poly (ethylene glycol)(PEG-PLA): Preparation and *in-vitro* characterization. *Iranian Journal of Pharmaceutical Research: IJPR, 18*, 142.

Kim, D. H., & Rossi, J. J., (2007). Strategies for silencing human disease using RNA interference. *Nature Reviews Genetics, 8*, 173–184.

Kim, J. W., & Cochran, J. R., (2017). Targeting ligand–receptor interactions for development of cancer therapeutics. *Current Opinion in Chemical Biology, 38*, 62–69.

Kozlovskaya, V., Alexander, J. F., Wang, Y., Kuncewicz, T., Liu, X., Godin, B., & Kharlampieva, E., (2014). Internalization of red blood cell-mimicking hydrogel capsules with pH-triggered shape responses. *ACS Nano, 8*, 5725–5737.

Kuai, R., Yuan, W., Li, W., Qin, Y., Tang, J., Yuan, M., Fu, L., Ran, R., Zhang, Z., & He, Q., (2011). Targeted delivery of cargoes into a murine solid tumor by a cell-penetrating peptide and cleavable poly (ethylene glycol) commodified liposomal delivery system via systemic administration. *Molecular Pharmaceutics, 8*, 2151–2161.

Kurrikoff, K., Aphkhazava, D., & Langel, Ü., (2019). The future of peptides in cancer treatment. *Current Opinion in Pharmacology, 47*, 27–32.

Kwak, H., Hwang, I., Kim, J. H., Kim, M. Y., Yang, J. S., & Jeong, S., (2009). Modulation of transcription by the peroxisome proliferator-activated receptor δ–binding RNA aptamer in colon cancer cells. *Molecular Cancer Therapeutics, 8*, 2664–2673.

Lai, P., Tarapacki, C., Tran, W. T., El Kaffas, A., Lee, J., Hupple, C., Iradji, S., et al., (2016). Breast tumor response to ultrasound mediated excitation of microbubbles and radiation therapy *in vivo*. *Oncoscience, 3*, 98.

Larrañeta, E., Stewart, S., Ervine, M., Al-Kasasbeh, R., & Donnelly, R. F., (2018). Hydrogels for hydrophobic drug delivery. Classification, synthesis, and applications. *Journal of Functional Biomaterials, 9*, 13.

Ledward, D., (2000). Gelatin. In: Phillips, G. O., & Williams, P. A., (eds.), *Handbook of Hydrocolloids*. Woodhead Publishing. 2nd edition.

Lee, Y. J., & Lee, S. W., (2012). Regression of hepatocarcinoma cells using RNA aptamer specific to alpha-fetoprotein. *Biochemical and Biophysical Research Communications, 417*, 521–527.

Lee, Y. J., Kim, I. S., Park, S. A., Kim, Y., Lee, J. E., Noh, D. Y., Kim, K. T., et al., (2013). Periostin-binding DNA aptamer inhibits breast cancer growth and metastasis. *Molecular Therapy, 21*, 1004–1013.

Li, X., Jin, Q., Chen, T., Zhang, B., Zheng, R., Wang, Z., & Zheng, H., (2009). LyP-1 ultrasonic microbubbles targeting to cancer cell as tumor bio-acoustics markers or drug carriers: Targeting efficiency evaluation in, microfluidic channels. In: *2009 Annual International Conference of the IEEE Engineering in Medicine and Biology Society* (463–466). IEEE.

Li, Y., Wang, P., Chen, X., Hu, J., Liu, Y., Wang, X., & Liu, Q., (2016). Activation of microbubbles by low-intensity pulsed ultrasound enhances the cytotoxicity of curcumin involving apoptosis induction and cell motility inhibition in human breast cancer MDA-MB-231 cells. *Ultrasonics Sonochemistry, 33*, 26–36.

Liotta, L. A., (1992). Cancer cell invasion and metastasis. *Scientific American, 266*, 54–63.

Lu, M., Qiu, Q., Luo, X., Liu, X., Sun, J., Wang, C., Lin, X., Deng, Y., & Song, Y., (2019). Phyto-phospholipid complexes (phytosomes): A novel strategy to improve the bioavailability of active constituents. *Asian Journal of Pharmaceutical Sciences, 14*, 265–274.

Lu, Z., Yeh, T. K., Wang, J., Chen, L., Lyness, G., Xin, Y., Wientjes, M. G., et al., (2011). Paclitaxel gelatin nanoparticles for intravesical bladder cancer therapy. *The Journal of Urology, 185*, 1478–1483.

Mancarella, S., Greco, V., Baldassarre, F., Vergara, D., Maffia, M., & Leporatti, S., (2015). Polymer-coated magnetic nanoparticles for curcumin delivery to cancer cells. *Macromolecular Bioscience, 15*, 1365–1374.

Meghani, N. M., Amin, H. H., Park, C., Park, J. B., Cui, J. H., Cao, Q. R., & Lee, B. J., (2018). Design and evaluation of clickable gelatin-oleic nanoparticles using fattigation-platform for cancer therapy. *International Journal of Pharmaceutics, 545*, 101–112.

Mirtič, A., & Grdadolnik, J., (2013). The structure of poly-L-lysine in different solvents. *Biophysical Chemistry, 175*, 47–53.

Mishra, P., Barik, A., Parhi, P., & Dey, R., (2016). Poly (ethylene glycol)-poly (lactic acid) Polymeric micelles as novel drug delivery system in cancer therapy: An overview. *Trends in Biomaterials & Artificial Organs, 30*.

Mitra, A. K., Agrahari, V., Mandal, A., Cholkar, K., Natarajan, C., Shah, S., Joseph, M., et al., (2015). Novel delivery approaches for cancer therapeutics. *Journal of Controlled Release, 219*, 248–268.

Naidoo, S., Daniels, A., Habib, S., & Singh, M., (2022). Poly-l-lysine–lactobionic acid-capped selenium nanoparticles for liver-targeted gene delivery. *International Journal of Molecular Sciences, 23*, 1492.

Nangare, S., Patil, P., Patil, A., Deshmukh, P., Powar, T., Pantwalawalkar, J., Khan, Z., et al., (2022). Pharmacokinetics of drug-in-polymer matrix-based nanoparticulate drug delivery system. *Pharmacokinetics and Pharmacodynamics of Nanoparticulate Drug Delivery Systems*. Springer.

Nasrolahi, S. A., Mandal, D., Tiwari, R. K., Guo, L., Lu, W., & Parang, K., (2013). Cyclic peptide-capped gold nanoparticles as drug delivery systems. *Molecular Pharmaceutics, 10*, 500–511.

Ng, E. W., Shima, D. T., Calias, P., Cunningham, E. T., Guyer, D. R., & Adamis, A. P., (2006). Pegaptanib, a targeted anti-VEGF aptamer for ocular vascular disease. *Nature Reviews Drug Discovery, 5*, 123–132.

Nomura, S., Niki, M., Nisizawa, T., Tamaki, T., & Shimizu, M., (2015). Microparticles as biomarkers of blood coagulation in cancer. *Biomarkers in Cancer, 7*, BIC. S30347.

Obayemi, J., Danyuo, Y., Dozie-Nwachukwu, S., Odusanya, O., Anuku, N., Malatesta, K., Yu, W., et al., (2016). PLGA-based microparticles loaded with bacterial-synthesized prodigiosin for anticancer drug release: Effects of particle size on drug release kinetics and cell viability. *Materials Science and Engineering: C, 66*, 51–65.

Oh, Y. K., & Park, T. G., (2009). siRNA delivery systems for cancer treatment. *Advanced Drug Delivery Reviews, 61*, 850–862.

Orive, G., Tam, S. K., Pedraz, J. L., & Hallé, J. P., (2006). Biocompatibility of alginate–poly-l-lysine microcapsules for cell therapy. *Biomaterials, 27*, 3691–3700.

Pantwalawalkar, J., Chandankar, S., Tade, R., Khan, Z., Shaikh, M., Powar, T., Patil, P., et al., (2022). Graphene quantum dot based ultrasensitive probe for biosensing of prostate cancer biomarkers: Current updates and future challenges. *Advances in Natural Sciences: Nanoscience and Nanotechnology, 13*, 013001.

Park, D., Jung, B., Lee, Y., Jang, J., Kim, M., Lee, J., Park, H., Seo, J., & Kim, C., (2015). Evaluation of *in vivo* antitumor effects of ANT2 shRNA delivered using PEI and ultrasound with microbubbles. *Gene Therapy, 22*, 325–332.

Patiño, T., Soriano, J., Barrios, L., Ibáñez, E., & Nogués, C., (2015). Surface modification of microparticles causes differential uptake responses in normal and tumoral human breast epithelial cells. *Scientific Reports, 5*, 1–12.

Peters, G. J., (2014). Novel developments in the use of antimetabolites. *Nucleosides, Nucleotides and Nucleic Acids, 33*, 358–374.

Phillips, G. D. L., Li, G., Dugger, D. L., Crocker, L. M., Parsons, K. L., Mai, E., Blättler, W. A., et al., (2008). Targeting HER2-positive breast cancer with trastuzumab-DM1, an antibody–cytotoxic drug conjugate. *Cancer Research, 68*, 9280–9290.

Pietersz, G., Kanellos, J., Smyth, M., Mckenzie, I., & Zalcberg, J., (1987). The use of monoclonal antibody conjugates for the diagnosis and treatment of cancer. *Immunol. Cell. Biol, 65*, 111–125.

Poppe, J., (1992). Gelatin. *Thickening and Gelling Agents for Food.* Springer.

Prakobvaitayakit, M., & Nimmannit, U., (2003). Optimization of polylactic-co-glycolic acid nanoparticles containing itraconazole using 23 factorial design. *Aaps Pharmscitech, 4*, 565–573.

Rajput, M. S., & Agrawal, P., (2010). Microspheres in cancer therapy. *Indian Journal of Cancer, 47*, 458.

Raucher, D., & Ryu, J. S., (2015). Cell-penetrating peptides: Strategies for anticancer treatment. *Trends in Molecular Medicine, 21*, 560–570.

Rautio, J., Kumpulainen, H., Heimbach, T., Oliyai, R., Oh, D., Järvinen, T., & Savolainen, J., (2008). Prodrugs: Design and clinical applications. *Nature Reviews Drug Discovery, 7*, 255–270.

Repta, A., Rawson, B. J., Shaffer, R. D., Sloan, K. B., Bodor, N., & Higuchi, T., (1975). Rational development of a soluble prodrug of a cytotoxic nucleoside: Preparation and properties of arabinosyladenine 5′-formate. *Journal of Pharmaceutical Sciences, 64*, 392–396.

Rudge, S., Kurtz, T., Vessely, C. R., Catterall, L., & Williamson, D., (2000). Preparation, characterization, and performance of magnetic iron–carbon composite microparticles for chemotherapy. *Biomaterials, 21*, 1411–1420.

Sabzichi, M., Hamishehkar, H., Ramezani, F., Sharifi, S., Tabasinezhad, M., Pirouzpanah, M., Ghanbari, P., & Samadi, N., (2014). Luteolin-loaded phytosomes sensitize human breast carcinoma MDA-MB 231 cells to doxorubicin by suppressing Nrf2 mediated signaling. *Asian Pacific Journal of Cancer Prevention, 15*, 5311–5316.

Samadian, H., Hosseini-Nami, S., kamrava, S. K., Ghaznavi, H., & Shakeri-Zadeh, A., (2016). Folate-conjugated gold nanoparticle as a new nanoplatform for targeted cancer therapy. *Journal of Cancer Research and Clinical Oncology, 142*, 2217–2229.

Segovia, N., Pont, M., Oliva, N., Ramos, V., Borrós, S., & Artzi, N., (2015). Hydrogel doped with nanoparticles for local sustained release of siRNA in breast cancer. *Advanced Healthcare Materials, 4*, 271–280.

Senter, P. D., Saulnier, M. G., Schreiber, G. J., Hirschberg, D. L., Brown, J. P., Hellström, I., & Hellström, K. E., (1988). Anti-tumor effects of antibody-alkaline phosphatase conjugates in combination with etoposide phosphate. *Proceedings of the National Academy of Sciences, 85*, 4842–4846.

Senter, P. D., Wallace, P. M., Svensson, H. P., Vrudhula, V. M., Kerr, D. E., Hellstroem, I., & Hellstroem, K. E., (1993). Generation of cytotoxic agents by targeted enzymes. *Bioconjugate Chemistry, 4*, 3–9.

Shahani, K., Swaminathan, S. K., Freeman, D., Blum, A., Ma, L., & Panyam, J., (2010). Injectable sustained release microparticles of curcumin: A new concept for cancer chemoprevention. *Cancer Research, 70*, 4443–4452.

Shakeri-Zadeh, A., Zareyi, H., Sheervalilou, R., Laurent, S., Ghaznavi, H., & Samadian, H., (2021). Gold nanoparticle-mediated bubbles in cancer nanotechnology. *Journal of Controlled Release, 330*, 49–60.

Sharma, R., (2020). An examination of colorectal cancer burden by socioeconomic status: Evidence from GLOBOCAN 2018. *EPMA Journal, 11*, 95–117.

Sharma, S., & Roy, R. K., (2010). Phytosomes: An emerging technology. *Int. J. Pharm. Res. Dev., 2*, 1–7.

Shirsath, N. R., & Goswami, A. K., (2019). Nanocarriers based novel drug delivery as effective drug delivery: A review. *Current Nanomaterials, 4*, 71–83.

Simon, D. D. A., & Parish, C. R., (2013). Heparan sulfate: A ubiquitous glycosaminoglycan with multiple roles in immunity. *Frontiers in Immunology, 4*, 470.

Stone, E. L., Citossi, F., Singh, R., Kaur, B., Gaskell, M., Farmer, P. B., Monks, A., et al., (2015). Antitumour benzothiazoles. Part 32: DNA adducts and double strand breaks correlate with activity; synthesis of 5F203 hydrogels for local delivery. *Bioorganic & Medicinal Chemistry, 23*, 6891–6899.

Sun, H., Zhu, X., Lu, P. Y., Rosato, R. R., Tan, W., & Zu, Y., (2014). Oligonucleotide aptamers: New tools for targeted cancer therapy. *Molecular Therapy-Nucleic Acids, 3*, e182.

Sung, H., Ferlay, J., Siegel, R. L., Laversanne, M., Soerjomataram, I., Jemal, A., & Bray, F., (2021). Global cancer statistics 2020: GLOBOCAN estimates of incidence and mortality worldwide for 36 cancers in 185 countries. *CA: A Cancer Journal for Clinicians, 71*, 209–249.

Surana, K. R., Ahire, E. D., Sonawane, V. N., & Talele, S. G., (2021). Biomolecular and molecular docking: A modern tool in drug discovery and virtual screening of natural products. In: *Applied Pharmaceutical Practice and Nutraceuticals* (pp. 209–223). Apple Academic Press.

Sworn, G., & Stouby, L., (2021). Gellan gum. *Handbook of Hydrocolloids.* Elsevier.

Tabatabaei, M. F. S., Nejati-Koshki, K., Akbarzadeh, A., Yamchi, M. R., Milani, M., Zarghami, N., Zeighamian, V., et al., (2014). PLGA-based nanoparticles as cancer drug delivery systems. *Asian Pacific Journal of Cancer Prevention, 15*, 517–535.

TAKAKURA, Y., & HASHIDA, M., (1995). Macromolecular drug carrier systems in cancer chemotherapy: Macromolecular prodrugs. *Critical Reviews in Oncology/Hematology, 18*, 207–231.

Talele, S. G., Ahire, E. D., Talele, G. S., & Derle, D. V., (2021). An innovative approach as self-emulsifying drug delivery system for phytoconstituents. In: *Enhancing the Therapeutic Efficacy of Herbal Formulations* (pp. 69–84). IGI Global.

Tan, L., Han, S., Ding, S., Xiao, W., Ding, Y., Qian, L., Wang, C., & Gong, W., (2017). Chitosan nanoparticle-based delivery of fused NKG2D–IL-21 gene suppresses colon cancer growth in mice. *International Journal of Nanomedicine, 12*, 3095.

Thombre, N. A., Niphade, P. S., Ahire, E. D., & Kshirsagar, S. J., (2022). Formulation development and evaluation of microemulsion based lornoxicam gel. *Biosciences Biotechnology Research Asia, 19*(1), 69–80.

Todorova, M., Agache, V., Mortazavi, O., Chen, B., Karshafian, R., Hynynen, K., Man, S., et al., (2013). Antitumor effects of combining metronomic chemotherapy with the antivascular action of ultrasound stimulated microbubbles. *International Journal of Cancer, 132*, 2956–2966.

Tommelein, J., De Vlieghere, E., Verset, L., Melsens, E., Leenders, J., Descamps, B., Debucquoy, A., et al., (2018). Radiotherapy-activated cancer-associated fibroblasts promote tumor progression through paracrine IGF1R activation. *Cancer Research, 78*, 659–670.

Tritton, T. R., (1991). Cell surface actions of Adriamycin. *Pharmacology & Therapeutics, 49*, 293–309.

Tsou, K., & Su, H. C., (1963). Synthesis of possible cancer chemotherapeutic compounds based on enzyme approach. v. tetrazolium nitrogen mustards. *Journal of Medicinal Chemistry, 6*, 693–696.

Veltkamp, S. A., Pluim, D., Van, T. O., Beijnen, J. H., & Schellens, J. H., (2008). Extensive metabolism and hepatic accumulation of gemcitabine after multiple oral and intravenous administration in mice. *Drug Metabolism and Disposition, 36*, 1606–1615.

Villar-Cheda, B., Dominguez-Meijide, A., Valenzuela, R., Granado, N., Moratalla, R., & Labandeira-Garcia, J. L., (2014). Aging-related dysregulation of dopamine and angiotensin receptor interaction. *Neurobiology of Aging, 35*, 1726–1738.

Villarreal-Otalvaro, C., & Coburn, J. M., (2021). Fabrication methods and form factors of gellan gum-based materials for drug delivery and anti-cancer applications. *ACS Biomaterials Science & Engineering.*

Vlad, R. M., Brand, S., Giles, A., Kolios, M. C., & Czarnota, G. J., (2009). Quantitative ultrasound characterization of responses to radiotherapy in cancer mouse models. *Clinical Cancer Research, 15*, 2067–2075.

Waller, D., & George, C., (1989). Prodrugs. *British Journal of Clinical Pharmacology, 28*, 497.

Wei, X., Gong, C., Gou, M., Fu, S., Guo, Q., Shi, S., Luo, F., et al., (2009). Biodegradable poly (ε-caprolactone)–poly (ethylene glycol) copolymers as drug delivery system. *International Journal of Pharmaceutics, 381*, 1–18.

Weinberg, R., & Hanahan, D., (2000). The hallmarks of cancer. *Cell, 100*, 57–70.

Wu, L., Zhang, F., Wei, Z., Li, X., Zhao, H., Lv, H., Ge, R., et al., (2018). Magnetic delivery of Fe_3O_4@ polydopamine nanoparticle-loaded natural killer cells suggest a promising anticancer treatment. *Biomaterials Science, 6*, 2714–2725.

Xu, L., Tang, W. H., Huang, C. C., Alexander, W., Xiang, L. M., Pirollo, K. F., Rait, A., & Chang, E. H., (2001). Systemic p53 gene therapy of cancer with immunolipoplexes targeted by anti-transferrin receptor scFv. *Molecular Medicine, 7*, 723–734.

Xue, P., Zhang, G., Zhang, J., & Ren, L., (2022). Synergism of ellagic acid in combination with radiotherapy and chemotherapy for cancer treatment. *Phytomedicine,* 153998.

Yan, F., Li, L., Deng, Z., Jin, Q., Chen, J., Yang, W., Yeh, C. K., et al., (2013). Paclitaxel-liposome–microbubble complexes as ultrasound-triggered therapeutic drug delivery carriers. *Journal of Controlled Release, 166*, 246–255.

Yan, F., LI, X., Jin, Q., Jiang, C., Zhang, Z., Ling, T., Qiu, B., & Zheng, H., (2011). Therapeutic ultrasonic microbubbles carrying paclitaxel and LyP-1 peptide: Preparation, characterization, and application to ultrasound-assisted chemotherapy in breast cancer cells. *Ultrasound in Medicine & Biology, 37*, 768–779.

Yang, D. H., Kim, H. J., Park, K., Kim, J. K., & Chun, H. J., (2018). Preparation of poly-l-lysine-based nanoparticles with pH-sensitive release of curcumin for targeted imaging and therapy of liver cancer *in vitro* and *in vivo*. *Drug Delivery, 25*, 950–960.

Yang, S. J., Lin, F. H., Tsai, K. C., Wei, M. F., Tsai, H. M., Wong, J. M., & Shieh, M. J., (2010). Folic acid-conjugated chitosan nanoparticles enhanced protoporphyrin IX accumulation in colorectal cancer cells. *Bioconjugate Chemistry, 21*, 679–689.

Yang, X. Z., Dou, S., Sun, T. M., Mao, C. Q., Wang, H. X., & Wang, J., (2011). Systemic delivery of siRNA with cationic lipid assisted PEG-PLA nanoparticles for cancer therapy. *Journal of Controlled Release, 156*, 203–211.

Yokoyama, M., Miyauchi, M., Yamada, N., Okano, T., Sakurai, Y., Kataoka, K., & Inoue, S., (1990). Characterization and anticancer activity of the micelle-forming polymeric anticancer drug Adriamycin-conjugated poly (ethylene glycol)-poly (aspartic acid) block copolymer. *Cancer Research, 50*, 1693–1700.

Yoshida, J., Mizuno, M., & Wakabayashi, T., (2004). Interferon-β gene therapy for cancer: Basic research to clinical application. *Cancer Science, 95*, 858–865.

Yu, L., Ni, C., Grist, S. M., Bayly, C., & Cheung, K. C., (2015). Alginate core-shell beads for simplified three-dimensional tumor spheroid culture and drug screening. *Biomedical Microdevices, 17*, 1–9.

Zawilska, J. B., Wojcieszak, J., & Olejniczak, A. B., (2013). Prodrugs: A challenge for the drug development. *Pharmacological Reports, 65*, 1–14.

Zelikin, A. N., (2020). Unique enzymatic repertoire reveals the tumor. *Nature Chemistry, 12*, 11, 12.

Zhang, G., Zhu, Y., Wang, Y., Wei, D., Wu, Y., Zheng, L., Bai, H., et al., (2019). PH/redox sensitive nanoparticles with platinum (iv) prodrugs and doxorubicin enhance chemotherapy in ovarian cancer. *RSC Advances, 9*, 20513–20517.

Zhang, L., Chen, Z., Wang, H., Wu, S., Zhao, K., Sun, H., Kong, D., et al., (2016). Preparation and evaluation of PCL–PEG–PCL polymeric nanoparticles for doxorubicin delivery against breast cancer. *RSC Advances, 6*, 54727–54737.

Zhang, L., Yang, N., Mohamed-Hadley, A., Rubin, S. C., & Coukos, G., (2003). Vector-based RNAi, a novel tool for isoform-specific knock-down of VEGF and anti-angiogenesis gene therapy of cancer. *Biochemical and Biophysical Research Communications, 303*, 1169–1178.

Zhang, R., Song, X., Liang, C., Yi, X., Song, G., Chao, Y., Yang, Y., Yang, K., et al., (2017). Catalase-loaded cisplatin-prodrug-constructed liposomes to overcome tumor hypoxia for enhanced chemo-radiotherapy of cancer. *Biomaterials, 138*, 13–21.

Zhao, Y. Z., Luo, Y. K., Tang, J., Zhang, Y., Lin, Q., & Mei, X. G., (2006). Comparison of sonoporation effect of liposomes and phospholipids-based microbubbles on cultured cell membrane. *Yao Xue Xue Bao= Acta Pharmaceutica Sinica, 41*, 1176–1179.

Zhou, G., Wilson, G., Hebbard, L., Duan, W., Liddle, C., George, J., & Qiao, L., (2016). Aptamers: A promising chemical antibody for cancer therapy. *Oncotarget, 7*, 13446.

Zidan, A. S., Sammour, O. A., Hammad, M. A., Megrab, N. A., Hussain, M. D., Khan, M. A., & Habib, M. J., (2006). Formulation of anastrozole microparticles as biodegradable anticancer drug carriers. *Aaps Pharmscitech, 7*, E38–E46.

Zunino, F., Giuliani, F., SAVI, G., Dasdia, T., & Gambetta, R., (1982). Anti-tumor activity of daunorubicin linked to poly-l-aspartic acid. *International Journal of Cancer, 30*, 465–470.

CHAPTER 8

Novel Drug Delivery Systems in HIV/AIDS and Associated Complications

DEEPAK S. BHAMBERE,[1] VARSHA CHAUDHARI,[2] SAPANA P. AHIRRAO,[1] and SANJAY J. KSHIRSAGAR[1]

[1]MET's Institute of Pharmacy, Bhujbal Knowledge City, Adgaon, Nashik, Maharashtra, India

[2]Department of Pharmacognosy, Matoshri College of Pharmacy, Eklahare, Maharashtra, India

ABSTRACT

Acquired immunodeficiency syndrome is a killer disease and to cure HIV/AIDS is one of the major challenges of the world in this 21st century. In the market, different antiviral drugs are available to reduce the worst condition and manage improved survival rate which are limited due to their severe side effects, remarkable incompatibilities, frequent drug administration, limited absorption, and reduced access reservoir sites of viruses like macrophages. In AIDS patients, if the drug is preferred in higher dosage, the drug leads to adverse effects and toxicity to patients and also increases resistance against HIV. But if the drug is preferred in small dose along with nano-carriers, it will reach the target area for beneficial and favorable effect. Failure of complete elimination of HIV from the infected cells and shortness of widespread preventive measures are the most significant challenges today which need to be overcome. The main aim in AIDS/HIV research is the development of a therapeutic strategy which can be effective in long-lasting remission of AIDS/HIV without daily drug treatment and ultimately a cure for AIDS/HIV. Novel drug delivery system (NDDS) provides an opportunity to avoid the limitations and drawbacks related to antiretroviral treatment. It

Novel Drug Delivery Systems in the Management of Chronic Diseases.
Raj K. Keservani, Umesh D. Laddha, Eknath D. Ahire, and Sanjay J. Kshirsagar (Eds.)

also helps in managing the dosage form development complexity such as insolubility, limited entrapment of the drug, and instability. Extended release or long-acting nanoproducts which are used in antiretroviral treatment and other AIDS-related treatments and are given on a less frequent or monthly basis. Nanotechnology-based delivery system Nanocarriers like dendrimers, liposomes, nanoparticles, nano-emulsion, nano-vesicles, polymeric micelles play an important role in the delivery of the drug to targeting tissue. Nanocarriers amended the field of Pharmacokinetics and Pharmaceutics in targeted drug delivery. Nano-based antiretroviral drug is superior because of fewer adverse effects and high efficiency to control AIDS.

8.1 INTRODUCTION

Human immunodeficiency virus (HIV) infection and AIDS (acquired immune deficiency syndrome), which are commonly referred to as HIV/AIDS, is one of the severe infectious diseases challenging to public health globally. There are currently two known species of HIV, viz., HIV-1 and HIV-2, with their respective subspecies. HIV-1 is the globally common infection while HIV-2 is more prevalent in West Africa and takes a longer time to develop into immunodeficiency from infection than HIV-1 (Ojewole et al., 2008). The first five cases of the acquired immunodeficiency syndrome were reported by the CDC and Centers for Disease Control and Prevention, and HIV was identified in 1983. HIV causes AIDS by infecting helper T cells of the immune system. HIV-1 is the most common serotype worldwide (Marina, 2008). HIV is a retrovirus that belongs to the lentivirus family and is responsible for the immune dysfunction which leads to AIDS and is responsible for 20 million deaths. Most of these are because of respiratory diseases with pathogens such as the fungus *Pneumocystis jirovecii*, bacteria (*Pneumococcus* and gram-negative organisms), and *M. tuberculosis*. The immune dysfunction induced by HIV also leads to a variety of non-infectious pulmonary complications (Lipman, 2006). From the mucosal layer, HIV invades and damages the immune system, because of which the host body gets infected easily with an extensive range of viruses, bacteria, protozoa, and fungi. HIV infection propagates to offspring from an infected mother, due to blood transfusion by the transfer of body fluids, sexual contact, organ transplant. Sexual transmission or intercourse is one of the major sources to enter into the mucosal surfaces. The initial and main targets of HIV infection are the immune cells, i.e., macrophages and dendritic cells, which are present in the subepithelial layer of the vagina or cervix mucosa. In HIV, a glycoprotein is present on the surface of the viral coat called gp120, which binds with the chemokine receptor CXCR4, CCR5 of the T-helper lymphocytes, or transmembrane protein receptor CD4 and infects the cell. In the present situation, multiple ARV (antiretroviral) medications are present, which renders the disease chronic rather than worse and helps to enhance the survival rate (Sailaja et al., 2021).

8.2 NOVEL DRUG DELIVERY SYSTEM (NDDS)

NDDS, i.e., novel drug delivery systems, provide an opportunity for formulation scientists to overcome the challenges associated with ARV (antiretroviral) drug therapy, thereby improving the management of patients suffering from HIV/AIDS. There are various antiretroviral delivery systems that have been developed to achieve sustained drug release kinetics, specifically for targeting drugs to the brain and gastric mucosa, macrophages, and to overcome formulation difficulties such as stability, drug entrapment, and poor solubility (Elizabeth et al., 2008). Developing an effective NDDS for the treatment of AIDS/HIV is a global challenge. Conventional drug delivery methods, including HAART (Highly Active Anti-Retroviral Therapy), have increased the lifespan of AIDS/HIV patients. However, complete elimination of HIV is still not possible due to some limitations. Non-polymeric and polymeric nanotechnological approaches provide opportunities in this direction. Polymeric carriers such as nanoparticles (NP) and dendrimers have been reported for the targeting of anti-HIV drugs (Gupta et al., 2010).

Semisynthetic and synthetic polymers have several desirable features, such as stability, sustained drug release, and manufacturing reproducibility, and hence have attracted great attention for nanotechnology-based drug delivery. A variety of materials have been used to manufacture nanoparticles, and the choice of material may depend on the application. Poly(lactic acid) (PLA), poly(lactic-co-glycolic acid) (PLGA), poly(alkyl)cyanoacrylates, poly(ethylene glycol-co-(lactic-glycolic acid), poly(caprolactone), and poly(methyl) methacrylate are widely used in the fabrication of nanoparticles for drug delivery. Out of these, PLGA and PLA have been granted by the FDA (Food and Drug Administration) for human trials. A variety of drugs with different hydrophobicity or hydrophilicity can be incorporated into these polymers; their release properties are modified to meet dosing requirements. The effect of the polymer can be tailored to the intended application by controlling the molecular weight or copolymer composition, which influences properties such as degradation rate, thermal sensitivity, and pH sensitivity. These polymers have the ability to sustain drug release for several weeks. Loading of the drug is generally achieved by entrapment, dissolution/dispersion, or encapsulation. Polymeric nanoparticles (PNPs) are biocompatible and biodegradable in nature, making them attractive candidates for tissue engineering, drug delivery, and biomedical applications (Du Toic et al., 2010).

8.3 NANOCARRIERS IN ANTI-HIV DRUG

Need of effective drug delivery is a major concern for treatment of several diseases such as cancer, HIV/AIDS, etc., which requires systemically administered therapies. Thus, Importance of nanotechnology is gaining globally. Nanomedicines is application of nano-scale technologies in the practice of medicines. These carrier systems are characterized with monomeric size which allows them to cross biological barriers, improve cellular uptake and transport which enables them to deliver therapeutic agents to target sites like brain, liver. There are variety of nanocarriers which are used in the anti-AIDS/HIV drug delivery (Gupta et al., 2010).

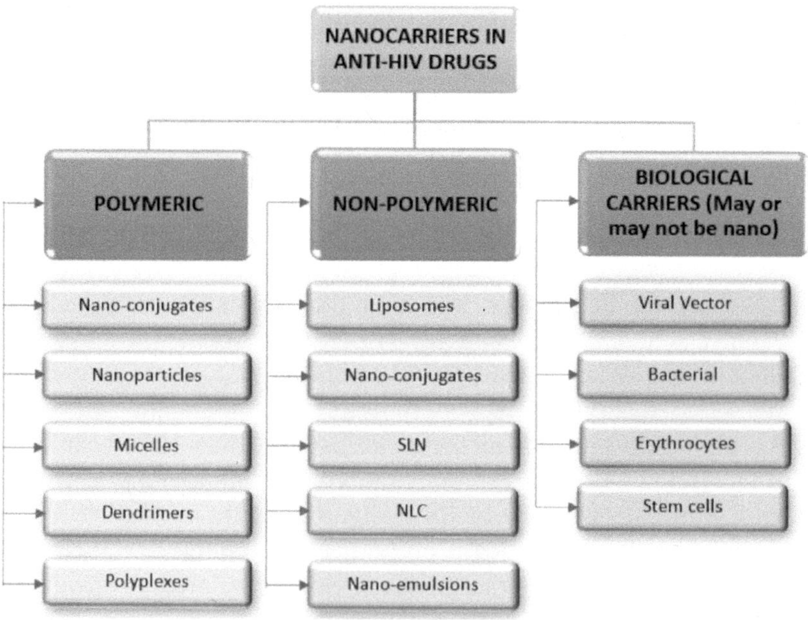

FIGURE 8.1 Nanocarriers in Anti-HIV Drugs

8.3.1 ANTIRETROVIRAL THERAPY BASED ON POLYMERIC NANOCARRIER

Polymeric nanocarriers are mostly consists of polymers such as gelatine, chitosan, poly(D,L-lactide-co-glycolide) (PLGA), poly(ε-caprolactone), etc. Nanoparticles and dendrimers are the important groups for drug delivery along with polymeric nanocarriers. Targeted delivery of ARV (antiretroviral) drugs which is nanoparticle based, has been studied to target the cells of MPS (mononuclear phagocytic system) such as macrophages/monocytes ((Mac/Mo) which acts as reservoir for HIV (Gupta et al., 2009).

8.3.1.1 NANO-CONJUGATES

Strategies of conjugation, which is based on a hydrophobic interaction, electrostatic interaction, hydrogen bonding, and host–guest interaction offer, the attachment of hydrophobic and hydrophilic fragments to create the polymer, which is amphiphilic in nature, which is then self-assembled by molecular associations between hydrophobic segments into nanoparticles in an aqueous environment. The hydrophilic segment which forms shows great impact on formation, drug interaction, drug encapsulation and stability of nano-conjugates. Drug which is poorly soluble in water are loaded into aquaphobic (hydrophobic) moieties to increase circulation time, improve its solubility, to prevent drug degradation and for targeting the tumor environment (Tran et al., 2019).

8.3.1.2 NANOPARTICLES

Nanoparticles are solid colloidal particles with size ranging from 10–1,000 nm. Depending upon polymeric composition and colloidal size, nanoparticles are able to target the drug to the specified sites in the body together with its sustained release (SR). They are explored for efficacy of drugs having physicochemical drawbacks like poor solubility and stability and improvement of formulation. Nanoparticles are increasingly investigated to accomplish sustained drug-release kinetics and for targeted delivery of the antiretroviral drugs to HIV infected cells. In such system drug encapsulation provides decreased drug resistance, improved efficacy, improvement in patient compliance, decrease in side effects and systemic toxicity (Kumar et al., 2015; Ahire et al., 2020a).

Nanoparticles are classified into three groups: Inorganic NP, polymer NP and SLNPs. Several nanoparticles such as magnetic and gold nanoparticles (AuNPs) are called inorganic nanoparticles. Solid lipid nanoparticles (SLNs) combine the advantages of another colloidal carriers such as liposomes, polymeric microparticles (MPs), nanoparticles, and emulsions but avoids the disadvantages. Polymer nanoparticles consist of several biocompatible synthetic or natural polymers (Lenjisa et al., 2014).

8.3.1.3 POLYMER NANOPARTICLES (NPS)

Semisynthetic and synthetic polymers due for some desirable properties, including manufacturing sustained drug release, stability, and reproducibility, have attracted great attention for drug delivery which is based on nanotechnology. Different materials have been employed in the preparation of nanoparticles, and material choice in some cases is driven by application. Synthetic or semi-synthetic polymers such as poly(lactic acid) (PLA), poly(lactic-co-glycolic acid) (PLGA), poly(alkyl) cyanoacrylates, poly(ethylene glycol-co-(lactic-glycolic acid)), poly(caprolactone), and poly(methyl) methacrylate, are frequently used for the manufacture of nanoparticles intended for drug delivery. Of these, PLGA and PLA were approved by the FDA for human use. Polymer performance can be tailored to the intended application by controlling the molecular weight or composition of the copolymer, which affects properties such as degradation rate, thermal sensitivity, and pH sensitivity. These polymers have ability to sustain release of drug for several weeks. Loading of drug is usually achieved by entrapment, encapsulation, dispersion or dissolution. PNPs are attractive candidates for biomedical applications due to its biodegradable and biocompatible properties. Inorganic materials such as titanium, gold, iron, silica, copper, zinc oxide and silver have been fabricated as nanoparticles for several applications in pharmaceutical field (Du Toic et al., 2010). Silver nanoparticles have proved great antimicrobial agents because of its effectiveness as bactericidal (Sondi et al., 2004) and antiviral (Lu et al., 2008; Ahirrao et al., 2022).

Polymer nanoparticles are the most extensively investigated materials for delivery of nucleic acid because of their favorable safety profile, capability to enhance immune responses and versatility. The release rate of these PNPs can be controlled by designing the chemical structure of the particle to behave specifically in different environments. For example, a pH-stimulated control system would include a structure that modulates permeability through the dissociation of various surface ligands as the pH changes. Based on the chemistry and biological effects inherent in different materials, the functionality and suitable applications for different polymeric NPs vary significantly. Chitosan is an abundant polysaccharide that is a biocompatible, biodegradable, and minimally toxic polymer suitable for biomedical applications (Lim et al., 2020).

8.3.1.4 MICELLES

Micelles are formed from amphiphilic surfactants that spontaneously assemble in an aqueous medium to form core (hydrophobic)-shell (hydrophilic) structures or vesicles. Polymeric micelles can reach parts of the body that are hard to reach. Loading of low drug and low drug incorporation stability leads to release of loaded drug before getting to the desired site of action (Lenjisa et al., 2014).

8.3.1.5 DENDRIMERS

Tomalia in 1985 proposed the word Dendrimer which originates from the Greek word 'dendron,' meaning tree and the structural shape yields its unique attributes (Tomalia et al., 1985). Dendrimers are different from conventional polymers that they are characterized with presence of multi-branched, 3-dimensional architecture with high functionality and minimal polydispersity. Dendrimers are comprised of 3 different topological components with chemical significance: (i) a multivalent surface; (ii) multi-functional core; and (iii) interior layers. Various chemical species can be encapsulated in the multi-functional focal core, and it exhibits unequaled properties because of the special nano-environment which is surrounded by extensive dendritic branching. Versatile class of regularly branched macromolecules which are characterized by unique topologic and structural features are Dendrimers. Narrow molecular weight distribution, the relative ease of incorporation of targeting ligands, and small size (usually less than 100 nm) make them attractive candidates for drug delivery. Dendrimers have high functionality and minimal polydispersity. Dendrimers are obtained by attaching numeral monomeric units, but unlike conventional polymers, they have highly branched 3-dimensional architecture (du Toic et al., 2010).

8.3.1.6 POLYPLEX

Formation of polymer-nucleic acid complex by combining non-viral polymeric carriers with the nucleic acids via an electrostatic interaction is called polyplexes. The tough electrostatic reactions between DNA and PEI transform the DNA in a condensed form and therefore a nanostructured positively charged complex called polyplex generated. Various cationic polymers are used as a nucleic acid carrier like PLL, PEI, and poly(2-dimethylaminoethyl methacrylate) (PDMAEMA) and polyamidoamine (PAMAM). Polyplex formation is mainly controlled by the kinetics. By the order of incorporation of reacting substances throughout the complexation process, the transfection capacity and the resulting size of polyplex can be influenced. Polyplexes are excellent medium for gene delivery without the serious immunological issues associated with viral systems (Aggarwal et al., 2019).

8.3.2 NON-POLYMERIC NANOCARRIERS

8.3.2.1 LIPOSOMES

Liposomes can be either MPs or colloidal carriers with diameters ranging from 20 nm to greater than 10 μm. There are many types of liposomes, e.g., small or large unilamellar vesicles and multilamellar vesicles, the conventional ones being composed of non-toxic and biodegradable zwitterionic phospholipids such as cholesterol, phosphatidylserine, phosphatidylcholine (PC). When lipids are subjected to hydration, spontaneous formation of liposomes occurs, leading to the formation of lipid bilayers surrounding the aqueous core (Gao et al., 2018). It is accepted that macrophages play a crucial role in HIV pathogenesis and acts as a reservoir for spread of virus throughout the immune system. As liposomes are taken up naturally by the cells of MPS, therapy which is based on liposome represents an advantageous approach to improve anti-HIV agent's delivery into infected cells and thereby decreasing their adverse side-effects and improving efficacy of drugs. A more specific targeting of HIV-infected cells could also be obtained by using liposomes bearing surface attached-antibodies (Désormeaux et al., 1998).

8.3.2.2 ETHOSOMES

Ethosomes are soft and malleable lipid vesicles mainly composed of water, alcohol (ethyl alcohol or isopropanol) in comparably high concentration (20–45%) and phospholipids. The phospholipids having numerous chemical structures includes hydrogenated PC, phosphatidyl choline (PC), phosphatidyl ethanolamine (PE), hydrogenated PC, etc. The non-aqueous phase of ethosome ranges in between 22% and 70%. The alcohol may be ethyl alcohol or isopropanol. This carrier presents interesting properties correlated with its ability to penetrate intact human skin due to its high deformability. The physicochemical properties of ethosomes allow vesicular phospholipids to be a component of the vesicle-forming ethosomal system. Some selected phospholipids are Phospholipon 90 (PL-90), Lipoid S100, soy phospholipid. A high alcohol concentration of about 20–45% in the formulation provides flexible and soft properties along with stability for the blisters and also disrupts the structure of the lipid bilayers of the skin, ultimately leading to an increase in membrane permeability.

Advantages of ethosomal drug delivery:

- Delivery of peptides, protein molecules (large molecules) are possible.
- Permeation enhancement of drug through the skin for transdermal drug delivery.
- Ethosomes contains raw material in the formulation which is nontoxic in nature.
- Ethosomal system is non-invasive, passive, and available for instant commercialization.
- Ethosomal drug is administered in gel or cream form (semisolid) because of producing high patient compliance.
- Ethosomal drug delivery system can be applied broadly in cosmetic, pharmaceutical, and veterinary fields (Nandure et al., 2013).

8.3.2.3 CUBOSOMES

Highly stable structures and are organized in curved bicontinuous lipid bilayers that forms the soft 3D honeycomb like structures. They are composed of two non-connected water channels and a continuous periodic bilayer. Main components of cubosomes are glyceryl monoolein and phytantriol (Faria et al., 2021).

8.3.2.4 SOLID LIPID NANOPARTICLES (SLNS)

Solid lipid nanoparticles (SLNs) are nanocrystalline structures made up of fatty acids. SLNs are colloidal carrier systems which are comparatively stable, and in them, melted lipid is dispersed in an aqueous surfactant by micro-emulsification or high-pressure homogenization. SLNs are characterized by the presence of unique properties like high drug loading, large surface area, small size, and interaction of the phases at the interfaces. Their hydrophobic core usually entraps the dissolved or entrapped drug. They have been shown to be taken up by the brain and due to the biodegradable nature of the constituent materials, exhibit the least toxicity. Their nano-scale dimensions allow them to cross various anatomical barriers and bypass the liver. They have a large efficiency for entrapment of drugs which render the drug more stable in their lipid matrix and also provide a controlled release of the entrapped drug. The surface coating of SLNs with surfactants or hydrophilic polymers, such as PEG (polyethylene glycol), minimizes their uptake in liver cells which results in improving the bioavailability (Thomas et al., 2018). SLNs are gaining importance nowadays due to various advantages over other colloidal DDSs, such as increased drug stability, increased drug protection against enzymatic metabolism, the possibility of controlled drug release, high drug loading capacity, biocompatibility, easy large-scale production and sterilization, less variability in release mechanisms and their kinetics, potential for increased permeability due to lipid and surfactant content, and therefore increased bioavailability and ligand-mediated or passive targeting due to their small size via dermal, ocular, pulmonary, oral, nasal, and parenteral routes of administration (Gupta et al., 2017).

8.3.2.5 NLC

NLCs are developed to solve potential difficulties that are arising with SLNs in order to prevent drug expulsion and increase the drug loading. High loading of drugs can be achieved by mixing solid lipids with a small quantity of liquid lipids (oils). This model is called imperfect type nanostructured lipid carriers (NLCs). Drugs which exhibit higher solubility in oils compared to solid lipids can be dissolved into oil, thus protecting them from the surrounding solid lipids. These types of NLCs are called multiple types of NLCs, which are similar to W/O/W type emulsions as it is an oil-in-solid lipid-in-water dispersion. Transformation of the solid lipids or continuous crystallization results in the expulsion of the drug, which can be avoided by the formation of amorphous types of NLCs. In this type, the particles are solid in nature, but crystallization after cooling can be prevented by mixing special types of lipids like isopropyl myristate, hydroxyl stearate, and hydroxyl octacosanyl. The NLCs are chiefly investigated in dermatological and topical formulations (Thomas et al., 2018).

8.3.2.6 NANOEMULSION (NE)

Nanoemulsions (NEs) or mini-emulsions are translucent or transparent water-in-oil (W/O) or oil-in-water (O/W) droplets with a mean diameter of droplet in between 100 and 500 nm. NEs are also called submicron emulsions and NEs are stable kinetically unlike the microemulsions which are thermodynamically stable. NEs have great stability in suspension because of their small droplet size. NEs are advantageous over coarse emulsions or macroemulsions because of free energy without the flocculation, higher surface area, inherent creaming, sedimentation, and coalescence associated with macroemulsions. They can be manufactured in a various formulations like sprays, liquids, creams, foams, gels, and ointments. Nowadays, NEs are of great interest in anti-HIV/AIDS drug targeting. O/W NE of Saquinavir is formulated to improve its brain localization and oral bioavailability (Gupta et al., 2010).

8.3.3 BIOLOGICAL CARRIERS

HAART (highly active antiretroviral therapy) has problematic side effects and is not totally effective. So, it is necessary to investigate new therapeutic approaches.

8.3.3.1 VIRUS VECTORS FOR HIV/AIDS

Animal retroviruses, lentiviruses of humans and animals, and simian and bovine spuma viruses have all been used for gene delivery. RNA viruses reverse their gene sequences in DNA and integrate this DNA into the genome of host. Among several virus-based gene vectors MLV (Murine Leukemia Virus) derived vectors is mostly used for clinical gene transfer research. In case of patients who are infected from HIV-1, gene therapy is not for elimination of HIV-1 from the body but for persistence of the therapeutic anti-HIV-1 gene is very important for long term treatment solutions. Lentivirus-based retroviral gene delivery vectors are of particular interest because they can genetically modify. HIV-1-based vectors expressing various anti-HIV-1 for delivery into lymphocytes, monocytes, stem cells and cells in the central nervous system (CNS) (Strayer et al., 2005).

8.3.3.2 STEM CELLS

Anti-HIV gene therapy which is based on hematopoietic stem cell aims to reconstitute immune system of patient by transplanting the genetically engineered hematopoietic stem cells with anti-HIV genes. Hematopoietic stem cells can multiply and differentiate into mature immune cells and able to self-renew. HIV resistant immune cells can continuously provide throughout the life of a patient by Anti-HIV gene modified hematopoietic stem cells. Hence, anti-HIV gene therapy based on hematopoietic stem cell has a big potential for providing a life-long remission of AIDS/HIV by single treatment (Pernet et al., 2016).

8.3.3.3 ERYTHROCYTES AS A DRUG CARRIERS

Erythrocytes (RBCs or Red Blood Cells) are the huge population in mammals like human. Main function of erythrocytes is to transfer oxygen to body cells and tissues. Most organelles and cell nucleus are absent in mature RBCs, but they contain large amount of hemoglobin (Hb), a special protein which is capable for oxygen binding. Lifetime of the erythrocytes is 100–120 days, in the bloodstream after that they are removed from the spleen. Erythrocytes are very flexible because of their biconcave shape and flexibility and pass through the narrow capillaries. Erythrocytes are used as carriers in two different methods: by incorporation of drug on the RBCs' surface (using specific association or non-specific adsorption, which involves several chemical cross-linking compounds or antibodies). For incorporation of drug into the erythrocyte, the cell requires to undergo some influences externally so that pores reversibly formed in its membrane, across which drug can enter. This unique property of erythrocyte allows them to load with substances of different molecular weight which are biologically active. For these reasons, RBCs are auspicious biocompatible cells for drug delivery (Koleva et al., 2020).

8.4 NOVEL DRUG DELIVERY SYSTEMS (NDDSS) FOR ANTI-RETROVIRAL DRUG

8.4.1 ENTERIC COATED MATRIX/BIOADHESIVE/SUSTAINED RELEASE (SR) TABLET

For achieving continuous drug delivery at reproducible and predictable kinetics over an extended time period, sustained DDSs are preferred and designed. Decrease in side effects related to drug because of improvement in patient compliance due to reduction in dosing frequency, controlled therapeutic blood levels rather than oscillating blood levels, and reduction in the total dose of drug are the main advantages of these system. The DDSs which are designed for retention of drug on the mucosa for prolonged period of time to facilitate the absorption of drug over a prolonged time period by interacting with mucin is called Bioadhesive system. Therefore, the combination of both bioadhesive and SR properties in DDS will further increase the therapeutic efficacy. ARVs like ddI (Didanosine) would be an ideal candidate for sustained drug release because of its short half-life of 1.3–1.6 h, necessitating frequent administration of doses, as well as its severe dose-dependent side effects (Ojewole et al., 2008).

8.4.2 NANOPOWDERS

To increase the dissolution rates for drugs which are poorly soluble, nanopowders are used as a delivery system for oral administration of such drugs. Poloxamer 188/Tween 80 stabilized nanosuspensions of loviride, hydrophobic antiretroviral drug is prepared by sucrose co-freeze-drying and media milling to obtain the solid nano-powders. Morphological characterization of loviride showed structures similar to plates in nano-powder which is different from the loviride crystals which was untreated (Ojewole et al., 2008).

8.4.3 TRANSDERMAL DRUG DELIVERY SYSTEMS (DDSS)

Generally, most of the anti-retroviral drugs are administered by oral route, hence it is necessary to recognize the major adversities of transdermal route of administration, along with those related to the hepatic first-pass metabolism, bioavailability, and frequency of administration. TDDS (transdermal drug delivery systems) are investigated to overcome those issues. Transdermal

DDSs present substitute to the oral anti-retroviral regimens since this system of administration allows to obtain continuous and controlled release of drug by avoiding the variations in pharmacokinetic results in oral route of administration. So, it can be possible to use drugs which have a short half-life and also to find an acceptable and simpler regimen for patients. Besides the drug and the reservoir, itself TDDS should have other components also which are essential – (i) pressure-sensitive adhesives (PSA) to keep contact between the skin and the device; (ii) a backing layer which should not allow the diffusion of any excipients; (iii) a release liner to cover the patch; and (iv) permeation enhancers for increasing the permeability through stratum corneum (e.g., solvents, surfactants, pyrrolidones, and phospholipids) (Cunha et al., 2021).

8.4.4 SUSPENSIONS

Nanosuspensions consists of finely dispersed solid dry particles in an aqueous vehicle for either topical or oral use for pulmonary and parenteral administration, which are made from a sterile or non-toxic, biodegradable or non-biodegradable polymer. Suspensions formulated with lipid-linked complexes for subcutaneous administration have been reported to increase localization in lymphoid tissue and also reduce viral load in macrophages of HIV-2287 (Vedha Hari et al., 2013).

8.4.5 VAGINAL RINGS

Vaginal ring is a delivery device of circular ring type which containing two layers was developed to be inserted into the vaginal cavity, which releases drugs at a controlled rate. There are systems made with a third layer of drug-free rate-controlling elastomeric membrane that plays an excellent role in minimizing drug loading and release. The production of such a device merits the use and control of the position by the patient in a comfortable way to prevent the disruption of intercourse and also to provide a steady drug delivery (Vedha Hari et al., 2013).

8.4.6 BIOADHESIVE INTRA-VAGINAL SYSTEMS

To minimize the failings of the traditional intra vaginal dosage forms such as leakage of the formulations, improper dose administration and poor retention, new Bioadhesive type DDSs are more superior. The bio-adhesive polymers that are used for intra-vaginal formulations includes polyacrylic acids, hydroxypropyl cellulose and polycarbophil. There are systems used for microbicide delivery using mucoadhesive microparticulate vaginal systems that must be multiphase liquid or semi-solid that have been designed to not slip out of the vagina (Vedha Hari et al., 2013).

8.4.7 NANOPARTICLE BASED IMMUNO-ENGINEERED APPROACHES FOR HIV

By suppressing viral replication, HAART serves as an effective approach to combat HIV infections in patients with AIDS/HIV, but HAART do not provide AIDS/HIV patients with a functional cure and also introduces several comorbidities. The virus is capable to remain within latent reservoirs, both unaffected by HAART and fails to detect by the immune system and hence increases the risk of a viral bounce. Immuno-engineering makes use of varied bioengineering approaches for interaction with the immune system and to enhance its therapeutic effects against HIV (Bowen et al., 2020; Thombre et al., 2022).

8.4.8 IMMUNOTHERAPY FOR HIV/AIDS

Immunotherapy involves use of immunomodulatory agents for modulation of immune response for treatment of disease. Like vaccines, it is based on individual's immunization with several immunologic formulations; for treating HIV-infected patient. There are several immunotherapy approaches based on antigens or delivering cytokines (such as IL-2, IL-7 and IL-15) for AIDS/HIV patients. For development of large degree humoral immunity and cellular immunity, it requires APCs (antigen-presenting cells) for dispensation and offering antigens to $CD4^+$ and $CD8^+$ T-cells (Mamo et al., 2010).

8.4.9 NANOVACCINES

Traditional vaccines for HIV-1 are problematic to progress, and clinical trials using HIV-1 vaccines have confirmed reduced efficiency. Vaccines fail because of various reasons such as poor delivery to dendritic cells; vaccine is too weak to help an immune reaction or if degeneration of a live attenuated virus to its lethal form. Nanoparticles overcome these limitations by processing by antigen presenting cells (APCs), promoting antigen uptake, antigen protection by proteolytic enzymes and along with being biodegradable and biocompatible (Bowen et al., 2020).

8.5MNPS TARGETING HIV VIRAL BLEND TO IMMUNE CELLS

The focus was to target the HIV replication cycle by inhibiting the ability of HIV-1 to fuse and/or enter the target cell. Inhibition of entry or fusion leads to inhibition of viral cytotoxicity and viral activity. Lara et al. showed that silver nanoparticles are antiviral and prophylactic against HIV-1 fusion with target cells. Silver nanoparticles exhibit anti-retroviral activity at the early stage of replication of virus, likely as viral entry inhibitor or as a virucidal agent. Silver nanoparticles bind with gp120 in such a manner that it prevents virion CD4-dependent binding, infectivity, and fusion acts as an efficient virucidal agent against cell associated and cell-free virus. SEVI (semen-derived enhancer of viral infection) is type of amyloid fibril present in the semen of human is another approach which increases viral fusion by enhancing target cells HIV-1 infection and capturing HIV-1 virions. Further, silver nanoparticles also inhibit the post-entry stages of life cycle of the HIV-1. SEVI due to its highly cationic nature serves as mediator for HIV-1 viral attachment. Sheik et al. synthesized hydrophobic PNPs to reduce infection mediated by SEVI fibrils (Ahire et al., 2018). Nanoparticle due to its hydrophobicity interferes with Aβ amyloid structure and forms amorphous aggregates which disrupts the amyloid HIV-1 trafficking protein to target cells. Thus, the nanoparticles because of their hydrophobic nature are capable for reduction in binding affinity of HIV-1 virion towards their target cells (Bowen et al., 2020).

8.6 NANOTECHNOLOGY FOR AIDS/HIV TREATMENT

In many areas for treatment of various diseases nanotechnology provides revolutionizing platform for drug delivery. With the use of nanotechnology, it become possible to achieve targeted drug delivery to specific tissues or cells and intracellular delivery of macromolecules, improvement in delivery of poor water-soluble drugs. Controlled-release systems of drug delivery, may improve their half-lives, keeps them in circulation for longer period of time with therapeutic concentrations. Nanoscale delivery systems because of their small size, modulate, and enhance the distribution of hydrophobic as well as hydrophilic drugs within and into different tissues. This distinct characteristic of nanoscale delivery systems makes them most promising for their use in prevention and clinical treatment of AIDs. Specifically, targeted of antiretroviral drugs delivery to macrophages, the brain, other organ systems and $CD4^+$ T-cells ensures that drugs reach latent reservoirs (Mamo et al., 2010; Ahiore et al., 2020b).

8.7 GENE THERAPY FOR TREATMENT OF AIDS/HIV

To improve existing antiretroviral therapy, there are continuous efforts to discover replacement approaches for the treatment of AIDS/HIV. One of the approaches is treatment by gene therapy, in that gene is imbedded into a cell, to interfere with viral replication or infection. Other compounds which are nucleic acid based, like RNA decoys, DNA, ribozymes siRNA, protein-based agents such as zinc-finger nucleases and fusion inhibitors and aptamers (Apts) are also be used to interfere with viral replication (Mamo et al., 2010).

8.8 COMPLICATIONS

NDDSs clearly provide an opportunity for the formulation scientists to defeat the challenges associated with anti-HIV drug therapy but the toxicity and complications associated with it also need to focus.

8.8.1 THE COMPLEXITY OF AIDS/HIV AS A DISEASE

If left untreated, HIV infection may cause great viral burden in the body that leads to progressive fall in the immune cells especially $CD4^+$ T-cells so, the eradication of virus completely from the body is the main challenge. This can be disrupted by treating HAART which must comprise at least three drugs' regimes of at least two antiretroviral agents. Instantly, after initial infection, the virus is capable to establish the pools where it discharges from the drug effects and continuously releasing the viral progeny to the blood until the patient lives. Because of this reason, HIV is one of the lifelong and chronic disease. There are two kinds of viral reservoirs present within the tissues which serve this role. Anatomical reservoirs; these are the tissues unreachable to optimal levels of anti-retroviral drugs because of the presence of various barriers, like the blood-testes barrier, blood cerebrospinal fluid barrier and BBB (blood-brain-barrier) and the others are cellular reservoirs; in such reservoirs virus remains latent in the cells and thus escaping the action of antiviral drugs because of the presence of efflux proteins like multidrug resistance protein and P-glycoprotein on the cell surface which prevents the drugs from attaining therapeutic intracellular concentrations (Lenjisa et al., 2014).

8.8.2 CHALLENGES TO WIDELY APPLY NANOMEDICINE FOR HIV/AIDS THERAPY

Although nanomedicines are the auspicious future of AIDS/HIV treatment and prevention, various obstacles remain unresolved, including but not limited to toxicity, undesirable biological interactions, and the difficulty and cost of large-scale synthesis of nano-pharmaceuticals. Other big challenge is targeted antiretroviral drugs delivery to viral reservoir sites by using nanopolymers may cause HIV drug resistance. Which may be due to the targeted delivery of anti-HIV drugs which if not achieved by systemic HAART administration will cause suboptimal doses of drug in the non-targeted tissues and with the potential to select out the drug resistant mutations there (Lenjisa et al., 2014).

8.8.3 TOXICITY

It is considered that PEI (polymer) is a secure carrier for gene-mediated therapy, but PEI complexes had revealed cell toxicity effects. The toxicity of polymer depends upon the size, molecular weight, zeta potential of derivatives prepared from PEI and degree of branching. Apoptotic pathways and Cell necrotic are also affected by the cytotoxicity. More compact nanoparticles can be produced by high ionic strength buffer and higher molecular weight polymer which affects cytotoxicity. Studies reported that the cytotoxicity is inversely related to size of polyplex, therefore, smaller polyplexes showed higher cytotoxicity and bigger polyplexes had lower cytotoxicity (Aggarwal et al., 2019).

Inorganic nanoparticles also exhibit potential toxicity. Multiple studies have revealed the toxicity of these inorganic nanoparticles. For mammalian cells, silver nanoparticles were found highly cytotoxic, based on the evaluation of abnormal cell morphologies, membrane leakage of the lactate dehydrogenase and mitochondrial function (du Toic et al., 2010).

8.8.4 MLIMITATIONS OF THE USE OF RBCS AS A DRUG CARRIERS

RBCs are more auspicious for use as a drug carrier but their use has numerous limitations. Blood is the source of RBCs; thus, use of the allogeneic blood can cause errors in selecting the right blood type and transmission of several infections. These are the common disadvantages for all transfusion of blood products. Also, the production of the carrier erythrocytes is associated with the requirement for sterile work and complexity of the large-scale production of such cells. Along with this, there are other restrictions also. Some compounds having low molecular weight can pass easily through the erythrocyte membrane are easy to enter as well as easy to leave the cells, which makes it unfeasible to create the long term the red blood cell-based depot form of these compounds in the bloodstream. To slow down the release of such substances from the red blood cells, the cells were treated with various cross-linking agents such as glutaraldehyde, which slows down the release of medicinal compounds from the cells, the membrane of such erythrocytes changes so much that they are quickly recognized by the RES (reticuloendothelial system) cells and removed from the blood circulation. Thus, there are many limitations to the use of red blood cells as drug carriers (Koleva et al., 2020).

Still, there are several challenges remaining in the anti-HIV drug prophylaxis/therapy, which include the following:

- Occurrence of viral resistance.
- Onset of severe adverse effects, which results in interruption or discontinuation of therapy or even prophylaxis.
- Troublesome regimens and/or drug delivery routes that lead to poor adherence by users/patients.
- Sub-optimal biodistribution and pharmacokinetics, particularly in reservoir sites or mucosae involved in sexual transmission.
- Reduced shelf-life and low stability of active molecules, which can be specifically challenging in the tropical climates and low-resource regions lacking sufficient refrigerated storage and distribution channels.
- Shortage of proper dosage forms for particular populations (e.g., children and women).
- Drug products which are costly are often not accessible to populations in need of prophylaxis/therapy.
- Legal and Social constraints resulting in poor access to and the discontinuation of anti-HIV therapy/prophylaxis (das Neves et al., 2019) (Table 8.1).

TABLE 8.1 Lipid Nanocarriers and Disadvantages for ARV Therapy

	Types			
Limitations	**Liposomes and Ethosomes**	**Cubosomes**	**SLN and NLC**	**Nanoemulsion (O/W)**
	• Most suitable for IV and nasal route. • Low polar drug loading. • Less stability.	• Difficult to load high molecular weight compound. • Difficult to scale-up. • Lack of reproducibility in terms of inner channel size.	• Low polar/semi-polar drug loading.	• Need high concentration of surfactant, which results in biological toxicity.

Source: Modified from: Faria et al. (2021).

8.9 CONCLUSION

Nanotechnology offers a groundbreaking platform for medication delivery in several therapeutic fields for the management of various ailments. With the help of nanotechnology, it is now possible to transport medications specifically to certain tissues or cells, deliver macromolecules inside of cells, and optimize the distribution of poorly water-soluble drugs. Drugs delivered using controlled-release techniques may have longer half-lives and remain in the bloodstream for longer at therapeutic concentrations. Due to their small size, nanoscale delivery methods modify and improve the distribution of both hydrophobic and hydrophilic medicines inside and into various tissues. Because of this distinctive quality, nanoscale delivery methods are the most promising for use in AIDS clinical therapy and prevention. Antiretroviral drug delivery specifically focused on macrophages, the brain, and other organ systems.

KEYWORDS

- **AIDS**
- **carriers**
- **cytotoxicity**
- **dendrimers**
- **HIV**
- **liposomes**
- **nanoparticles**
- **polymers**
- **retrovirus**

REFERENCES

Ahire, E. D., Talele, S. G., & Shah, H. S., (2020). Nanoparticles as a promising technology in microbial pharmaceutics. In: *Applied Pharmaceutical Science and Microbiology* (pp. 133–158). Apple Academic Press.

Ahire, E., Thakkar, S., Borade, Y., & Misra, M., (2020). Nanocrystal based orally disintegrating tablets as a tool to improve dissolution rate of Vortioxetine. *Bulletin of Faculty of Pharmacy, Cairo University*, *58*(1, 2), 11–20.

Ahire, E., Thakkar, S., Darshanwad, M., & Misra, M., (2018). Parenteral nanosuspensions: A brief review from solubility enhancement to more novel and specific applications. *Acta Pharmaceutica Sinica B*, *8*(5), 733–755.

Ahirrao, S. P., Sonawane, M. H., Bhambere, D. S., Udavant, P. B., Ahire, E. D., & Kanade, R., (2022). Cocrystal formulation: A novel approach to enhance solubility and dissolution of etodolac. *Biosciences Biotechnology Research Asia*, *19*(1), 111.

Allan, B., Elizabeth, E. S., & Rohan, F., (2020). Nanoparticle-based immunoengineered approaches for combating HIV. *Frontiers in Immunology*, *11*, 789.

André, D., & Michel, G. B., (1998). Liposomes as drug delivery system: A strategic approach for the treatment of HIV infection. *Journal of Drug Targeting*, *6*(1), 1–15,.

David, S. S., Ramesh, A., Bruce, A. B., Boro, D., Vicente, P., Roger, J. P., John, J. R., & John, A. Z., (2005). Current status of gene therapy strategies to treat HIV/AIDS. *Molecular Therapy*, *11*(6).

Elizabeth, O., Irene, M., Panjasaram, N., & Thirumala, G., (2008). Exploring the use of novel drug delivery systems for antiretroviral drugs. *European Journal of Pharmaceutics and Biopharmaceutics*, *70*, 697–710.

Hiraman, P. N., Prashant, P., Prabhanjan, G., & Vidya, L., (2013). Ethosome: A novel drug carrier. *International Journal of Pharmaceutical Research and Allied Sciences*, *2*(3), 18–30.

Jimma, L. L., Minyahil, A. W., & Gizaw, D. S., (2014). New hope for eradication of HIV from body: The role of polymertic nanomedicines in HIV/AIDS pharmacotherapy. *Journal of Nanobiotechnology*, *12*, 9.

Josc, D. N., (2019). Novcl approachcs for thc dclivcry of anti-IIIV drugs—What is ncw? *Pharmaceutics*, *11*, 554.

Jose, D. N., Mansoor, M. A., Maria, F., & Bruno, S., (2010). Nanotechnology-based systems for the treatment and prevention of HIV/AIDS. *Advanced Drug Delivery Reviews*, *62*, 458–477.

Joshy, K. S., Snigdha, S., Anne, T., Anne, G., Nandakumar, K., Laly, A. P., & Sabu, T., (2018). Nanotechnology in HIV/AIDS: Lipids and lipid hybrid systems- novel delivery approaches on antiviral drugs. *JSM Nanotechnology and Nanomedicine*, *6*(3), 1073.

Lalit, K., Shivani, V., Deo, N. P., Ankur, B., Bhuvaneshwar, V., & Amit, K. J., (2015). Nanotechnology: A magic bullet for HIV AIDS treatment. *Artificial Cells, Nanomedicines and Biotechnology*, *43*, 71–86.

Larisa, K., Elizaveta, B., Fazoil, A., & Elena, S., (2020). Erythrocytes as carriers: From drug delivery to biosensors. *Pharmaceutics*, *12*, 276.

Lipman, M., (2006). *Human Immunodeficiency Virus, Encyclopedia of Respiratory Medicine* (pp. 284–292). Academic Press.

Lisa, C. Du. T., Viness, P., & Yahya, E. C., (2010). Nano-microbicides: Challenges in drug delivery, patient ethics and intellectual property in the war against HIV/ AIDS. *Advanced Drug Delivery Reviews*, *62*, 532–546.

Lu, L., Sun, R. W., Chen, R., et al., (2008). Silver nanoparticles inhibit hepatitis B virus replication. *Antivir. Ther., 13*(2), 253–262.

Maria, J. F., Carla, M. L., Jose, D. N., & Marlene, L., (2021). Lipid nanocarriers for anti-HIV therapeutics: A focus on physicochemical properties and biotechnological advances. *Pharmaceutics, 13*, 1294.

Marina, C. M. D., (2008). *Human Immunodeficiency Virus, Adolescent Medicine*, 220–228.

Olivier, P., Swati, S. Y., & Dong, S. A., (2016). Stem cell-based therapies for HIV/AIDS. *Advanced Drug Delivery Reviews, 103*, 187–201.

Rita, F. C., Sandra, S., Manuela, C., José, M. A. P., Quirina, C., & Andreia, A., (2021). Novel antiretroviral therapeutic strategies for HIV. *Molecules, 26*, 5305.

Rohan, A., Monika, T., Bhumika, K., Sahoo, P. K., & Meenakshi, K. C., (2019). Polyplex: A promising gene delivery system. *International Journal of Pharmaceutical Sciences and Nanotechnology, 12*(6).

Saialja, I., Manoj, B., & Ivvala, A. S. (2021). *Nanotechnology Based Drug Delivery For HIV-AIDS Treatment*. doi: Http://dx.doi.org/10.5772/intechopen.97736.

Shweta, G., Rajesh, K., Narendra, C., Ambikanandan, M., & Abdelwahab, O. (2021). Systematic approach for the formulation and optimization of solid lipid nanoparticles of efavirenz by high pressure homogenization using design of experiments for brain targeting and enhanced bioavailability, Hindawi, *BioMed Research International, 2017*, Article ID 5984014.

Sondi, I., & Salopek-Sondi, B., (2004). Silver nanoparticles as antimicrobial agent: A case study on E. coli as a model for gram-negative bacteria. *Journal Colloid Interface Sci., 275*(1), 177–182.

Talele, S. G., Ahire, E. D., Talele, G. S., & Derle, D. V., (2021). An innovative approach as self-emulsifying drug delivery system for phytoconstituents. In: *Enhancing the Therapeutic Efficacy of Herbal Formulations* (pp. 69–84). IGI Global.

Tewodros, M., Ashley, M. E., Nagesh, K., Carolina, S. M., Jinjun, S., Daniel, R. K., Robert, L., et al., (2010). Emerging nanotechnology approaches for HIV/AIDS treatment and prevention. *Nanomedicine (Lond), 5*(2), 269–285.

Thao, T. D. T., & Phuong, H. L. T., (2019). Nanoconjugation and encapsulation strategies for improving drug delivery and therapeutic efficacy of poorly water-soluble drugs. *Pharmaceutics, 11*(7), 325.

Thombre, N. A., Niphade, P. S., Ahire, E. D., & Kshirsagar, S. J., (2022). Formulation development and evaluation of microemulsion based lornoxicam gel. *Biosciences Biotechnology Research Asia, 19*(1), 69–80.

Tomalia, D. A., Baker, H., Dewald, J., Hall, M., Kallos, G., Martin, S., Roeck, J., Ryder, J., & Smith, P., (1985). A new class of polymers: Starburst-dendritic macromolecules. *Polymer Journal, 17*(I), 117–132.

Umesh, G., & Narendra, K. J., (2010). Non-polymeric nano-carriers in HIV/AIDS drug delivery and targeting. *Advanced Drug Delivery Reviews, 62*(4, 5), 478–490.

Vedha, H. B. N., Devendharan, K., & Narayanan, N., (2013). Approaches for novel drug delivery system for anti-HIV agents. *International Journal of Drug Development and Research, 5*(4), 16–24. ISSN 0975-9344.

Yong, G., Chanuka, W., & Jamie, F. S. M., (2018). Virus-like particle, liposome, and polymeric particle-based vaccines against HIV-1. *Frontiers in Immunology, 9*, Article 345.

CHAPTER 9

Novel Drug Delivery Systems for Diabetes: An Immense Hope

GOWTHAM MENON,[1] SACHIN GAIKWAD,[1] DEEPAK PISOTE,[2] ROHIT CHOURE,[1] GIRISH A. KASHID,[3] and RAOSAHEB S. SHENDGE[1]

[1]*Department of Pharmaceutics, Sanjivani College of Pharmaceutical Education and Research, Kopargaon, Maharashtra, India*

[2]*Department of Pharmaceutical Biotechnology, Sanjivani College of Pharmaceutical Education and Research, Kopargaon, Maharashtra, India*

[3]*Department of Pharmaceutical Chemistry, Sanjivani College of Pharmaceutical Education and Research, Kopargaon, Maharashtra, India*

ABSTRACT

Diabetes mellitus (DM) is a chronic hyperglycemic condition marked by lackadaisical and sluggish metabolism. Despite the myriad of medications on the market, exhaustive, and competent DM solutions remain elusive owing to side effects. Finding a stable/non-invasive drug delivery system (DDS) for these medications, as well as organized and monitored drug administration, might be more advantageous. Novel drug delivery methods for antidiabetic drugs have been increasingly popular in recent years as a result of demonstrable benefits such as lower dosage frequency, greater bioavailability, protection against degradation in the hostile stomach environment, site-specificity, and fewer adverse effects. The purpose of this chapter is to give information on the innovative drug delivery methods for antidiabetics, as well as potential problems. Various innovative medication delivery techniques have been implemented. The unique potential of catering physical stability, sustained, and site-specific drug delivery for a calculated period

Novel Drug Delivery Systems in the Management of Chronic Diseases.
Raj K. Keservani, Umesh D. Laddha, Eknath D. Ahire, and Sanjay J. Kshirsagar (Eds.)

can open up new vistas for precise, safe, and effective DM treatment in the current scientific landscape, where novel drug delivery systems (NDDSs) are being recognized for their tangible benefits. The purpose of this chapter is to give information on the innovative drug delivery methods for antidiabetics, as well as potential problems.

9.1 INTRODUCTION

9.1.1 *WHAT IS DIABETES MELLITUS (DM)?*

Diabetes mellitus (DM), universally established as Diabetes, had already emerged as a global epidemic in recent decades (Misra et al., 2011), ranking as the fifth leading cause of mortality in most industrialized and developing countries. DM is a kind of hyperglycemia that lasts for a long time. This is considered a multidimensional metabolic disease etiology characterized by poor carbohydrate, lipid, and protein intake, etc. (Rai et al., 2016; Unnikrishnan et al., 2016). DM is caused mostly by hereditary and environmental factors. Infections, immune system malfunction, and mental issues all contribute to inadequate nutrition. Insulin production and insulin resistance are two different things (Petersmann et al., 2019). People with this condition have had to deal with a variety of issues for a long time. The life-threatening consequences might reduce their standard of health and put their lives in jeopardy. DM is a derivative of the Greek words 'diabetes,' which implies "to pass through," and 'Mellitus,' which stands for "sweet" which is exemplified both in blood and urine. The medical fraternity widely began to recognize that dietary modifications may aid in accomplishing diabetic management in the 1700s and 1800s, and they endorsed their people to do things like have an abundance of animal fat and meat in their routine diet (International Diabetes Federation, 2021).

9.1.2 *WHAT ARE THE SYMPTOMS OF DIABETES MELLITUS (DM)?*

Diabetes is associated with symptoms like thirst, polyuria, blurred eyesight, and weight loss. A menacing and life-threatening condition like Ketoacidosis or a non-ketotic hyperosmolar condition is the most serious clinical sign, which can lead to other issues like dehydration, unconsciousness, and, in the absence of satisfactory treatment, death. Symptoms are often moderate or non-existent in DM due to the slow pace at which hyperglycemia progresses. As an outcome, in the absence of accurate evidence of biochemical analysis, hyperglycemia-induced pathological and functional abnormalities may remain undiagnosed for a long period, causing complications at the time of diagnosis (International Diabetes Federation, 2021).

9.2 DIABETES ETIOLOGY AND PATHOLOGY

The malfunction or death of pancreatic beta cells is now universally endorsed as the fundamental characteristic component common to all diabetic classes. A drop in productivity can be caused by a variety of factors. Many convoluted mechanisms can precipitate a decline in performance or the full death of beta cells (Singh et al., 2016). After the age of 30, it appears to be unable to renew beta cells. These mechanisms include the following: autoimmunity, genetic propensity, and anomalies, epigenetic step-by-step mechanisms, insulin resistance, coeval ailments, inflammation, as well as environmental variables are all considerations to consider. Distinguishing between beta-cell malady and dysfunction as well as reduced beta-cell mass might have major ramifications for healing attempts to keep cells alive or boosting our glucose tolerance. Understanding beta-cell status can aid in the classification of diabetes subtypes as well as guiding therapy.

9.3 STATISTICS OF DIABETES MELLITUS (DM)

- The entire population with diabetes is forecasted to reach 643 million by 2030, which may further increase at an alarming rate to 783 million by 2045.
- In worldwide, a low- or mid-income country is home to 3 out of every 4 diabetics.
- One-quarter of the population has undiagnosed diabetes (240 million) and about 6.7 million individuals die as a result of diabetes.
- In the criteria of health expenses, diabetes accounted for approximately USD 966 billion, or statistically 9% of all adult spending.
- Almost 1.2 million children and teenagers aged 0 to 19 are stated to be vulnerable to type 1 diabetes.
- One in every six live babies is affected by diabetes (21 million).
- 541 million people are liable or likely to develop type 2 diabetes.
- 10.6% of the global adult population exhibit a condition wherein blood glucose levels accelerate over normal but not to the point where he/she can be called diabetic. In due course of time, these subjects are more inclined to develop diabetes and cardiovascular disease.

9.4 CLASSIFICATION OF DIABETES

In 1980, the World Health Organization (WHO) communicated the generally unquestioned and universally adopted classification for diabetes, which was subsequently amended in 1985. Recent demands have been made to examine and revise the diabetes categorization system because many persons with diabetes do not fall into a single group; recent breakthroughs in pathophysiological pathway understanding and developing technology to assess pathology and disease have made this possible. Treatments that target particular pathways are becoming more popular, and customized therapy is becoming more popular in this modern era (Singh et al., 2016).

A single categorization system for a diabetic would be ideal since it would help with three main goals: clinical treatment, etiopathology, and epidemiology. Nevertheless, given our modern and prevailing knowledge and the available scientific data in remarkably all worldwide countries, this is not conceivable. The WHO has formulated certain guidelines for the categorization of diabetes, which were circulated in 2019 and are as follows:

1. **Type-1 Diabetes:** The former WHO categorization of diabetes defined 2 subgroups amid type 1 diabetes patients: 1a which is considered autoimmune and 1b which is regarded as non-immune. However, because the pathophysiology of non-immune diabetes is uncertain, the new classification doesn't contain subtypes that are omitted. The detection of specific autoantibodies directed targeting glutamic acid decarboxylase (GADA), anti-islet antigen-2 (IA-2), anti-zinc transporter 8 (ZnT8), and anti-insulin (IAA) can prove the autoimmune basis of the illness in the majority of individuals (Diabetes Mellitus Type 2 – StatPearls – NCBI Bookshelf, n.d.; DiMeglio et al., 2018; International Diabetes Federation, 2021; Mahesh et al., 2016; Rai et al., 2016; Singh et al., 2016; WHO, 2019).
2. **Type-2 Diabetes:** Pancreatic β-cell malfunction is required in the development of diabetes, which is categorized as type 2, according to research. The majority of people with this kind of diabetes have a comparative insulin shortage due to varying characteristics of insulin resistance. There is frequently a rise in absolute insulin levels in the initial steps of the illness. Diabetes of type 2 is by far the most prevalent form of the illness, attributed to 90–95% of all cases. It's linked to being overweight or obese, not getting enough exercise, and leading an unhealthy lifestyle that includes eating processed foods and drinking sugary drinks (Chatterjee et al., 2017; *Diabetes*

Mellitus Type 2 – StatPearls – NCBI Bookshelf, n.d.; International Diabetes Federation, 2021; Mahesh et al., 2016; Olokoba et al., 2012; Rai et al., 2016; Singh et al., 2016; WHO, 2019).

3. **Hybrids Forms of Diabetes:** In an attempt to separate T1DM from T2DM in adults, additional disease classifications and nomenclatures have been proposed, including steadily progressing immune-mediated diabetes as well as type 2 diabetes which is ketosis-prone (Rai et al., 2016; Singh et al., 2016).

   Immune-interceded diabetes is slowly progressing. Among individuals who have pancreatic autoantibodies that really can interact with quasi cytoplasmic antigens in islet cells but do not have type 2 DM, GAD, protein tyrosine phosphatase IA-2, insulin, or ZnT8 are all enzymes that decarboxylate glutamic acid. This kind of "Latent autoimmune diabetes in adults" is a term that has been used to describe diabetes (LADA). In the case of such patients, initial management in the form of lifestyle changes along with oral medications is preferred rather than an initial insulin therapy but eventually, they may need it more quickly than other patients. Obesity, metabolic syndrome traits, preserving more – cell functionality, expressing a single autoantibody and having the transcription factor 7-like 2 (TCF7L2) gene polymorphism are all distinctions between slowly growing immune-mediated diabetes and T1DM.

   Type 2 diabetes which is ketosis-prone (also known as "Flatbush diabetes") has characteristics of both type 1 and 2 diabetes and is quite common in African communities. The sudden development of acute hyperglycemia with ketoacidosis necessitated hospitalization and treatment with insulin, as well as water and electrolyte replacement.

4. **Other Types of Diabetes**: Like monogenic diabetes, drug or chemical caused diabetes, Diabetes caused by endocrine disorders, etc. (Singh et al., 2016; WHO, 2019).

9.5 THERAPY FOR TREATMENT OF DIABETES

9.5.1 SULFONYLUREAS

The drug sulfonylurea is a hypoglycemic agent which was used since the mid-1950. These drugs are classified into four generations. The drugs such as acetohexamide, chlorpropamide, tolbutamide, tolazamide, etc., are classified under the first generation, gliclazide, glibenclamide, glipizide, glisoxepide, etc., are classified under the second generation, glimepiride included in third-generation drugs, and JB253 and JB558 are classified under fourth-generation drugs. Fourth-generation drugs are also known as light-dependent drugs. Sulfonylureas act as direct stimulation of insulin from pancreatic β-cells, and it may help to maintain blood glucose levels (Grover & Utreja, 2014; Mahesh et al., 2016).

9.5.2 GLINIDES

The class glinides includes the drug classes such as meglitinide and D-phenylalanine analogs. The drug Repaglinide in 1998, was introduced by the United States, and Nateglinide in late 2000, was introduced by the United States. These are short-acting and quick-developing drugs. In advancement, the plasma half-life for these drugs is about 1 hour. The directions such as it is administered before the meal to avoid postprandial hyperglycemia. These are used as insulin-releasing medicines (Grover & Utreja, 2014; Mahesh et al., 2016).

9.5.3 BIGUANIDES

Biguanides are the drugs for the treatment of diabetes. The first drug in the class biguanides such as phenformin was introduced in late 1950, and it was used till 1977 in the United States. After this, it was banned because of the deaths of the patients due to its relationship with the lactic acidosis higher risks. When it was removed from many markets then in India also it was banned in 2003. The drug metformin was introduced by the United States into the market after the tracking of its effectiveness and safety in 1995, and currently, it is a widely used drug in diabetic patients. These drugs increase the action of insulin (Grover & Utreja, 2014; Mahesh et al., 2016).

9.5.4 THIAZOLIDINEDIONES (TZDS)

A pharmacologically important class of heterocyclic compounds known to be thiazolidinediones (TZDs) is used for the treatment of T2DM and its complications. These were introduced in the form of glitazone. The drugs in this class are ciglitazone, troglitazone, pioglitazone, rosiglitazone, englitazone, etc. These are acts by increasing the action of the insulin (Grover & Utreja, 2014; Mahesh et al., 2016).

9.6 PROBLEMS POSED BY THE CURRENT ANTIDIABETIC DRUG THERAPY REGIMEN

The world drug market is flooded with antidiabetic drugs in the form of tablets or capsules, which have gained wide acceptance because of their ease of administration. The most prevalent issues that are related to antidiabetic drug therapy are that these dosage forms have several side effects and limitations, including gastric hyper irritation followed by discomfort, metal taste, diarrhea, loss of appetite, acute pancreatitis, lactic acidosis in people with abnormal kidney or liver activity due to gastrointestinal (GI) degradation, swelling of legs or ankles, weight gain and do not meet the FDA's safety and efficacy requirements. Anti-diabetic medications in traditional dose forms have variable/low bioavailability and a $t_{1/2}$, necessitating continual administration and creating significant adverse effects, ensuing in therapeutic ineffectuality and patient non-compliance. These problems possess a serious limitation to the availability of conventional dosage forms to the site of action, limited therapeutic index, and non-adherence to antidiabetic treatment in long term. To be effective, the drugs must have their ideal level at the targeted spot and sustain that proportion for a more course of time. After rapid dispensation, oral antidiabetic drugs penetrate through biological membranes and move to the location of pharmacological action (Souto et al., 2019).

Nonetheless, it takes around 2 to 3 doses to obtain the ideal blood concentration for the whole day. Some antihyperglycemic drugs because of their sensitivity to proteolytic degeneration in the gastric surroundings and shorter plasma $t_{1/2}$ are avoided to be used for oral administration. Exenatide injections induce nausea, vomiting, and antibody formation when given two times a day, whereas liraglutide injections cause moderate, transitory nausea and vomiting when given once daily. Another stumbling block is injection site discomfort, which causes injection anxiety. As a consequence, a thorough analysis of the area of drug distribution is required, which might be facilitated by the use of artificial intelligence (AI). All these factors have stressed the need for and importance of the development of innovative drug delivery systems (DDSs) for antidiabetic drugs (Grover & Utreja, 2014; Meng et al., 2019; Zhao et al., 2020).

9.7 NOVEL DRUG DELIVERY SYSTEMS (NDDSS) FOR ANTIDIABETIC DRUGS – A NEED IN THE CURRENT SCENARIO

Investigators were able to create delivery strategies with greater treatment efficacy due to the restrictions associated with antidiabetics. Novel treatment methods for delivering anti-diabetic drugs are being developed, which help in delivering drugs to the right location in a healthier and more precise manner. A set period is set aside for better a result which leads to enhanced treatment effectiveness and, as a consequence, better control of diabetes. Furthermore, changes in antidiabetic physicochemical and pharmacological features have abhorred/altered biological barricades that restrict active access to diverse therapeutic targets. Many investigators worldwide are keenly working on incorporating antidiabetic drugs into innovative DDSs and many of them have been successful in it. So, more work has to be done on such core areas including clinical trials so that such systems can be commercialized and launched into the market.

In diabetes, the development of novel drug delivery methods has permitted enhanced distribution of small molecule medications, which might significantly improve diabetic quality of life. Given the incessant condition of diabetes, novel technology-based techniques hold more promise in terms of delivering medications with improved bioavailability and a lower dose regimen to particular sites. Novel medicines work at the molecular and cellular levels to improve drug absorption or prevent efflux processes, allowing the medication to stay within the cell for longer. It might be accomplished by adjusting formulation variables such as procedure or concentration (independent variables), increasing to end-user. Because each drug transmittal system has its own set of pharmacological as well as physiological properties, careful taking into account independent factors is essential in the end product development. It is also necessary to be familiar with the physical, chemical, and pharmacological properties of each ingredient. Oral treatment, proteolytic and physical stabilities in a hostile biological environment are all factors that have been tried to increase antidiabetic delivery. Because each delivery strategy has its unique combination of biochemical and therapeutic features, independent elements must be carefully considered when designing the final product.

NDDSs have also proved successful in circumventing the limits of high dose frequency by allowing the medicine to be released continuously. Most notably, novel formulations upgrade the pharmacodynamic as well as the pharmacokinetic profile of antidiabetic drugs by acting at the molecular level to boost drug absorption by the cells or hinder outflow mechanisms. Encapsulating drugs into novel delivery systems has several advantages like improved bioavailability and outbreaking first-pass metabolism, eventually lowering the dosage and hence decreasing the lethal effects of the drug on cells that are non-target for the medication. A novel-based targeted medicine delivery lowers negative effects and improves efficacy. The drug's concentration is localized, which allows for a faster beginning of the action. Other obstacles to the conventional dosage forms were the solubility in lipids and minimal water solubility of several oral hypoglycemics, such as glibenclamide. Enhancement of dissolving rate is one of the formulation scientist's goals which aimed to enhance these medicines' solubilization by modifying several processes in mimicked biological fluids with different variables. Controlled release, on the other hand, helps by better permeability and steady medication plasma concentration which is beneficial.

9.8 A FOCUS ON CURRENT AND NEW TRENDS IN INSULIN ADMINISTRATION

Insulin is required for the therapy of Type 1 diabetes mellitus (T1DM) in addition to a variety of Type 2 diabetes mellitus patients (T2DM). Improved insulin administration through the oral is advantageous in imitating the normal pathway of insulin and reducing patient pain. Currently, a variety of nanoparticle vesicles for oral and alternative insulin administration methods are being researched. Future research will focus on improving existing treatments through pharmacokinetic adjustments, the development of a fully automated approach to replicate pancreatic insulin supply, as well as eliminating invasiveness upon intake.

9.8.1 SUBCUTANEOUS

Insulin delivery throughout the twentieth century necessitated the use of big needles and glass syringes, as well as less pure insulin, which might result in infection and lipoatrophy at the injection site. Insulin seems to be the only therapeutic option that is realistic. Insulin is presently administered by injections and infusions. Multiple daily injections (MDI) are now used to deliver it, which comprises many shots of brief enacting prandial insulin and 1–2 dosages of long-acting basal preparation (Rys et al., 2018). Insulin is now delivered mostly by the use of a conventional metal needle, which necessitates a high level of professionalism. Microneedles were created to provide for safe intradermal administration without disrupting a nerve, allowing for the most precise targeting of macromolecules into the dermis (Bariya et al., 2012).

The insulin pen, which uses disposable insulin cartridges, provides an option for active patients and those who are unable to deliver injections. A continual supply of insulin is provided by a continuous subcutaneous insulin infusion (CSII) pump with superimposed meal-related boluses (Atkinson et al., 2014). Micron Jet is a microneedle system that contains four 0.45 mm long microneedles fitted to a regular syringe rather than a traditional needle. In patients with T2DM, the phosphomimetic nature of insulin administration utilizing micron jet needles is improved, as seen by the early achievement of systemic levels of insulin, lesser late insulin levels, and greater late glucose levels, implying that postprandial hypoglycemia is reduced. Since the introduction of insulin treatment, there have been numerous breakthroughs. Novel insulin delivery routes have been developed in response to the desire for a more convenient and less painful method of delivery. Buccal or oral, transdermal, ophthalmic, nasal and pulmonary, and rectal administration systems were among those used (Kochba et al., 2016).

9.8.2 OCULAR

The clinical use of ophthalmic insulin administration is defined by the quantity of insulin which can be absorbed via the eyes and reach the systemic circulation. An absorption booster was added to the ocular solution at the ideal pH to boost insulin absorbed into the systemic circulation, and variations in the pH of the insulin ocular solution were explored to maximize the systemic absorption of insulin via the ocular route. Insulin can be administered systemically via the eyes, especially when surfactant enhancers like saponin are used. When compared to parenteral delivery and/or minipump implantation, this innovative technique is more cost-effective and uncomplicated. It is believed that this innovative technique would increase patient adherence to insulin therapy for diabetes (Chiou et al., 1989).

Nanoparticles allow for regulated drug release into intraocular tissues without the use of invasive methods. The amount of medicine absorbed, as well as its bioavailability, increases when the retention time is increased. Nanoparticles can be utilized to target specific sites, reducing the likelihood of negative side effects. Indeed, advanced nanoparticles will be used for drug carriers because, due to their unique physicochemical properties, they can bypass eye barriers and boost therapeutic efficacy when compared to standard pharmacotherapy. Nanoparticles can be applied topically to the sclera of the eye and remain on the surface, giving protection to conjunctival liquid and blinking as well as functioning as medication storage. If cationic nanoparticles are used, this is particularly true (Souto et al., 2019).

9.8.3 BUCCAL

Thanks to its elevated vasculature and low content of proteolytic enzymes than the digestive tract (GIT) mucosa, the buccal mucosa is an ideal location for medication administration (Campisi et al., 2010). Poor diffusivity, instability, quick metabolism, aggregation, and immunogenicity are all linked to high molecular mass and hydrophilicity. Insulin that will be supplied by buccal mucosal factors should be evaluated for structural characteristics, immunological competence, and physicochemical qualities, as well as buccal mucosal properties and possible transport mechanisms (Henkin, 2010; Kraan et al., 2014). Attempting to avoid pre-systemic metabolic activity in the GI and liver; good ease of access; no loss in other liquids; and the emergence of an increased drug gradient of concentration from across mucosa preferring drug diffusion into the deeper tissues are all potential benefits of insulin administration through the buccal mucosa (Sonia & Sharma, 2015).

9.8.4 NASAL

Insulin Delivery through the Nasal Mucosa and oral insulin formulations are appropriate for people who require daily insulin therapy. Insulin's impermeability across epithelial membranes, as well as local digestion and enzymatic degradation, must be overcome to accomplish mucosal absorption by oral or nasal delivery (Kamei et al., 2013). Nasal insulin is a suitable method to investigate brain insulin effects in people since it is comparatively safe and stable after repacking in spray flasks (Kullmann et al., 2018). A nasal application, on the other hand, frequently causes local symptoms such as a burning feeling in the nose (Benedict et al., 2007; Born et al., 2002; Schöpf et al., 2015). Insulin therapy caused local side effects in the nasal region, and infections appeared after long-term use of basal insulin (Schmid et al., 2018).

9.8.5 PULMONARY

This enormous size and the absence of stomach peptidases in the lungs make along with the absence of stomach peptidases in the lungs, making the pulmonary an appealing target for drug administration (Chan & Cheng-Lai, 2017). In comparison to the buccal, nasal, ocular, and rectal routes, pulmonary insulin administration increased bioavailability by 4 to 40 times. Earlier many studies showed that insulin aerosolized and nebulized solutions lowered blood glucose levels (Mastrandrea, 2010).

9.8.6 COLON AND ILEUM INSULIN DELIVERY

Insulin transport across the ileum and colon has been explored, and numerous approaches such as tablets, intestinal patches, microparticles (MPs), and nanoparticles have been developed (Li et al., 2002). Insulin is delivered via a method that includes substances and excipients that prevent insulin from aggregation and promote absorption via the digestive tract (Krauland et al., 2004). Due to its poor solubility in acidic pH values at the stomach up to pH 6, anionic polymeric Eudragit S1000 was selected for intestinal distribution (Khan et al., 1999).

9.8.7 NANOTECHNOLOGY AS A NEW TREND IN INSULIN DRUG DELIVERY

Nanotechnology is an emerging scientific subject concerned with the manufacture and use of substances, devices, and networks on the nanoscale level, or between 1 and 100 nanometers in size. Potential applications in varied sectors such as electronics, environmental cleanup, energy storage, and so on, interest in this area is growing due to biological applications such as medicine administration, target-specific drug treatment, imaging, biomarkers, and biosensors (Zaric et al., 2019; Ahire et al., 2018, 2020a, b; Thombre et al., 2022).

Biomedical nanotechnology has resulted in technological advancements in drug delivery efficiency. The use of nanoparticulate DDSs to formulate medications has shown to have different advantages over standard dosage forms. Encapsulating medicines in nano-based DDSs for oral drug administration should enhance dissolution, stability, and oral absorption competence substantially. Furthermore, the oral absorption mechanisms of nanoparticulate DDSs differ significantly from those of free medicines in traditional formulations. In the oral absorption of medicines into the blood, lymph absorption and endocytosis of nanoparticles (NPs) play a larger role (Chen et al., 2020; Talele et al., 2021).

Many researchers have designed and investigated the rise and promise of nanotechnology in DDSs throughout the last few years. The fast development of nanotechnology will surely enable a new approach to overcome restrictions in current traditional antidiabetic drug delivery, hence enhancing diabetes treatment for patients. Liposomes (Keservani & Gautam, 2020), niosomes (Keservani et al., 2010; Keservani & Gautam, 2022).

Nanoparticles (Ahirrao et al., 2022), dendrimers (Gautam et al., 2015), and micelles are among the nano-carrier-based delivery methods discussed here. With greater bioavailability, dosage proportionality, lower toxicity, and reduced dosing frequency, those nanocarriers (Keservani et al., 2018) have demonstrated higher capability in drug delivery. Additionally, a protein-based oral delivery nano-system has been proposed for effective drug transfer to the biological circulation while avoiding protease degradation in the GI environment (Subramani, 2006; Zaric et al., 2019) (Figure 9.1).

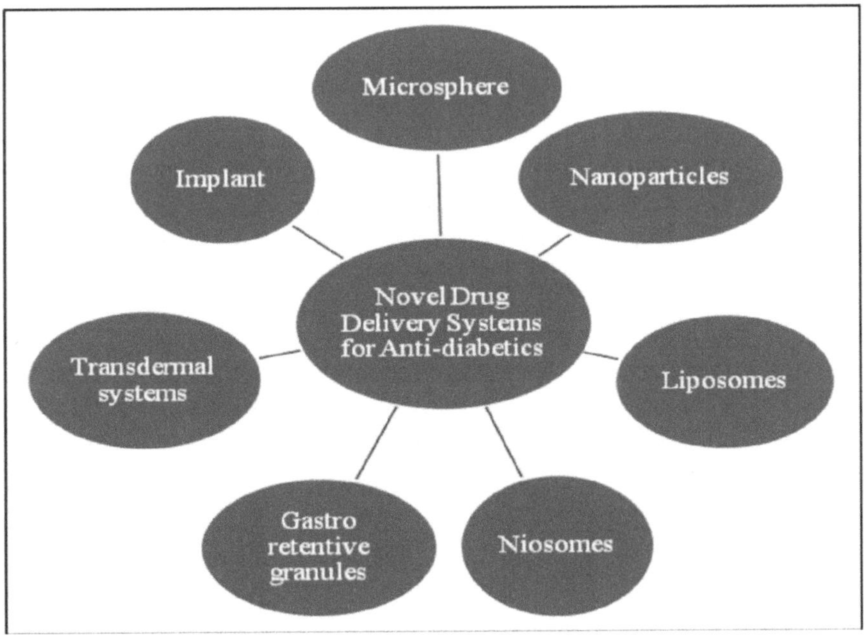

FIGURE 9.1 Novel drug delivery systems for anti-diabetics.

9.9 RECENT ADVANCEMENTS IN THE DRUG DELIVERY SYSTEMS (DDSS) FOR ANTI-DIABETICS

9.9.1 *LIPOSOMES*

The microscopic cavities are made up of bilayer of one or more phospholipids produced from hypoallergic naturally obtained cholesterol and phospholipids known to be "liposomes." They're used in medication administration seeing as they're compostable, bioabsorbable, hypoallergic, entrap hydrophilic and lipophilic pharmaceuticals, and deliver therapeutics to specific locations. There have been numerous approaches to using liposomes to lessen drug allergy and target determining cells to improve potency and safety (Rai et al., 2016; Souto et al., 2019).

For oral delivery, Zhang et al. modified insulin-loaded liposomes with the specific ligand biotin. Biotin-1,2-distearoyl-sn-glycero-3-phosphatidylethanolamine (DSPE) was added to the lipid bilayer of liposomes to form liposomes. In internal aqueous compartments, insulin leakage was minimized by a 3:1 lipid: cholesterol ratio. The biotin-DSPE ratio and the carrier's average size were shown to affect the insulin-loaded liposomal hypoglycemic impact. Liposomes were found to have a greater absorption via endocytosis via intestinal epithelia mediated by receptors due to their smaller size. Liposomes with biotin and DSPE outperformed ordinary liposomes in terms of insulin bioavailability (Zhang et al., 2014).

For oral insulin administration, Agrawal et al. created liposomes with folic acid functionalization with polyacrylic acid anion and polyallylamine hydrochloride cation. Functionalized liposomes showed increased absorption in Caco-2 cells and gastroenteric absorption by *ex-vivo* as compared to regular liposomes. Pharmacodynamic investigations of orally delivered liposomes with folic acid demonstrated greater hypoglycemic impact and higher bioavailability of insulin when compared with subcutaneous insulin administration. More sustained hypoglycemia was observed in recombinant human insulin coated functionalized liposomes by oral delivery with fewer adverse effects than subcutaneous insulin. Liposomes may thus be an excellent candidate for delivering peptides (GLP-1, insulin) and additional blood-sugar-lowering drugs, allowing for improved hyperglycemic control via different administration methods (Souto et al., 2019).

9.9.2 *NIOSOMES*

Non-ionic surfactant vesicles are known as niosomes, made by combining an alkyl or dialkyl polyglycerol ether, non-ionic surfactant family with cholesterol and hydrating it in an aqueous medium (Mujoriya et al., 2011). Because surfactants have better chemical stability than phospholipids, which are utilized to make liposomes, they've been discovered to be more substantial as compared with the liposomal DDSs. The facile hydrolysis of phospholipids is the cause of their instability (A & B, 2014). Because they are non-ionic, they are less harmful, and they increase a drug's therapeutic index by limiting its effect on currently available drugs by novel anti-diabetic DDSs. Apart from that, niosomes are stable and osmotically active, and they exist on a nanometric size. In terms of structural properties (composition, fluidity, and size), these are adjustable, allowing them to be tailored to the needs of the scenario. Niosome surfactants are biodegradable and biocompatible (Mujoriya et al., 2011). Depending on the method required to produce them, niosomes can be unilamellar or multilamellar. They have an infrastructural structure made up of hydrophilic, amphiphilic, and lipophilic components. As a consequence, they can accommodate drug molecules with a wide range of solubility, which enhances the oral bioavailability of poorly absorbed medications, increases drug penetration via the skin, and so on. Furthermore, surfactants utilized in the synthesis of niosomes do not require specific handling or storage (Kauslya et al., 2021). More research into these structures could lead to a new medication delivery system. Niosomes have recently been discovered to be valuable in focused drug administration (Rajera et al., 2011). The number of bilayers, size, aqueous phase entrapment effectiveness, and vesicle membrane permeability is all affected by how niosomes are prepared.

Tamizharasi et al. (2009) used a lipid film hydration approach to generate and analyze gliclazide carrying niosomes utilizing non-ionic surfactants such as cholesterol and Span 60 in various quantities. Several formulas were developed and tested. All of the formulations were put through an *in vitro* release test. The best formulation had a long release time of up to 24 hours, hence it was chosen for *in vivo* testing. Finally, the niosomal formulation might be a potential gliclazide delivery strategy with enhanced bioavailability and a longer drug release profile (S et al., 2011).

Hasan et al. (2013) used Span 40 and cholesterol to construct and evaluate metformin hydrochloride-loaded niosomes. The reverse-phase evaporation process was employed in the synthesis of niosomes. The main goal of the research was to increase the bioavailability of metformin hydrochloride when taken orally. Negatively and positively charged vesicles were made with dicetyl phosphate and 1,2-dioleoyl-3-trimethylammonium-propane chloride salt, respectively. In the instance of positively charged niosomes, high entrapment efficiency and long-term drug release were achieved. Wistar rats were used to test bioavailability. The bioavailability of metformin hydrochloride solution was shown to be superior to that of free metformin hydrochloride solution (Hasan et al., 2013).

9.9.3 MICROSPHERE

Microspheres, also known as microscopic particles, are solid spherical particles with molecular sizes ranging from 1–1,000 μm. Drug particles are disseminated in structures made up of continuous phases of one or more miscible polymers at the molecular or microscopic level. They may be made from a variety of polymers and have a wide range of uses in DDSs. They are commonly utilized to prolong drug release while also reducing GIT discomfort. When compared to traditional dosage forms, these are more consistent. Microspheres can help to prevent unwanted polymeric material retention in the intestine, which can occur with matrix tablets when used on a long-term basis (Mathew Sam et al., 2008). Similarly, floating microspheres are utilized as a drug delivery method to extend the drug's duration in the stomach. Microspheres are a useful way to extend or maintain the DDS, allowing for less frequent dosage, which improves patient compliance. Several research in the literature detailed the development of microspheres for delivering drugs to the body and sustaining pharmacological activity. In this part, we'll go through the microspheres that were used to test several oral hypoglycemic medications.

Using the porous carrier, Jain et al. (2006) developed gastro-retentive floating microspheres of repaglinide. Eudragit S was the polymer utilized to make the microspheres. *In vitro* floating and drug release properties of the improved formulation were excellent. A gastric retention time of over 6 hours was obtained (Jain et al., 2006).

Glipizide microspheres were investigated and optimized by Phutane et al. (2010). Ethyl cellulose and Eudragit S100 were combined to make five major microsphere formulations. For all formulations, fragments size, density, flow properties, surface structure pattern, drug content, and *in-vitro* release of drug properties were examined. Ethyl cellulose and eudragit were combined in a 2:3 ratio in the improved formulation. Glipizide was encapsulated to maximum capacity and released continuously for 12 hours (Phutane et al., 2010). Patel et al. (2005) used chitosan loaded with glipizide to develop and test mucoadhesive microspheres. It was found that the optimum batch of microspheres had a more entrapment efficiency (75%). The release of the drug was also for up to 12 hours. In comparison to the control group, optimal formulation demonstrated a strong hypoglycemic impact (Patel et al., 2005).

9.9.4 TRANSDERMAL DRUG DELIVERY

An innovative way of administering a drug to the body through the skin is the transdermal DDS (Keservani et al., 2020a; Gujarathi et al., 2022). This approach provides several benefits over the oral route of drug administration, including the capacity to prevent the first-pass effect, maintain drug release, and adjust the features of biological barriers to absorption (Pathan & Setty, 2009). A transdermal patch is a medicated adhesive patch that is put to the skin and delivers a controlled amount of medication over time. There are three types of reservoir patches: reservoir patches, micro-reservoir patches, and matrix patches. Transdermal patches are made from a variety of polymers, including vinyl polymers, acrylic polymers, and cellulosic polymers (Kandavilli et al., 2002).

Anitha et al. (2011) developed and tested a glibenclamide and atenolol transdermal patch combination. Patches were made using various polymer ratios, including Carbopol, HPMC, and PVP. *In vitro* drug release, investigations, and physicochemical characterization were carried out. For additional *in vivo* experiments, the patch formulation, i.e., optimized (PVP 0.5%, polymer ratio HPMC 1%, and Carbopol 0.5%) was used. The interaction between *in-vitro* and *in-vivo* characteristics was found to be good. In the biological setting, it also generated repeatable outcomes. Many studies have reported on the development of transdermal patches for drugs. Furthermore, various patches for anti-diabetic drugs have been observed (Anitha et al., 2011).

Microcapsules and transdermal patches of glibenclamide were developed, evaluated, and reviewed by Mishra et al. (2009). Different physicochemical characteristics, as well as *in vitro* and *in vivo* tests, were assessed for both types of formulations. Transdermal patches outperformed oral microcapsules in terms of improved management of hyperglycemia, longer plasma half-life (9.6 ± 1.2 hours), and greater bioavailability, according to the findings of the testing (Mishra et al., 2009).

Transdermal patches containing the medication rosiglitazone maleate were developed and characterized by Hasan et al. (2010). A polymer combination of HPMC and PVA was used to make these patches. Characteristic parameters were investigated both *in vitro* and *in vivo*. *In vitro* experiments showed that 81% of the medication was released in 12 hours. Normal rats developed hypoglycemia upon using the patch, whereas diabetic rats' blood glucose levels restored to normal after 16 hours (Hasan et al., 2011).

9.9.5 NANOPARTICLES

Drugs can be loaded into nanoparticles (Keservani et al., 2017a; Sharma et al., 2018) to increase their stability and protect them against enzymatic and/or chemical destruction in the GI tract. Nanoparticles also promote interaction with the GI epithelium, resulting in higher absorption of medicine and bioavailability. Drugs can be encapsulated in nanoparticles or bonded to their surfaces, and they must be released near the absorption site. Nanoparticles may be employed to boost drug absorption outside of the cell. For epithelial endocytosis, nanoparticles having hydrophobic surfaces are favorable. Cationic nanoparticles interact with the negatively charged mucus layer, preventing them from being absorbed. By creating mucins and antimicrobial peptides, the secretory goblet and Paneth cells protect and cure the intestine. Insulin nanoparticles taken orally must penetrate through mucus and epithelial barriers. While neutral and hydrophilic surface nanoparticles are better for overcoming mucus, their interface with epithelial cells may be compromised (Souto et al., 2019).

9.10 CONCLUSION

The cases of DM are extensively increasing annually, and it affects the economy as well as the extensive health of the people in the world. It is ranked as the fifth leading cause of mortality. There are so many drugs available for the treatment of DM, but these drugs are having some drawbacks such as lesser bioavailability, stability, etc., therefore, there is a need to develop an NDDS to overcome the problems that are associated with conventional drug delivery. Recent advancements in drug delivery such as liposomes, niosomes, nanoparticles, microspheres, etc., play a major role in increasing bioavailability and stability and minimizing the problems of conventional drug delivery. The exceptional potential of providing physical stability, sustained, and site-specific drug delivery for a predetermined period can introduce new opportunities for precise, safe, and effective treatment of diabetes in the current scientific landscape. NDDSs are being acknowledged for their tangible benefits, which indeed validate the statement 'NDDSs for diabetes: an immense hope' as true and meaningful.

ACKNOWLEDGMENT

The authors are thankful to respected Management, SRES Sanjivani College of Pharmaceutical Education and Research, Kopargaon, and Principal Dr. Kishor S Salunkhe for their timely support and encouragement.

CONFLICT OF INTERESTS

All authors declare that they have no conflicts of interest.

KEYWORDS

- **conventional drug delivery**
- **diabetes mellitus**
- **gestational diabetes**
- **liposomes**
- **microsphere**
- **monogenic types**
- **nanoparticles**
- **nanotechnology**
- **niosomes**
- **novel drug delivery systems**
- **targeted drug delivery**

REFERENCES

Ahire, E. D., Talele, S. G., & Shah, H. S., (2020). Nanoparticles as a promising technology in microbial pharmaceutics. In: *Applied Pharmaceutical Science and Microbiology* (pp. 133–158). Apple Academic Press.

Ahire, E., Thakkar, S., Borade, Y., & Misra, M., (2020). Nanocrystal based orally disintegrating tablets as a tool to improve dissolution rate of Vortioxetine. *Bulletin of Faculty of Pharmacy, Cairo University*, *58*(1, 2), 11–20.

Ahire, E., Thakkar, S., Darshanwad, M., & Misra, M., (2018). Parenteral nanosuspensions: A brief review from solubility enhancement to more novel and specific applications. *Acta Pharmaceutica Sinica B*, *8*(5), 733–755.

Ahirrao, S. P., Sonawane, M. H., Bhambere, D. S., Udavant, P. B., Ahire, E. D., & Kanade, R., (2022). Cocrystal formulation: A novel approach to enhance solubility and dissolution of etodolac. *Biosciences Biotechnology Research Asia*, *19*(1), 111.

Anitha, P., Ramkanth, S., Saleem, M. T. S., Umasankari, K., Reddy, B. P., & Chetty, M., (2011). Preparation, *in-vitro*, and *in-vivo* characterization of transdermal patch containing glibenclamide and atenolol: A combinational approach. *Pakistan Journal of Pharmaceutical Sciences, 24*(2), 155–163.

Atkinson, M. A., Eisenbarth, G. S., & Michels, A. W., (2014). Type 1 diabetes. *Lancet (London, England), 383*(9911), 69–82. https://doi.org/10.1016/S0140-6736(13)60591-7.

Bariya, S. H., Gohel, M. C., Mehta, T. A., & Sharma, O. P., (2012). Microneedles: An emerging transdermal drug delivery system. *The Journal of Pharmacy and Pharmacology, 64*(1), 11–29. https://doi.org/10.1111/J.2042-7158.2011.01369.X.

Benedict, C., Hallschmid, M., Schmitz, K., Schultes, B., Ratter, F., Fehm, H. L., Born, J., & Kern, W., (2007). Intranasal insulin improves memory in humans: Superiority of insulin aspart. *Neuropsychopharmacology: Official Publication of the American College of Neuropsychopharmacology, 32*(1), 239–243. https://doi.org/10.1038/SJ.NPP.1301193.

Born, J., Lange, T., Kern, W., McGregor, G. P., Bickel, U., & Fehm, H. L., (2002). Sniffing neuropeptides: A transnasal approach to the human brain. *Nature Neuroscience, 5*(6), 514–516. https://doi.org/10.1038/NN849.

Campisi, G., Paderni, C., Saccone, R., Fede, O., Wolff, A., & Giannola, L., (2010). Human buccal mucosa as an innovative site of drug delivery. *Current Pharmaceutical Design, 16*(6), 641–652. https://doi.org/10.2174/138161210790883778.

Chan, J., & Cheng-Lai, A., (2017). Inhaled insulin: A clinical and historical review. *Cardiology in Review, 25*(3), 140–146. https://doi.org/10.1097/CRD.0000000000000143.

Chatterjee, S., Khunti, K., & Davies, M. J., (2017). Type 2 diabetes. *The Lancet, 389*(10085), 2239–2251. https://doi.org/10.1016/S0140-6736(17)30058-2.

Chen, Y., Shan, X., Luo, C., & He, Z., (2020). Emerging nanoparticulate drug delivery systems of metformin. *Journal of Pharmaceutical Investigation, 50*(3), 219–230. https://doi.org/10.1007/s40005-020-00480-1.

Chiou, G. C. Y., Chuang, C. Y., & Chang, M. S., (1989). Systemic delivery of insulin through eyes to lower the glucose concentration. *Journal of Ocular Pharmacology, 5*(1), 81–91. https://doi.org/10.1089/JOP.1989.5.81.

Diabetes Mellitus Type 2 – StatPearls – NCBI Bookshelf. (n.d.). from https://www.ncbi.nlm.nih.gov/books/NBK513253/ (accessed on 6 December 2023).

DiMeglio, L. A., Evans-Molina, C., & Oram, R. A., (2018). Type 1 diabetes. *Lancet (London, England), 391*(10138), 2449. https://doi.org/10.1016/S0140-6736(18)31320-5.

Gautam, S. P., Keservani, R. K., Gautam, T., Gupta, A. K., & Sharma, A. K., (2015). An alternative approach for acetylation of amine terminated polyamidoamine (PAMAM) dendrimer. *ARS Pharmaceutica, 56*(3), 155–159.

Grover, M., & Utreja, P., (2014). Recent advances in drug delivery systems for anti-diabetic drugs: A review. *Current Drug Delivery, 11*(4), 444–457. https://doi.org/10.2174/1567201811666140118225021.

Gujarathi, N. A., Patil, T., Rane, B., Babu, A., & Keservani, R. K., (2022). Micro- and nano-systems IN transdermal drug delivery system; Chapter 2. In: Nayan, A. G., Juliana, P. A., Raj, K. K., & Anil, K. S., (eds.), *Topical and Transdermal Drug Delivery Systems: Applications and Future Prospects* (pp. 41–78). Apple Academic Press, Taylor & Francis. ISBN: 9781774910702.

Hasan, A. A., Madkor, H., & Wageh, S., (2013). Formulation and evaluation of metformin hydrochloride-loaded niosomes as controlled release drug delivery system. *Drug Delivery, 20*(3, 4), 120–126. https://doi.org/10.3109/10717544.2013.779332.

Hasan, Md. K., Rahman, Md. A., Shahin, S., & Islam, Md. A. U., (2011). *In Vitro and In Vivo Evaluation of a Rosiglitazone Maleate-loaded HPMC-PVA Blend Patch. 3*(3), 2011, 11–15.

Henkin, R. I., (2010). Inhaled insulin-intrapulmonary, intranasal, and other routes of administration: Mechanisms of action. *Nutrition (Burbank, Los Angeles County, Calif.), 26*(1), 33–39. https://doi.org/10.1016/J.NUT.2009.08.001.

International Diabetes Federation, (2021). *IDF Diabetes Atlas* (10th edn.). International Diabetes Federation. www.diabetesatlas.org (accessed on 6 December 2023).

Jain, S. K., Agrawal, G. P., & Jain, N. K., (2006). A novel calcium silicate-based microspheres of repaglinide: *In vivo* investigations. *Journal of Controlled Release: Official Journal of the Controlled Release Society, 113*(2), 111–116. https://doi.org/10.1016/J.JCONREL.2006.04.005.

Kamei, N., Nielsen, E. J. B., Khafagy, E. S., & Takeda-Morishita, M., (2013). Noninvasive insulin delivery: The great potential of cell-penetrating peptides. *Therapeutic Delivery, 4*(3), 315–326. https://doi.org/10.4155/TDE.12.164.

Kandavilli, S., Nair, V., & Panchagnula, R., (2002). Polymers in transdermal drug delivery systems. *Pharmaceutical Technology North America, 26*(5), 62–80. https://doi.org/10.5360/membrane.19.75.

Kauslya, A., Borawake, P. D., Shinde, J. V., & Chavan, R. S., (2021). Niosomes: A novel carrier drug delivery system. *Journal of Drug Delivery and Therapeutics, 11*(1), 162–170. https://doi.org/10.22270/jddt.v11i1.4479.

Keservani, R. K., & Gautam, S. P., (2020). Formulation and evaluation of baclofen liposome vesicles using lecithin. *ARS Pharmaceutica, 61*(3), 175–180.

Keservani, R. K., & Gautam, S. P., (2022). Skeletal muscle relaxant activity of different formulation of span 60 niosome. *ARS Pharmaceutica, 63*(1), 32–44.

Keservani, R. K., Bandopadhyay, S., Bandyopadhyay, N., & Sharma, A. K., (2020a). Design and fabrication of transdermal/skin drug-delivery system. In: *Drug Delivery Systems* (pp. 131–178). Academic Press.

Keservani, R. K., Kesharwani, R. K., & Sharma, A. K., (2017a). Introduction to nanotechnology in drug delivery. In: *Drug Delivery Approaches and Nanosystems* (pp. 1–19). Apple Academic Press. ISBN: 9781771885836.

Keservani, R. K., Sharma, A. K., & Kesharwani, R. K., (2018). Nanoparticulate nanocarriers in drug delivery. In: *Nanobiomaterials* (pp. 3–22). Apple Academic Press. ISBN: 9781771885911.

Keservani, R. K., Sharma, A. K., & Ramteke, S., (2010). Novel vesicular approach for topical delivery of baclofen via niosomes. *Lat. Am. J. Pharm., 29*, 1364–1370.

Khan, M. Z. I., Prebeg, Ž., & Kurjaković, N., (1999). A pH-dependent colon targeted oral drug delivery system using methacrylic acid copolymers. I. Manipulation of drug release using Eudragit L100-55 and Eudragit S100 combinations. *Journal of Controlled Release: Official Journal of the Controlled Release Society, 58*(2), 215–222. https://doi.org/10.1016/S0168-3659(98)00151-5.

Kochba, E., Levin, Y., Raz, I., & Cahn, A., (2016). Improved insulin pharmacokinetics using a novel microneedle device for intradermal delivery in patients with type 2 diabetes. *Diabetes Technology & Therapeutics, 18*(9), 525–531. https://doi.org/10.1089/DIA.2016.0156.

Kraan, H., Vrieling, H., Czerkinsky, C., Jiskoot, W., Kersten, G., & Amorij, J. P., (2014). Buccal and sublingual vaccine delivery. *Journal of Controlled Release: Official Journal of the Controlled Release Society, 190*, 580–592. https://doi.org/10.1016/J.JCONREL.2014.05.060.

Krauland, A. H., Guggi, D., & Bernkop-Schnürch, A., (2004). Oral insulin delivery: The potential of thiolated chitosan-insulin tablets on non-diabetic rats. *Journal of Controlled Release: Official Journal of the Controlled Release Society, 95*(3), 547–555. https://doi.org/10.1016/J.JCONREL.2003.12.017.

Kullmann, S., Veit, R., Peter, A., Pohmann, R., Scheffler, K., Häring, H. U., Fritsche, A., et al., (2018). Dose-dependent effects of intranasal insulin on resting-state brain activity. *The Journal of Clinical Endocrinology and Metabolism, 103*(1), 253–262. https://doi.org/10.1210/JC.2017–01976.

Li, J., Yang, L., Ferguson, S. M., Hudson, T. J., Watanabe, S., Katsuma, M., & Fix, J. A., (2002). *In vitro* evaluation of dissolution behavior for a colon-specific drug delivery system (CODES) in multi-pH media using United States pharmacopeia apparatus II and III. *AAPS PharmSciTech, 3*(4). https://doi.org/10.1208/PT030433.

Mahesh, D., Laxman, K., Vandana, N., & Rohini, D., (2016). Review on diabetes mellitus. *Inter. J. of Pharmacotherapy, 6*(1), 24–45.

Mastrandrea, L. D., (2010). Inhaled insulin: Overview of a novel route of insulin administration. *Vascular Health and Risk Management, 6*(1), 47–58. https://doi.org/10.2147/VHRM.S6098.

Mathew, S. T., Devi, G. S., Prasanth, V. V., & Vinod, B., (2008). NSAIDs as microspheres. *The Internet Journal of Pharmacology, 6*(1), 332–338.

Meng, Q., Hu, H., Zhou, L., Zhang, Y., Yu, B., Shen, Y., & Cong, H., (2019). Logical design and application of prodrug platforms. *Polymer Chemistry, 10*(3), 306–324. https://doi.org/10.1039/C8PY01160E.

Mishra, M. K., Ray, D., & Barik, B. B., (2009). Microcapsules and transdermal patch: A comparative approach for improved delivery of antidiabetic drug. *AAPS PharmSciTech, 10*(3), 928–934. https://doi.org/10.1208/S12249–009–9289-Z.

Misra, P., Upadhyay, R. P., Misra, A., & Anand, K., (2011). A review of the epidemiology of diabetes in rural India. *Diabetes Research and Clinical Practice, 92*(3), 303–311. https://doi.org/10.1016/j.diabres.2011.02.032.

Mujoriya, R. Z., & Bodla, R. B., (2014). *Niosomes – the Challenge in Preparation for Pharmaceutical Scientist Academic Sciences. 13*(2), 60–63.

Mujoriya, R., Bodla, R. B., Dhamande, K., Singh, D., & Patle, L., (2011). Niosomal drug delivery system: The magic bullet. *Journal of Applied Pharmaceutical Science, 1*(9), 20–23.

Olokoba, A. B., Obateru, O. A., & Olokoba, L. B., (2012). Type 2 diabetes mellitus: A review of current trends. *Oman Medical Journal, 27*(4), 269–273. https://doi.org/10.5001/omj.2012.68.

Patel, J. K., Patel, R. P., Amin, A. F., & Patel, M. M., (2005). Formulation and evaluation of mucoadhesive glipizide microspheres. *AAPS PharmSciTech, 6*(1). https://doi.org/10.1208/PT060110.

Pathan, I. B., & Setty, C. M., (2009). Chemical penetration enhancers for transdermal drug delivery systems. *Tropical Journal of Pharmaceutical Research, 8*(2), 173–179. https://doi.org/10.4314/tjpr.v8i2.44527.

Petersmann, Müller-Wieland, D., Müller, U. A., Landgraf, R., Nauck, M., Freckmann, G., Heinemann, L., & Schleicher, E., (2019). Definition, classification, and diagnosis of diabetes mellitus authors definition of diabetes mellitus diagnostic criteria of diabetes mellitus. *German Diabetes Association, 127*(Suppl 1), 51–57.

Phutane, P., Shidhaye, S., Lotlikar, V., Ghule, A., Sutar, S., & Kadam, V., (2010). *In vitro* evaluation of novel sustained release microspheres of glipizide prepared by the emulsion solvent diffusion-evaporation method. *Journal of Young Pharmacists: JYP*, *2*(1), 35–41. https://doi.org/10.4103/0975-1483.62210.

Rai, V. K., Mishra, N., Agrawal, A. K., Jain, S., & Yadav, N. P., (2016). Novel drug delivery system: An immense hope for diabetics. *Drug Delivery*, *23*(7), 2371–2390. https://doi.org/10.3109/10717544.2014.991001.

Rajera, R., Nagpal, K., Singh, S. K., & Mishra, D. N., (2011). Niosomes: A controlled and novel drug delivery system. *Biological and Pharmaceutical Bulletin*, *34*(7), 945–953. https://doi.org/10.1248/bpb.34.945.

Rys, P. M., Ludwig-Slomczynska, A. H., Cyganek, K., & Malecki, M. T., (2018). Continuous subcutaneous insulin infusion vs multiple daily injections in pregnant women with type 1 diabetes mellitus: A systematic review and meta-analysis of randomized controlled trials and observational studies. *European Journal of Endocrinology*, *178*(5), 545–563. https://doi.org/10.1530/EJE-17–0804.

S, T., A, D., V, R., & JC, R., (2011). Development and characterization of niosomal drug delivery of gliclazide. *Journal of Young Pharmacists*, *1*(3), 205. https://doi.org/10.4103/0975–1483.57065.

Schmid, V., Kullmann, S., Gfrörer, W., Hund, V., Hallschmid, M., Lipp, H. P., Häring, H. U., et al., (2018). Safety of intranasal human insulin: A review. *Diabetes, Obesity & Metabolism*, *20*(7), 1563–1577. https://doi.org/10.1111/DOM.13279.

Schöpf, V., Kollndorfer, K., Pollak, M., Mueller, C. A., & Freiherr, J., (2015). Intranasal insulin influences the olfactory performance of patients with smell loss, dependent on the body mass index: A pilot study. *Rhinology*, *53*(4), 371–378. https://doi.org/10.4193/RHINO15.065.

Sharma, A. K., Keservani, R. K., & Keseharwani, R. K., (2018). *Nanobiomaterials: Applications in Drug Delivery.* CRC Press. ISBN:9781771885911.

Singh, N., Kesharwani, R., Tiwari, A., & Patel, D., (2016). A review on diabetes mellitus. *The Pharma Innovation*, *5*(7), 36–40.

Sonia, T. A., & Sharma, C. P., (2015). *Oral Delivery of Insulin*, 1–345. https://doi.org/10.1016/C2013-0-18177-9.

Souto, E. B., Souto, S. B., Campos, J. R., Severino, P., Pashirova, T. N., Zakharova, L. Y., Silva, A. M., et al., (2019). Nanoparticle delivery systems in the treatment of diabetes complications. *Molecules*, *24*(23), 1–28. https://doi.org/10.3390/molecules24234209.

Subramani, K., (2006). Applications of nanotechnology in drug delivery systems for the treatment of cancer and diabetes. *International Journal of Nanotechnology*, *3*(4), 557–580. https://doi.org/10.1504/IJNT.2006.011179.

Talele, S. G., Ahire, E. D., Talele, G. S., & Derle, D. V., (2021). An innovative approach as self-emulsifying drug delivery system for phytoconstituents. In: *Enhancing the Therapeutic Efficacy of Herbal Formulations* (pp. 69–84). IGI Global.

Thombre, N. A., Niphade, P. S., Ahire, E. D., & Kshirsagar, S. J., (2022). Formulation development and evaluation of microemulsion based lornoxicam gel. *Biosciences Biotechnology Research Asia*, *19*(1), 69–80.

Unnikrishnan, R., Anjana, R. M., & Mohan, V., (2016). Diabetes mellitus and its complications in India. *Nature Reviews Endocrinology*, *12*(6), 357–370. https://doi.org/10.1038/nrendo.2016.53.

World Health Organization, (2019). *Classification of Diabetes Mellitus*. htttp://www.who.int/health-topics/diabetes (accessed on 6 December 2023).

Zaric, B. L., Obradovic, M., Sudar-Milovanovic, E., Nedeljkovic, J., Lazic, V., & Isenovic, E. R., (2019). Drug delivery systems for diabetes treatment. *Current Pharmaceutical Design, 25*(2), 166–173. https://doi.org/10.2174/1381612825666190306153838.

Zhang, X., Qi, J., Lu, Y., Hu, X., He, W., & Wu, W., (2014). Enhanced hypoglycemic effect of biotin-modified liposomes loading insulin: Effect of formulation variables, intracellular trafficking, and cytotoxicity. *Nanoscale Research Letters, 9*(1), 185. https://doi.org/10.1186/1556-276X-9-185.

Zhao, R., Lu, Z., Yang, J., Zhang, L., Li, Y., & Zhang, X., (2020). Drug delivery system in the treatment of diabetes mellitus. *Frontiers in Bioengineering and Biotechnology, 8*, 1–16. https://doi.org/10.3389/fbioe.2020.00880.

CHAPTER 10

Novel Drug Delivery Systems in the Management of Asthma and Pulmonary Diseases

MANGESH A. BHUTKAR,[1] DHEERAJ S. RANDIVE,[1]
SOMNATH D. BHINGE,[2] GANESH H. WADKAR,[3] and
CHANDRAKANTSING V. PARDESHI[4]

[1]*Department of Pharmaceutics, Rajarambapu College of Pharmacy, Kasegaon, Maharashtra, India*

[2]*Department of Pharmaceutical Chemistry, Rajarambapu College of Pharmacy, Kasegaon, Maharashtra, India*

[3]*Department of Pharmacognosy, Rajarambapu College of Pharmacy, Kasegaon, Maharashtra, India*

[4]*Department of Pharmaceutics, R. C. Patel Institute of Pharmaceutical Education and Research, Shirpur, Maharashtra, India*

ABSTRACT

Novel pulmonary drug delivery systems (DDSs) are of remarkable concern in the field of medicine and drug delivery owing to the fact that a greater number of drugs can be effectively used via the lung; both for localized effect and for its systemic action. Owing to their unique properties, these DDSs have acquired considerable interest over the past decades. Furthermore, the shortcomings of conventional drug therapy, such as extreme drug clearance exhibited by the macrophages of the alveoli, drug deposition of inhaled formulations, rapid drug degradation, and limited control over the particle deposition, can be effectively overcome by employing the principles

Novel Drug Delivery Systems in the Management of Chronic Diseases.
Raj K. Keservani, Umesh D. Laddha, Eknath D. Ahire, and Sanjay J. Kshirsagar (Eds.)

of novel pulmonary drug delivery. The present chapter critically focuses on the significant contributions to enhance the efficiency of drug for its prompt and targeted release to the lungs so as to achieve effective treatment of several respiratory diseases.

10.1 CHRONIC RESPIRATORY DISORDERS

Chronic respiratory diseases are one of the most prevalent noncommunicable diseases of the airways and other structures of the lung internationally. Inhalational exposures from work and behavioral patterns are common due to the proliferation of toxic and harmful environmental, occupational, as well as other factors (James et al., 2017). In addition to chronic obstructive pulmonary disease (COPD) and asthma, it includes interstitial lung disease, respiratory sarcoidosis, and pneumoconiosis like silicosis and asbestosis. Economies of several nations are adversely affected owing to their ever-rising expenditures for respiratory diseases and disorders.

In 2019, the treatment of people with chronic respiratory diseases solely cost the EU's 28 member states approximately €380 billion per year (https://www.erswhitebook.org/chapters/the-economic-burden-of-lung-disease/). Chronic respiratory diseases, on the other hand, have received very little public attention and funding for research than other disease categories such as heart disease, stroke, cancer, Alzheimer's disease (AD), and diabetes (Boehm et al., 2019; Gross et al., 1999). As a result, knowing the regional and international prevalence, morbidity, and mortality of chronic respiratory disease is crucial for particular events preventive, treatment screening, along with research activities.

10.2 COMMON DISEASES RESPONSIBLE FOR RESPIRATORY IMPAIRMENT AND ILLNESS

Any disease or disorder of the airways and lungs that affects human respiration is referred to as a respiratory disease and it includes acute respiratory distress syndrome (ARDS) and COPD. Out of them, ARDS can be a life-threatening condition. Injuries, infections, as well as other exposures cause inflammation and fluid accumulation inside the alveolar air sacs, obscuring gas exchange and resulting in hypoxemia. A low percentage of oxygen in the blood hampers organ functioning, especially the brain, kidneys, and heart, being the most vulnerable. Patients suffering from ARDS frequently require ventilators, which can further aggravate the inflammatory reaction (Uhlig et al., 2004). COPD is a group of disorders that includes inflammation (swelling) and irritation of the bronchial tubules (chronic bronchitis (CB)), damage to the alveoli that leads to their rupture (emphysema), and asthma, which together weaken ventilation. COPD is frequently associated with inflammation of the small airways that also prevents air from reaching the alveoli and the successive gas exchange from occurring. It is a serious ailment marked by irreparable airway obstruction and persistent inflammation in response to toxic and harmful environmental stimuli, most widely cigarette smoke. COPD comprises CB on one end and emphysema on the other, with most patients having some properties of both. CB has a number of clinical implications, including an enhanced exacerbation rate, a rapid decline in lung function, a lower health-related quality of life (HRQoL), and possibly an increased risk of mortality (Vestbo et al., 1996).

Emphysema is a lung condition associated with abnormal, irreversible expansion of air sacs followed by wall destruction, but without apparent fibrosis. Destruction is defined as a lack of uniformity in the structure of respiratory airspace expansions, such that the organized appearance of the acinus and its components has been disrupted and may have been lost (Karlinsky & Snider, 1978).

Asthma, a condition that causes narrowing and swelling of the airways, involves many cells and cellular elements, including mast cells, T lymphocytes, eosinophils, macrophages, epithelial cells, and neutrophils. This inflammation causes frequent episodes of wheezing, breathlessness, coughing, and chest tightness in the affected individuals, especially during bedtime or in the early hours of the day. These incidences of attack are typically accompanied by numerous but variable airflow obstruction, which is frequently reversible, whether instantaneously or with the aid of drug(s) (https://www.nhlbi.nih.gov/sites/default/files/media/docs/EPR).

10.2.1 ASTHMA

Asthma is among the most prevalent diseases in the world, resulting in a significant disease burden. Although asthma-related deaths have decreased significantly over the last 25 years, no potential treatment regimens can heal asthma, and the burden of asthma will proceed to be led by growing incidence. Because the causes of the rise in prevalence of asthma have not been identified, opportunity to build, targeted primary preventative measures are limited. Although relationships have been reported between a range of different risk factors and childhood asthma, documentation of causality from observational studies is inherently difficult, and very few risk factors have been evaluated in primary prevention studies. In the meantime, public health efforts should continue to focus on measures that have the potential to increase lung and overall health such as reduce tobacco smoking and environmental tobacco smoke exposure; reduce interior and exterior air pollution and occupational exposures; decrease childhood obesity and promote a diet high in vegetables and fresh fruits (Beasley et al., 2015).

It is the most common chronic lung disease, with Great Britain exhibiting a record number of cases considered universally. It affects one in every five children and one in every 12 adults. Exacerbation, in which disease control deteriorates over several days, is a significant component of asthma, accounting for the majority of deaths. Following the development of PCR-based detection systems for human rhinoviruses (RV), we used translational cohorts to first determine the role of respiratory viruses in causing asthma symptoms in both children (>85%) and adults (>65%), with RV major and minor subtypes predominating (Holgate, 2013).

Asthma diagnosis is dependent on what we imply by the term. Its definition is still debatable because there is no single environmental or genetic reason. Inflamed airways, symptoms, variable airflow limitation, and chronic airflow limitation are the disease components addressed from a descriptive standpoint (Hargreave & Nair, 2009).

Before analyzing the addition of biologics, the patient's asthma must be detected. Although this appears to be an extremely simple requirement, when careful evaluations are performed, a substantial percentage of patients considered to have severe asthma do not have severe asthma or even asthma. Medication non-adherence is popular in any disease, such as asthma, and a lack of consistent medication use gives rise to ongoing indications and further threat of worsening of the condition (Busse, 2019).

10.2.2 COPD

Airflow congestion is a symptom of COPD that is associated with inflammation (swelling) and irritation of the bronchial tubules (CB), damage of the alveoli that leads to their rupture (emphysema). Other symptoms include airway hyperreactivity, persistent cough, sputum formation, shortness of breath (SOB), intolerable exercise, wheezing, and poor quality of life (Liu et al., 2015). Neutrophilic inflammation occurs in the airways, mostly in COPD. Parenchymal damage is a vital irreversible characteristic and prone to airflow obstruction through vibrant compression. The corticosteroids can be used to suppress eosinophilic inflammation in asthma, but they lack significant effect in case of COPD, consistent with a failure of long treatment of corticosteroids to modify the development of COPD (Peter, 2000).

10.2.2.1 PATHOPHYSIOLOGY OF COPD

It is a treatable and preventable disease, clinically heterogeneous syndrome with leading symptoms of respiratory tract and different systemic consequences. In general, COPD is a lingering inflammatory reaction of a genetically prone subject to long-term inhalation exposure to, injurious particles, air pollution and various gases, which progresses in time. The pulmonary element is connected with respiratory airflow constraint, which is not completely reversible. The airflow restriction in this disease develops progressively as an effect of chronic, mainly non-infectious inflammation of the airways and lung parenchyma, to which a contributing factor is both the innate as well as acquired immunity. Resistance along with impending collapse of peripheral airways is because of elastic characteristics of lungs. Residual volume (RV) is characteristically augmented at rest in patients with COPD and furthermore dynamic hyperinflation occurs during physical activities, which is connected with amplified expiratory effort.

10.2.2.2 EPIDEMIOLOGY

COPD is a recurrent cause of death, rise morbidity and decreased standard of life. The disease's worldwide incidence and prevalence are increasing as the population ages and air pollution worsens. COPD affects at least 4–6% of the EU population. In the Czech Republic, the prevalence has recently been estimated to be 7–8%; nearly 16,000 patients are hospitalized every year as a result of COPD, and over 2,500 die each year.

10.2.2.3 RISK FACTORS

Cigarette smoking remains the main prominent cause for COPD. COPD smokers have more severe symptoms, a higher mortality rate, and faster degradation of respiratory function than COPD nonsmokers. Besides active smoking, which is able to account for 70–80% of COPD cases, the following are also generally regarded as risk factors: environmental tobacco smoke – ETS, cigar, smoking, cannabis, pipe or water pipe smoking; long time exposure to pollution from industry, long term inhalation of vehicle smoke, biomass fumes as well as burnt fossil fuel, low weight at birth and regular infections throughout childhood. The disease's progression is determined by the particular interaction of genetic and epigenetic factors, as well as environmental effects (Vladimir et al., 2013). COPD is primarily caused by smoking cigarettes and is characterized by restricted airflow and chronic inflammation of the lungs. In COPD, the most common inflammatory cells in the lungs are alveolar macrophages, neutrophils, and cytotoxic T-lymphocytes (CD-8+ cells). An increased incidence of the disease, which eventually result in bursts of systemic inflammatory markers released into the systemic circulation, is a possible significant source of systemic inflammation in COPD. Airflow obstruction causes air entrapment and hyperinflation, which leads to dyspnea, functional limitation, and poor HRQoL. Dyspnea can then lead to anxiety and distress, which increases tachypnea and, as a result, air trapping. Systemic inflammation in COPD is a reason of concern in osteoporosis. COPD has been linked to higher circulating levels of IL-6 and TNP-alpha, both of which are potent promoters of osteoclast generation and activity (Decramer et al., 2008).

10.2.3 TUBERCULOSIS (TB)

Tuberculosis (TB) is an infectious disease spread through the air for which the acid-fast pathogenic Mycobacterium TB bacteria are responsible. Even though *M. tuberculosis* is mainly a pulmonary pathogen, it can cause infection in nearly every other body part. *M. tuberculosis* infection can progress from a state of restraint in the host wherein, the pathogenic microorganism remains alienated within granulomas which is referred as latent TB infection, to an infectious state which is characterized by symptoms such as persistent cough, sweating in the nights and reduction in body weight of patients. Treatment typically necessitates a lengthy course of multiple antimicrobials, prompting efforts to create shorter drug regimens. Even though the Bacillus Calmette–Guérin (BCG) vaccine is widely used around the world, primarily to prevent life-threatening TB in babies and young children, it's been ineffective in attempting to control the global TB epidemic. Isoniazid, rifampicin, pyrazinamide, and ethambutol are the four first-line antimicrobials used to treat TB. Resistance to every drug is entirely feasible.

Indeed, multidrug-resistant tuberculosis (MDR-TB), defined as *M. tuberculosis* isoniazid and rifampicin resistance, is a well-known matter of serious concern confronted by nearly every country across the globe. Extensively drug-resistant TB, which causes more severe disease resistance to at least manifestations, is resistant not only to isoniazid and rifampicin, and also to any fluoroquinolone or any other of the three injectable second-line amino glycosides. Diagnostic and therapeutic options differ between latent TB infection and active TB, as well as between drug-sensitive and drug-resistant TB disease (Pai, 2020).

According to the WHO, an approximate 9.6 million people have developed active TB in 2014, with almost 2.0 million patients succumbed to the disease. The burden of TB is unevenly distributed. For example, the incidence of TB in South Africa is more than 250 times higher (834 cases per lakh population annually) than those in the US (reportedly exhibits a mere of only 3 cases per lakh of population every year). Rates of developing active TB are indeed very significant in exposed infants, but so much lower in children aged 2–10 years; risk then rises during teenage years and plateaus around the age of 25, remaining high all through adult life. Active TB disease affects men roughly twice as much as women, and children account for roughly 10% of all new cases in the world. Globally, TB disease is caused by the continuing spread of pathogenic mycobacterium and reactivation of LTBI. Greater number of TB incidences are reported by *M. tuberculosis* (sensustricto) or the closely linked pathogen *Mycobacterium africanum*; whereas the spread of disease caused by zoonotic *M. tuberculosis* complex members is not much common. Mycobacterium bevies and *Mycobacterium caprae* are two examples.

M. tuberculosis has no significant environmental reservoir; its only known reservoir is humans. Due to which, *M. tuberculosis* is a pathogen as well as a symbiont, which has profound consequences of host–pathogen interactions (McShane, 2011). To facilitate development of an efficient anti-TB vaccine, it is necessary to know the nature of protective immunity. To provide an optimum protection against mycobacterium an integral and strong cellular immunity is required and thus, the new TB vaccines under development are aimed at imparting greater degrees of cellular immune effect (Bussi & Gutierrez, 2019). The risk of development to infection and disease are two distinct factors that must be considered when developing TB control techniques. Exogenous factors primarily govern the risk of infection following TB exposure, which is ascertained by an intrinsic mixture of the infectiousness of the source case, proximity to contact, and social and behavioral risk factors such as smoking, alcohol, and indoor pollutants that deteriorate quality of air. Transmission will be high in settings where there is a greater chance of social mixing (along with overcrowding) (Narasimhan et al., 2013).

10.2.4 LUNG CANCER (LC)

Lung cancer (LC), also recognized as lung carcinoma, is a type of malignant tumor characterized by uncontrolled cell growth in lung tissues. It is essential to treat this in order to prevent it from spreading to certain other areas of the body via metastasis. The majority of LCs are carcinomas. Tobacco use over a long period of time is the primary cause of 85% of LCs. About 10–15% of the cases occur in persons who have never smoked but are exposed to secondhand smoke, asbestos, and radon gas as a result of increased pollutants in the air, thereby causing deterioration of its quality. Computer tomography (CT) and radiographs are some of the conventional techniques employed for detection of carcinomas of the lung (Asuntha et al., 2016). Lung carcinomas are categorized as small-cell lung carcinomas and non-small-cell lung carcinomas amongst which the latter form is more common which accounts for more than 85% of the entire cases and are further classified as squamous-cell carcinoma, adenocarcinoma, and large-cell carcinoma. Squamous cell carcinoma accounts for 25–30% of all LC cases and develops from squamous cells in the airway epithelium of the bronchial tubes in the center of the lungs. Cigarette smoking is strongly linked to this subtype of non-small cell lung carcinoma.

Adenocarcinoma is the most frequent kind of LC, accounting for roughly 40% of all cases. It is made up of type II alveolar cells, which are tiny airway epithelial cells that release mucus and other chemical substances. Adenocarcinoma is the most frequent type of LC in men and women of all ages, including smokers, passive smokers, and non-smoking persons. In contrast to other carcinomas, this type of lung carcinoma tends to transpire on the peripheral region of the lung. Adenocarcinoma grows more slowly and is more likely to be discovered before it spreads beyond the lungs. LCs that are large cell (undifferentiated) account for 5–10% of all malignancies. Large cell carcinoma usually starts in the middle area of the lungs and spreads to adjacent lymph nodes, the chest, and the other organs and tissues (Zappa & Mousa, 2016).

The majority of current LC research is focused on management of non-small cell lung carcinomas. Different somatic mutations have been discovered in LCs that can be utilized as molecular targets and biomarkers of anticancer drug response. Mutations in the epidermal growth factor receptor (EGFR), Kirsten rat sarcoma virus oncogene homolog (KRAS), and anaplastic lymphoma kinase (ALK) have all showed promising effects in LC treatment. The tyrosine kinase receptor EGFR is frequently changed in epithelial malignancies. Many malignancies have enhanced EGFR activity, which leads to amplification of the EGFR signaling pathway, which promotes cell proliferation and tumor formation. EGFR mutations are more frequent in adenocarcinoma than in other histologies of LC (30% vs. 2%), and they are more common in never-smokers than in ever-smokers (45% compared with 7%). Around 60 women are more likely than men to get LC with EGFR mutations. After surgery, patients in stage IIA, IIB, and IIIA NSCLC typically get chemotherapy to eradicate any leftover cancer cells. Adjuvant therapy reduces the chance of LC recurrence. Radiation, chemotherapy, and targeted therapy may all be used (Zappa & Mousa, 2016).

Since the early 1950s, LC has been the main cause of cancer deaths among men, and in 1987, it overtook breast cancer (BC) as the top cause of deaths due to cancer in females in the US (Ridge et al., 2013).

10.2.5 COVID-19

In the Latin terms, the meaning of the word Corona is "Crown." The crown is surrounded by an irregularly shaped light. Corona is the term given to the virus because of its shape. On the corona surface, there is a crown of glycoprotein protein spikes made up of glycoprotein 120. Coronavirus is a zoonotic type of virus, which means it infects both humans and animals and can cause illnesses ranging from the common cold to diseases like SARS (severe acute respiratory syndrome) and MERS (multiple sclerosis) (Middle East respiratory syndrome) (Wu & McGoogan, 2020). Coronaviruses (CoV) are enclosed viruses having a positive sense RNA genome. They are members of the Coronovirinae subfamily of the Coronaviridae family, which is member of the Nidovirales order. Within the beta-CoV genus, there are four lineages (alpha, beta, gamma, and delta) and four genera (alpha, beta, gamma, and delta) (A, B, C, and D) In mammalian and avian species, CoVs cause a variety of respiratory and digestive illnesses (Tortorici & Veesler, 2019; Ahire et al., 2020a, b).

10.2.5.1 TYPES OF CORONA VIRUSES

Corona viruses come in seven different varieties, four of which are common corona viruses and three of which are other human corona viruses. MERS-CoV (transmitted from dromedary camels to people), SARs-CoV (transmitted from civet scats to humans), and SARs-CoV-2 or 2019-nCoV (Novel Corona Virus). Fever, cough, difficulty breathing, weariness, headache, runny nose, congestion, sore throat, and diarrhea are clinical features-signs and symptoms. COVID-19 symptoms are comparable to colds and flu in mild cases. Pneumonia, acute renal failure, multiple organ failure, and health are significant respiratory symptoms in severe COVID-19 instances.

10.2.5.2 Mode of Transmission

The virus is primarily disseminated from infected to healthy people by droplet infection caused by sneezing and coughing with close contact without the use of a mask. It's possible that it's transferred by touching things like doorknobs, pens, and papers, then touching a healthy person's eyes, nose, and mouth.

10.2.5.3 INVESTIGATION

In COVID-19 instances, CT scans reveal opacities in the lungs as well as bilateral pulmonary infiltrates. The PCT test can detect a decrease in lymphocytes (lymphocytopenia) and leukopenia COVID-19.

10.2.5.4 PHARMACOLOGIC THERAPIES

Antiviral medicines such as molnupiravir, paxlovid, and remdesivir are currently accessible as therapeutic options. Tocilizumab and baricitinib are immunomodulatory drugs (Tanuja, 2020; Ahire et al., 2021; Talele et al., 2021; Gunjal et al., 2021).

10.3 SIGNIFICANCE OF NOVEL PULMONARY DRUG DELIVERY SYSTEMS (DDSS)

Continually, the pulmonary tract has been explored with intent of delivering and targeting of drugs specifically to the lung cell populations and for non-parenteral systemic delivery of macromolecular drug moieties. For the purpose of achieving both, local as well as systemic effects, lungs as the chief component of the respiratory system provide a fascinating means for noninvasive administration of the drug (Bivas-Benita et al., 2005).

Certainly, pulmonary drug delivery has gathered remarkable attention of the researchers in the last two decades. But some prime concerns which need to be addressed are less amount of drug per puff, inappropriate dosing reproducibility, less effectiveness of the inhalation system, poor formulation stability for drug amongst several others (Javadzadeh et al., 2017). So as to get rid of the aforesaid hurdles, there is a vital necessity to launch the studied novel pulmonary DDS with their scientific utilization. Also, a novel pulmonary DDS should necessarily facilitate delivery of drug to the desired either to the desired part of the body or to its targeted site (Deb et al., 2018). Moreover, it provides targeted therapy for effective treatment of several respiratory conditions such as asthma, lung carcinomas, CB, emphysema amongst several others (Mishra & Singh, 2020).

10.3.1 APPROACHES FOR PULMONARY DRUG ADMINISTRATION

Pulmonary delivery of drug is often considered to be a lucrative target and of great scientific and biomedical significance owing to the fact that the lung exhibits tremendous potential of absorbing several drug moieties either for systemic delivery or for local deposition (Patil & Sarasija, 2012). Usually, the following two strategies are employed to deliver a drug via pulmonary route namely intranasal delivery and oral inhalative delivery.

10.3.1.1 INTRANASAL DELIVERY

Intranasal delivery means administration of drug moieties to the lungs via the nasal tract which indicates nose-to-lungs aerosol delivery (Mishra & Singh, 2020). It's a non-invasive means for drug delivery, wherein drug can get absorbed into the systemic circulation via nasal mucosa and is considered to be beneficial in an array of acute or chronic condition(s) that warrant substantial systemic exposure and recurrent dosing (Grassin-Delyle et al., 2012). Typically, the formulations preferred for drug administration via intranasal route include solutions, suspensions, gels, emulsions, and powders (Xu et al., 2020). But these traditional formulations face many challenges, basically, anatomical limitation, namely narrow airway lumen, and several others that include deficient in dose precision, greater particle size, greater viscosity and issues related to drug stability (Xu et al., 2020). So as to improve patient convenience and compliance there is a need to explore use of novel formulations such as nanosolutions, nanosuspension, nanogel, nano-emulsion (NE), and nanoparticles, etc., that will provide greater residence time, greater comfort and enhanced therapeutic efficacy (Xu et al., 2020). At present a greater part of intranasal formulations available in the market are meant to provide local relief or aim at prevention of nasal symptoms. In the next few years, the drift toward the development of intranasal products for systemic absorption will rise significantly with the development of peptides, proteins, recombinant products, and vaccines.

10.3.1.2 ORAL INHALATION DELIVERY

As greater tissue levels can be attained soon after application, the oral inhalation is the favored route for delivery of small molecules to the pulmonary tract and the lungs (Frohlich & Salar-Behzadi, 2021). Oral inhalation is achieved by two means, namely Intratracheal inhalation of the drug and Intratracheal instillation. Intratracheal instillation is employed for the exposure of animals to particles, both soluble and insoluble and involves delivery of drug solution or dispersion by using of a syringe (Mishra & Singh, 2020).

Whereas the intratracheal inhalation utilizes aerosol technology for delivering drug carriers to the lungs with improved penetration and consistent distribution of the drug. Drug deposition into the lungs and airways by use of aerosol inhalation techniques usually take place by three general means namely inertial impaction, sedimentation of the drug and Brownian diffusion (Mishra & Singh, 2020). Generally, the particles of the drug undergo deposition in the various regions of the respiratory tract by complex mechanisms and the process is dependent on several parameters such as size of the drug particles, flow of air and the existence of mucus material in the respiratory passage (Mishra & Singh, 2020). Conventionally, small drug molecules administered by oral inhalation is the commonest way for treatment of several diseases of the lung such as asthma, CB, emphysema, and cystic fibrosis (CF). Also, type of inhaler plays a vital role for delivery of several biological substances. It has been reported that 75% of inhaled protein formulations in clinical research are produced as liquids. Inhalation therapy offers many benefits over other drug delivery routes such as large surface area of alveolar sacs and relatively greater vascularization make the lungs a suitable site of drug absorption (Javadzadeh et al., 2017). Inhalable NPs in the solid form presented as powders for inhalation to achieve delivery of the drug in the pulmonary region provide exclusive benefits and are a novel topic of interest (Yurdasiper et al., 2018). Also, novel nebulizers will certainly enhance delivery by producing less shear stress and administration as dry powder appear suitable for delivery of drugs (Frohlich & Salar-Behzadi, 2021). Inhalation therapy can be successfully used for local inflammatory respiratory diseases such as asthma, CB and COPD. Nanocarriers possess the ability to interrelate with the biomolecules present on the cell surface as well as within the cell. Furthermore, owing to outstanding biocompatible nature and options available for their surface modification(s) the biodegradable polymeric nanoparticles (PNPs) have been utilized as efficient drug carriers for the sustained effect of the encapsulated drug. These nanocarriers have been investigated considerably owing to their capacity to achieve delivery of several drug moieties used against pulmonary diseases, COPD, asthma, carcinomas, and TB (Passi et al., 2020). Pulmonary delivery of drugs in the form of nanomedicines improves LC treatment by transportation of the encapsulated hydrophobic, cytotoxic drugs straight to their proposed target site thereby achieving targeted drug delivery and release of drug in a controlled fashion at the tumor regions (Hitzman et al., 2006). In case of drug delivery by inhalation, the administered drug gets distributed evenly throughout the lungs, and both the normal as well as affected body cells are exposed to chemotherapeutic agent which produce deleterious effects on the healthy cells (Anderson et al., 2019). Nanomedicine presents a new advancement that helps in the development of an effective formulation strategy for the treatment of LC using the inhaled anticancer therapy. In the 1990s, liposomes were tried for the first time as a nanocarrier developed for the delivery of chemotherapy by inhalation (Rosière et al., 2019). Beclomethasone-loaded lipidic Nanocarriers were developed by Jaafar-Maalej et al. Also, the *in vitro* aerodynamic performance of the nebulized formulas was effectively assessed so as to find out the prospective of the developed Lipidic Nanocarriers as a means for delivery of inhaled-steroids for the treatment of asthma (Jaafar-Maalej et al., 2011). Along with delivery of various pulmonary drugs, these carriers are useful for inhalation delivery nucleic acids like siRNA (Passi et al., 2020). In case of chronic inflammatory lung diseases such asthma and COPD, where inflammation and oxidative stress coexist, budesonide and resveratrol inhalable microparticles (MPs) dried co-spray when used together can produce beneficial effect as per studies performed by Trotta et al. (2016). Rojanarat et al. formulated nanocarrier proliposome powder of Isoniazid, for inhalation therapy in the treatment of TB (Rojanarat et al., 2011). The aerosolization performance of the Isoniazid proliposomes was assessed by ACI and FPF. As compared to pure Isoniazid, proliposomes of Isoniazid were observed to be unharmful and safe for the cells of the airway and produced greater action against the *Mycobacterium* Bovis-infected alveolar macrophages (Rojanarat et al., 2011). Where the conventional drugs have failed to exhibit satisfactory responses in adults suffering from CF and several other lung diseases, a patented formulation Arikayce® exhibited good response (Huang et al., 2021).

10.3.2 PARTICLE-BASED PULMONARY DELIVERY SYSTEMS

To accomplish prompt pulmonary drug delivery, it is necessary to overcome certain limitations like surpassing the pulmonary clearance mechanisms together with mucociliary transport, avoiding the engulfment by alveolar macrophages, prevention of degradative spoilage of drug by enzymes, providing adequate control on the rate of drug deposition and its retention in the airway, etc. (Dabbagh et al., 2017). Thus, to overcome the various pulmonary clearance mechanisms, to improve the therapeutic efficacy of the drug and good precision over its release pattern the particle-based delivery systems have been utilized in the recent times (Dabbagh et al., 2017). The absorption of the administered drug is quicker in the lungs owing to availability of relatively greater amount of drug and the abolition of its pre-systemic metabolism. Though, several factors such as like morphology, compositions, size, solubility, targeting moiety, state (dispersion/agglomeration) and pH are vital and needs to be critically monitored to achieve the desired action of the drug when administered *via* pulmonary route.

Specifically, the particle-based delivery systems have emerged as a hopeful option to rise above the various physiological barriers that exist in the pulmonary system, thereby increasing the bioavailability of the therapeutic moieties at lower regions of the lung. Dabbagh et al. have elaborated that the particle-based delivery systems must efficiently surpass the pulmonary clearance mechanisms, shield their contents from undergoing metabolic degradation, enhance the localized delivery to the affected area, improve the sustained release (SR), reduce the dosage frequency, and decrease the systemic toxicity of the drug. Nanoparticles have been studied at length for their prospects for pulmonary drug delivery. Based on the results of the *in vitro*, *ex vivo*, and *in vivo* studies, several researchers have suggested SLNs, solid lipid microparticles (SLMs), PNPs, and liposomes as potential drug carriers for pulmonary drug delivery.

1. **Solid Lipid Nanoparticles (SLNs):** Matthias et al. studied use of murine precision-cut lung slices (PCLS) and human alveolar epithelial cell line (A549) in order to assess the tolerable doses of formulated SLN for lung cells (Nassimi et al., 2009). SLN20 and SLN50 exhibited comparable cytotoxic action with respect to LDH release with EC_{50} values of 3,431 μg/ml and 1,253 μg/ml, respectively. Numerous types of *in vitro*, *ex vivo*, and *in vivo* work has been carried out to demonstrate the utility of the formulated SLN for the treatment of various diseases and disorders related to the pulmonary tract.
2. **Solid Lipid Microparticles (SLMs):** Silva et al. prepared and evaluated quercetin loaded SLMs intended to be used by inhalation for the treatment of pulmonary DDS (Silva et al., 2013). A multi-path model of particle deposition approach was employed to assess the regional deposition of inhaled drug particles. Study results revealed that MGTL primarily gets deposited in the tracheobronchial and pulmonary regions and also a major portion of MGBL gains entry into deeper regions of the lung.
3. **Polymeric Nanoparticles (PNPs):** Modification of the polymer is achieved in terms of its chemical and surface properties to facilitate enhanced engulfment of the drug moiety and offer optimum protection for the contents from undergoing degradation and to prolong the drug delivery, longer shelf life so as to impart them the desired biodegradability. For therapeutic applications, the usually preferred polymeric materials comprises of poly(lactic acid) (PLA), poly(lactic-co-glycolic acid) (PLGA), poly(ε-caprolactone) (PCL), Chitosan, chitin, alginates along with several gelatin bases. Trapani and coworkers studied the utility of chitosan and glycol chitosan nanoparticles comprising of Lipoid S100 as a surface-active agent. The developed system was utilized to achieve systemic action of low molecular weight heparin after administration via pulmonary route (Trapani et al., 2013). It was observed by *in vivo* studies that systems formulated in the acidic conditions, administered as pulmonary delivery did not raise the time required for coagulation in the mice. On the contrary, Lipoid S100-LMWH GCS NPs formulated at neutral pH displayed better results. In general, these results demonstrate a number of useful features of CS-based nanocarriers in order to achieve effective delivery of LMWH via pulmonary route.

4. **Liposomes:** For pulmonary uses, Liposomes are pretty DDS, particularly, because it is formulated chiefly from phospholipids, which are intrinsic in the lungs. At present, two liposomal products are in the last stage of clinical development as dry powder liposomes that is Arikace® and Pulmaquin™, comprising of Amikacin and Ciprofloxacin, respectively for their use in the treatment of lung infections as inhalation therapy.

Mitsopoulos and colleagues have reported efficiency of prepared liposomal-N-acetylcysteine system against LPS-induced lung injuries in rodents (Mitsopoulos et al., 2008). It was concluded that L-NAC ameliorated the LPS-induced lung injuries by decreasing the effects of ROS and inflammation owing to the enhanced antioxidant delivery and retention of NAC in the lung.

The aforesaid NPs and other forms are delivered to the pulmonary site by use of drug delivery devices namely dry powdered inhalers (DPI), nebulizers, and pressurized metered-dose inhalers for nanoparticles. Also, the metered dose inhalers (MDI), dry powder inhalers (DPIs), and nebulizers are the means of aerosol drug delivery that are employed for treatment of respiratory problems namely asthma, CF, CB, emphysema, obstructive lung disorders, pulmonary arterial hypertension, infectious pulmonary diseases and others (Sorino et al., 2020).

10.3.2.1 PRESSURIZED METERED-DOSE INHALERS

MDI is a pocket-sized hand-held drug delivery device which is equipped with a mouthpiece to provide reliable, consistent dose of medication after spraying. It is a pressurized canister of medicine mounted in a plastic holder. On pressing the canister in to the actuator the system provides a fine atomized spray over 100–200 ms of medicament usually with a particle size smaller than 5 μm. On actuation the medicament is expelled out via nozzle at an elevated velocity to form a spray. After spraying, it offers a reliable, steady dose of drug in each puff. MDI are the basis of asthma and COPD therapy universally. The device comprises of the drug moiety in the form of a suspension or a dilute solution formulation with a volatile propellant. The suspension provides 10% respirable fraction, whereas a dilute solution formulation with a volatile propellant may provide around 40% respirable fraction. The drug substance(s) are either uniformly suspended or dissolved completely in a liquefied propellant system. MDI is more general than nebulizers (Preedy et al., 2013). MDI is available as a valved holding chamber or without a chamber. However, a valved holding chamber is the more favored design as it facilitates prompt spraying of medication and ensures effective entry of medicament in to the lung. Delivery of drug with the aid of pressurized MDI has been effectively used since many years to achieve treatment of several respiratory diseases (Myrda et al., 2014). Tan et al. have investigated the stability and aerosolization of prepared pressurized metered dose inhalers (pMDIs) containing thymopentin nanoparticles (Tan et al., 2012). Authors had fabricated NPs by a bottom-up process and suspended them in hydrofluoroalkane (HFA) 134a along with cineole and/or n-heptane to produce pMDI formulations. Lastly, it was concluded that the pMDIs containing thymopentin NPs developed in this study were stable and suitable for inhalation therapy for systemic action.

Regardless of some environmental footprint and require an excellent practice by the patient to accomplish consistent remedial actions, these devices are still considered as fundamental tools by the Pulmonologists and the medical practitioners. The continuing research, to advance the fundamental features of these devices, bring in environmentally acceptable and safe propellants and unite these devices with modern functions of telemedicine and artificial intelligence (AI), make new path for the incessant use of these devices in day-to-day clinical services. More work is required to explore safer pMDI formulations containing nanoparticles for pulmonary drug delivery.

10.3.2.2 Dry Powdered Inhalers

Since 1970, the DPIs have been made available, though the earliest prototypes were described several decades earlier, are clinically appropriate and cost-effective alternatives. It is dry free-flowing powdered formulations of the drug, usually prepared by spray-drying or drug micronization.

Quite a lot of medicines against chronic respiratory diseases are available in the market in the form of DPI's. As the drugs are powdered substances with inert carriers, their delivery requires unique inhalers.

During rapid inhalation, sufficient medicament gets released. The principle for use of DPI's is to exhale away from the device then put mouthpiece in your mouth and breathes in quickly. Gradually, a new generation of novel DPIs became available with extensively different designs, operating characteristics and improved delivery of drug substances to the pulmonary tract and to the lung. It is evident from several studies that inhaler design exhibits a major impact on the aerosolization of the dry powder and therefore so many devices are designed to inhale the dry powder. The following are the common dry powdered inhalers namely Ellipta™, Aerolizer®, Handihaler®,

Diskus®, Rotahaler®, Flexhaler®, Twisthaler®, Neohaler®, Turbuhaler®, Pressair™ (Figure 10.1). It comprises of a plastic device to facilitate inhalation of the medicament. It's a breath activated device which indicates that on inhalation the device readily releases the medicament and prompt inhalation offers greater chance for the medicine to gain entry in to the respiratory passage.

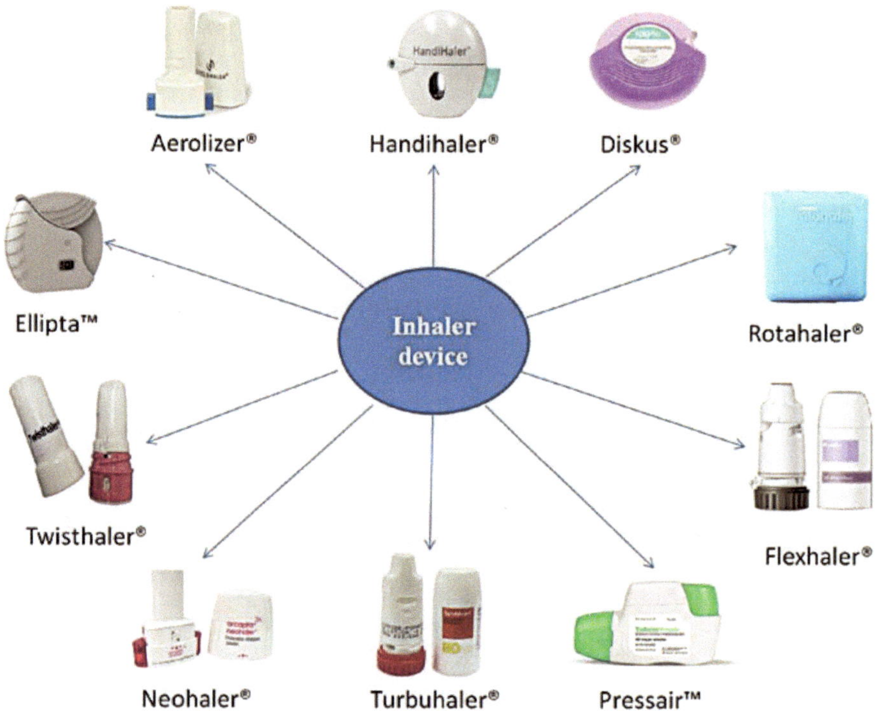

FIGURE 10.1 Different design of marketed inhaler devices.

10.3.2.3 NEBULIZERS

Nebulizers are characteristic devices utilized to achieve pulmonary delivery of drug substances are the oldest of the three main types of aerosol delivery devices (Labiris & Dolovich, 2003). It turns a liquid medicament in to a mist which can thereafter inhaled into the respiratory passage through a mouthpiece or a mask. It requires liquid preparations to create polydisperse particles of suitable size which can be easily inhaled. Nebulizers are generally used along with a suitable mouthpiece or a mask, which facilitates prompt delivery of the drug substance from a reservoir solution. Following are the types of nebulizers:

1. **Jet Nebulizers:** These small-volume jet nebulizers are considered to be the principal strongholds in the delivery of several bronchodilators especially in the older patients and bed riddens (Broaddus et al., 2022). To achieve breakup of the formulation in to droplets or thin films, the traditional pneumatic, or jet, nebulizers often utilize compressed air or oxygen, and these fine droplets are then suitable for inhalation therapy (Burks et al., 2020). Jet nebulizer work on the principle of Bernoulli theorem by which the compacted gas is allowed to pass through a fine orifice by generation of a lesser pressure area at the adjoining liquid feed tube (Mukherjee et al., 2017). The presence of a baffle within the inhaler helps to catch greater and non-inhalable particles. Lastly, fine inhalable drug particles will gain entry into the respiratory tract to reach the targeted site or undergo the systemic absorption (Javadzadeh et al., 2017).
2. **Ultrasonic Nebulizers:** These nebulizers use ultrasonic power so as to yield micron-scaled inhalable particles by break down of the drug solution into smaller droplets by piezoelectric vibration and do not need gas flow therefore, these nebulizers are more portable and largely preferred (Javadzadeh et al., 2017). It utilizes is a piezoelectric transducer crystal as a source of particle generation which facilitate conversion of electrical energy into ultra-high-frequency oscillations that generate aerosol particles from the surface of the liquid (Miller et al., 2009). The presence of baffle within the inhaler that eliminates large droplets and returns them to the principal chamber of device (Javadzadeh et al., 2017). Despite, of the fact that Ultrasonic nebulizers generate greater consistent size of the particles than as compared to the jet nebulizers they are not much preferred owing to their relatively higher costs.

10.3.3 NOVEL DELIVERY SYSTEMS TARGETING RESPIRATORY DISEASES

Various methods have been successfully utilized to achieve pulmonary drug delivery for effective treatment of several respiratory disease(s). Moreover, the potential of lungs to effectively absorb the administered drug *via* pulmonary route has developed a keen interest amongst the researchers and biomedical scientists to increase the utility of the existing drugs by converting them into nanoformulations(s) (Mishra & Singh, 2020). The principal mechanisms for absorption of drug via pulmonary route are transcellular diffusion, paracellular diffusion, vesicle-based transport, and carrier-mediated transport (CMT) (Mishra & Singh, 2020). A novel pulmonary system offers various advantages, as depicted in Figure 10.2.

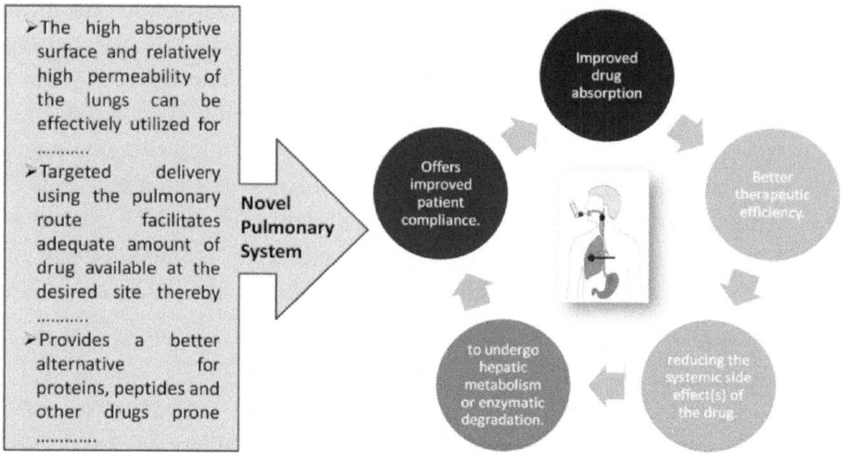

FIGURE 10.2 Advantages of novel pulmonary drug delivery system.

10.3.3.1 NANOPARTICLES

Both, the dispersed liquid droplet dosage forms such as MDIs, nebulizers, and dry powder formulations are offered by advent of nanoparticles for pulmonary drug delivery (Bailey & Berkland, 2009). Several benefits such as improved dissolution properties and effective intracellular drug delivery are offered by nanoparticle formulations as compared to their conventional counterparts (Bailey & Berkland, 2009).

In particular, pure drug nanoparticles, PNPs, polyelectrolyte complexes, and drug-loaded liposomes provide fruitful results for delivering drugs to and through the lungs.

10.3.3.1.1 Polymeric Nanoparticles (PNPs)

Pulmonary delivery of drug using PNPs provide various benefits such as overcoming drug clearance, targeted action of the drug, sustained action of the drug, avoidance of drug inactivation by enzymes, greater efficacy of the drug and its lowered adverse effect(s) (Mishra & Singh, 2020). Deacon et al. proposed the use of Tobramycin-loaded alginate/chitosan-based polymeric

NPs against *P. aeruginosa*. It was observed that the prepared NPs showed almost comparable antibacterial effect as that of Tobramycin, whereas after functionalization with dornase alfa, there was a marked increase in the penetration ability of the aforesaid functionalized NPs into the CF sputum. Thus, the research revealed that the challenge of crossing the hurdle of the mucus in the instance of several infectious diseases can be successfully achieved by use of antibiotic-loaded functionalized NPs, which will contribute to increase the drug efficacy (Deacon et al., 2015). Costa-Gouveia et al. carried out a study with an intent to enhance the efficacy of Ethionamide, a second-line agent against TB. The drug along with its booster was loaded in to nanoparticles of β-cyclodextrin. To resolve the problems arising out of drug-like crystallization behavior and less solubility of the booster in water the use of NP's was recommended. *In vivo* study results in mice revealed that the prepared NPs showed a reduction of the mycobacterial load in the pulmonary region of the mice by 3-log. Therefore, this study involving coencapsulation of anti-TB drugs intended for pulmonary administration offers a better option for the effective treatment of the mycobacterial disease (Costa-Gouveia et al., 2017; Ahire et al., 2018). Drug can be targeted to the infected lungs by use of the drug loaded PNPs, wherein the drug substances can be loaded on the surface of the polymers or can be encapsulated in the matrix (Hadinoto et al., 2007). The drug loaded PNPs after being administered, when reach to the humid conditions of the infected airway regions/lungs undergo dissolution of the polymeric matrix and facilitates prompt release of the loaded nanoparticulate drug moiety (Sung et al., 2007; Ahire et al., 2020c, d). Dawson et al. highlighted on the probable lowering of the target specificity of the functionalized NPs in the biological milieu. The researchers studied the uptake of fluorescent silica NPs (SiO_2-PEG) conjugated with human transferrin (Tf), a transferrin receptor (TfR)-binding glycoprotein, in different types of serum. It was observed that TfR-non-specific and specific uptakes were lowered in the presence of fetal bovine serum (FBS). It was thus concluded that the reduced targeting capability of Tf-conjugated NPs was attributed to their interactions with other proteins in the media that led to creation of a protein corona. The work importantly focused on need of critical evaluation of several factors such as identification of interactions, and fortune of the bioconjugated NPs in the complex biological environment (Salvati et al., 2013). Irrespective of various benefits of polymeric NPs for pulmonary delivery, they suffer from certain shortcomings that require prompt attention in order to facilitate translation of these formulations to the market. Often, the natural polymers used for the preparation of NP's are prone to undergo degradation in short period of time that causes immediate release of the drug moiety after its administration. This pitfall can be encountered by use of biodegradable synthetic polymers to achieve sustained effect of the drug. On repeated dosing, the polymeric NPs tend to accumulate in the airways and may precipitate unwanted effect(s) over a period of time. Thus, this issue of toxicity and ability of polymers to undergo degradation must be addressed promptly (Mishra & Singh, 2020).

10.3.3.1.2 Solid Lipid Nanoparticles (SLNs)

Solid lipid nanoparticles (SLNs) have been known to possess ample applications for pulmonary drug delivery and therefore have been widely focused. Nassimi et al. showed the utility of phospholipid- and triglyceride-based SLN in the (30:70) ratio for their varied pulmonary drug delivery uses (Nassimi et al., 2010). For the treatment of asthma, Wang et al. prepared Curcumin loaded stearic acid- and lecithin-based SLN using the solvent injection method. The stability of the prepared SLN was assessed and was found to be stable in the size ranges of 190–200 nm. *In vivo* study which comprised of ovalbumin-induced asthma, demonstrated that the cytokine levels got reduced in the group treated with Curcumin in contrast with the group left untreated (Wang et al., 2012).

Varshosaz et at. prepared SLN of Amikacin, with the aid of cholesterol as the lipid by employing high pressure homogenization technique. The researchers also focused on biodistribution of the prepared SLN of Amikacin after administration by pulmonary route. It was studied by using radioisotope technetium (99mTc) labeled Amikacin. The *in vivo* experiments were carried out in rats which were administered with and without radiolabeled Amikacin by using the pulmonary route and the IV route. Gamma scintigraphic studies revealed that 99mTc-amikacin SLN stayed for longer duration of time in the lungs when administered by pulmonary route as compared to the IV route. Moreover, it was found that 99mTc-amikacin SLN got deposited in the greater proportion in the lungs as compared to the kidneys (Varshosaz et al., 2013). Mussi et al. used a polyunsaturated fatty acid, Docosahexaenoic acid (DHA) and Triethanolamine with an intent to increase the encapsulation and solubility of doxorubicin (DOX), respectively in the lipid melt. The researchers prepared DOX SLN and studied its anti-cancer effect. It was found that encapsulation efficacy of DOX got increased from 36% to 99%. The results of the studies revealed that solubility enhancers had a positive effect to enhance efficacy of the formulation, thereby contributing to maximize uptake and, improvement in the cytotoxic activity of the drug (Mussi et al., 2013). Rosière et al. studied the effect of paclitaxel-loaded solid lipid NP's for treating tumors of lung. Study results revealed that paclitaxel loaded SLNs when administered pulmonary route *in vivo* showed improved retention of the drug in the lungs. Thus, it resulted in reduced systemic delivery of drug, and greater antitumor effect (Rosière et al., 2018). Ji et al. studied the encapsulation of poorly water-soluble drug Naringenin into SLN's. The prepared SLNs after administration by pulmonary route showed marked increase in the bioavailability of Naringenin as compared with the suspension formulation of the same drug. The study results concluded that SLN's thus provide a lucrative alternative to achieve increase in the bioavailability of hydrophobic drugs (Ji et al., 2016). Makled et al. performed studies on targeted delivery of sildenafil citrate loaded SLNs by inhalation as a substitute for conventional oral route for the treatment of pulmonary hypertension. The encapsulation efficiency, colloidal stability of the developed formulation was found to be agreeable. Sildenafil citrate loaded SLNs showed SR of the drug over a period of 24 h (Makled et al., 2017). Gaspar et al. prepared Rifabutin loaded SLNs and further encapsulated them in to microspheres with an intent to improve their stability profile and facilitate the administration via pulmonary route as a DPI. The results of the *in vivo* studies carried out in mice revealed that the developed formulation efficiently delivered the antibiotic to the lungs. Also, a significant quantity of active form of Rifabutin was found to be deposited in the vital organs namely liver and spleen of the mice. The group of mice that undergone treatment with the prepared formulation exhibited marked decline in the growth index of mycobacteria, in contrast with the control group. It was therefore concluded that administration of antimycobacterial loaded SLNs via pulmonary route provides immense benefit for effective treatment of TB (Gaspar et al., 2017). Silva et al. prepared quercetin-loaded SLMs intended for inhalation therapy. A multipath model of particle deposition was used to study the regional deposition of the inhaled particles. It was revealed that the drug loaded MPs prepared from a mixture of glyceryl trimyristate and soy lecithin exhibited prime deposition in the tracheobronchial and pulmonary regions, whereas drug loaded MPs prepared from a mixture of glyceryl behenate and soy lecithin successfully reached the lungs at depths (Silva et al., 2013). Numerous research articles have highlighted the utility of SLNs for drug delivery via pulmonary route. However, their successful clinical translation and launch of the drug product in to the market have been much sluggish. The prime reasons for this slow clinical translation include issues related to the stability of the formulations in the solution state and in the aerosolized/nebulized form, lack of precision in the delivery of drug and concerns over safety and compatibility facets. Therefore, it is the need of time to locate the reasons of failure of SLN-based systems in clinical trials and initiate further research and improvements in these systems to explore their potential for pulmonary drug delivery (Mishra & Singh, 2020).

10.3.3.1.3 Inorganic Nanoparticles

Several materials such as silver, gold, silica, platinum, and others are used for the preparation of Inorganic nanoparticles (Mishra & Singh, 2020). Richa Singh et al. synthesized phytogenic silver, gold, and gold-silver bimetallic nanoparticles from plants namely *Barleria prionitis, Plumbago zeylanica,* and *Syzygium cumini* and evaluated for their antimycobacterial potential. *In vitro* and *ex vivo* macrophage infection model methods exhibited that Gold-silver bimetallic nanoparticles promoted inhibition of both active and dormant stage of the mycobacteria.

It was observed that these nanoparticles effectively gained entry into the macrophage cells and showed 45% cytotoxicity at a concentration of 30 μg/mL (10 times MIC concentration) at the end of 48 h. Also, the selectivity index of the bimetallic nanoparticles was found in the range of 94–108 (Sing et al., 2016). Ramalingam et al. studied the effect of polyvinylpyrrolidone (PVP)-gold nanoparticles (AuNPs) conjugated with cytotoxic drug DOX against human lung carcinoma, wherein the release of the drug from the conjugate was observed to be dependent on pH. Decrease in the pH caused increased detachment of the drug from PVP-gold NP-conjugate. The prepared formulation exhibited outstanding cytotoxic effect as compared to DOX in free form and carbon nanoparticles. This result suggested that PVP-gold nanoparticle when conjugated with DOX produced greater drug concentration in the A549 cells and led to greater cytotoxic action. Moreover, DOX conjugated PVP-gold nanoparticle caused greater generation of reactive oxygen species thereby stimulating the process of apoptosis in LC cells along with enhancement of expression of tumor suppressor genes. Thus, the drug conjugated PVP-gold nanoparticle approach will provide a beneficial means to target the cancer cell of the lung (Ramalingam et al., 2018). The ability of superparamagnetic iron oxide NPs conjugated to anti-IL4Rα blocking antibodies (SPION-anti-IL4Rα NPs) to inhibit the inflammatory response in case of asthma was studied by Halwani et al. OVA-sensitized mice when administered with SPION-anti-IL4Rα NPs via pulmonary route caused marked reduction in the proinflammatory cytokines in the lung tissues. It also caused reduced proliferation with a marked decrease in its occurrence of the inflammatory cells of lung. The NPs offered effective and extended control of lung inflammatory cytokines for a period of 07 days after administration which contributed to decrease the levels of lung inflammation. Some SPION-based systems have been granted approval and a few have attained the stage of clinical trials (Halwani et al., 2016). An attempt was made by Wu et al. to achieve concurrent delivery of Paclitaxel and DOX by employing hollow calcium phosphate NP's which were coated with phospholipids. The prepared NPs were evaluated against the LC cells A549. The NPs showed profound cell inhibition rate, effective suppression of tumor proliferation, apoptosis of A549 tumor cells and a decline in tumor volume after treatment. The results of *in vitro* and *in vivo* studies indicated the utility of calcium phosphate nanoparticles for drug delivery to the lungs (Wu et al., 2017).

10.3.3.2 LIPOSOMES

Treatment of several pulmonary diseases, namely CF, pulmonary infections, asthma, and lung carcinomas, require local delivery of the drug moieties to the pulmonary tract and the lungs and often poses a challenge to the formulation scientists owing to the hydrophobic nature of the drug moieties used for the treatment which creates hurdles in formulation development and bioavailability related issues. Inspite of prompt local drug delivery to the lung, the deposited insoluble drug substances may produce toxic effect(s).

Therefore, to achieve targeted treatment of several pulmonary infectious diseases, Liposomes have been extensively studied. Lung surfactants comprise of 80% of phospholipids and neutral lipids around 8%, the composition resembles to that of Liposome and therefore they are considered to be a worthwhile option for pulmonary delivery. Moreover, their biodegradability, biocompatibility, and scare chances to produce local lung irritation further favor their use for pulmonary delivery. Also, the intracellular delivery to alveolar macrophages is fundamentally achievable owing to the phospholipid nature of liposomes. Additionally, surface modifications facilitate targeted delivery of drug with the aid of ligand or antibody binding (Mishra & Singh, 2020). Currently, Amikacin loaded liposomal suspension, Arikayce intended to be used as inhalation therapy for the treatment of several lung diseases produced due to Mycobacterium avium complex has been approved by FDA.

The controlled trials for this preparation showed promising results, wherein about 29% of patients provided with Arikayce treatment witnessed absence of Mycobacteria in the cultures of the collected samples of their sputum for a period of 3 months. Moreover, the formulation on its administration once a day proved to be useful against Mycobacterium abscesses in patients with CF (Caimmi et al., 2018). Dual release ciprofloxacin for inhalation comprising of liposomal Ciprofloxacin was prepared by Serisier et al. with an intent to optimize airway antibiotic delivery. It was observed that once-daily inhaled Dual release ciprofloxacin for inhalation exhibited strong antipseudomonal activity in patients with non- CF, bronchiectasis, and ciprofloxacin sensitive *P aeruginosa* (Serisier et al., 2013). Also, another liposomal formulation of ciprofloxacin for inhalation therapy against non-cystic fibrosis bronchiectasis (nontuberculosis mycobacterial infections) has successfully entered the Phase 2 and Phase 3 of clinical trials (Haworth et al., 2019). Verschraegen et al. developed liposomal formulations of 9-nitrocamptothecin to resolve the issues of slight aqueous solubility and reduced bioavailability. The liposomal 9-nitrocamptothecin intended to be administered by nebulizer for the treatment of pulmonary malignancies was found to be harmless and viable for use in the patients during Phase II clinical trials (Verschraegen et al., 2004). Several formulations of Liposomes comprising of chemotherapeutic agents such as Cisplatin and DOX have already been brought up in the laboratory scale, however, they have not undergone the clinical transition due to their toxicity or lack of prospective benefits over the existing formulations. To name a few, Lipoplatin, Li-PlaCls, and L-NDDP are some of the liposomal formulations containing Cisplatin that possess lesser toxic effect(s) in contrast to free cisplatin, however, were turned down during the stage of clinical trials because they lack greater efficacy (Ventola et al., 2017). Along with enhanced bioavailability, several aerosolized formulations of liposomes have offered numerous advantages such as greater drug concentration at the localized site, minimized adverse and cytotoxic effects that were evident in clinical trials conducted so far. However, for successful translation of these formulations to the market the key component is that enhanced drug efficacy profiles are necessary to be evident as compared with the parent drugs (Mishra & Singh, 2020).

10.3.3.3 MICELLAR SYSTEMS

Micelles possess an ability to encapsulate and deliver hydrophobic drugs. Moreover, it imparts several benefits such as enhanced stability profile *in vitro* and *in vivo*, changes and modifications in size as desired, greater encapsulation efficiency to incorporate hydrophilic and hydrophobic drug moieties at the same time. Numerous micellar systems have been studied for their utility in pulmonary drug delivery (Mishra & Singh, 2020). Kim et al. prepared cholesterol-conjugated polyamidoamine (Chol-PAM)-based polymeric micelles comprising of coencapsulated Resveratrol and gene. *In vitro* studies revealed anti-inflammatory activity of the drug loaded micelles. In acute lung infections model, the delivery of resveratrol and heme oxygenase-1 gene combination by Chol-PAM micelles by inhalation therapy resulted in increased anti-inflammatory effects. The study proved utility of micelles as drug carriers and their use in inhalation-based drug delivery (Kim et al., 2018). Hu et al. studied the effect of curcumin acetate and Nile red-loaded PEG-PLGA micelles on their intratracheal use in rats. It was observed that the intratracheal administration of micelles resulted in increased drug retention time in the pulmonary region as compared with IV route of administration and showed localized SR of the drug up to 24 h. Furthermore, the micelles promoted the uptake to pulmonary vascular endothelium which suggested their utility in treatment of pulmonary arterial hypertension. Also, it was observed that micelles bear the ability to cross the alveolar-blood barrier and gain entry into other organs without any disruption, thereby providing the option for such systems for noninvasive systemic delivery (Hu et al., 2014; Thombre et al., 2022). To achieve effective treatment of lung tumors, paclitaxel-loaded polymeric micelles of Pluronic, in combination with succinylated gelatin, intended to be administered by inhalation were prepared by Wang et al. The characteristic feature of the developed system was that release of the drug was activated by matrix metalloproteinase 2/9 (MMP 2/9) and the amount of drug release was dependent on the concentration of MMP 2/9. Studies indicated that drug-loaded micelles were able to attain alike drug concentrations in the lungs as that attained by IV route. However, the systemic concentration of paclitaxel in case of pulmonary administration was 7-fold less as compared to IV administration. Thus, the drug-loaded micelles effectively targeted the tumor cells thereby, reducing exposure to the healthy cells. Such types of environment responsive systems may be considered in the future to achieve targeted delivery of the drug and minimize side effect(s), which are desired in the treatment of LCs (Wang et al., 2018). Regardless of several benefits of polymeric micelle-based systems namely good choice for delivery of hydrophobic drugs, lesser side effects, environment responsive drug delivery and several others, the successful advent of this system in to the market has not been possible. It is attributed to some matters related to lesser options of polymers for micelle preparation, stability of the systems, burst release of drugs in the instances of instability and difficulty in scale up operations. Considering the benefits offered by the systems its shortcomings there is a need to further modify it to make it suitable for clinical use (Mishra & Singh, 2020).

10.3.3.4 DENDRIMER

It is a kind of polymeric nanostructure that is dissimilar from traditional polymers. It possesses an extremely branched monodisperse three-dimensional structure. As a nanocarrier, its versatility and biocompatibility are enhanced owing to the presence of multiple functional groups on its surface (Mehta et al., 2019). Moreover, the betterment of its functional groups is possible by other charged compounds by electrostatic interaction. Numerous drug substances with varied solubility's can be easily delivered by employing dendrimers possessing both hydrophobic and hydrophilic functional groups (Ahmad et al., 2015). In case of use dendrimers for drug delivery it is possible to combine the loaded drug with the functional groups on its surface or the drug can be enclosed in its cavity (Mendes et al., 2017). Rajabnezhad et al. prepared inhalable nanodrug formulation of Rifampicin for the treatment of TB, by encapsulation of the said drug in different generations of polyamidoamide dendrimers. Interestingly, the third generation polyamidoamide dendrimers exhibited SR of Rifampicin, its enhanced absorption and greater bioavailability as compared with IV dosing (Rajabnezhad et al., 2016). Bellini et al. performed the studies on the 4th generation polyamidoamide dendrimers containing Rifampicin. The results of the study indicated that the formulation exhibited high stability under the physiological pH conditions. Also, the polyamidoamide dendrimers facilitates a favorable change in pH that resulting in quick release the drug moiety in the acidic region of the macrophages (Bellini et al., 2015). Conti et al. utilized an amine-terminated G4-PAMAM dendrimer (G4NH2) loaded with siRNA for the treatment of LC. The formulation successfully diminished the appearance of enhanced green fluorescent protein in a model of A549 cells, which signify that G4NH2-siRNA can effectively target the alveolar epithelial cells, and also efficiently quiet the target gene (Conti et al., 2014). Zhong et al. highlighted the utility of a complex comprising of DOX carboxyl-terminated G4-PAMAM dendrimers in lung metastasis. The study results exhibited that the aforesaid complex improved the drug accumulation and its retention time at the targeted site. Moreover, it contributed significantly to decrease the systemic drug toxicity and achieved greater efficiency of the drug on melanoma lung metastasis, thereby mounting the survival rate of mice. Additionally, dendrimers have been employed against inflammatory respiratory diseases and disorders associated to asthma (Zhong et al., 2016). Using a mouse lung inflammation model, Inapagolla et al. explored that that G4-PAMAM conjugated with methylprednisolone can possibly increase the capability of the drug to slow down inflammation through improvement of the deposition time of the drug substance in the lungs (Inapagolla et al., 2010). Khan et al. highlighted the effect of PAMAM and poly(propylene imine) (PPE) dendrimers with lipid substitution that were encapsulated with siRNA. It was observed that siRNA enclosed in chemically modified dendrimers possessed stronger affinity for the endothelial cells of the lungs. Out of various combinations studied, PAMAM-conjugated with C_{15} lipid tails, PPE dendrimers with C_{15}, and C_{14} lipid chains initiated strong potency dependent on the diameter. Also, several issues such as composition of the modified dendrimers, pKa values of the nanoparticles greatly influence the targeting window of the dendrimers. Similar systems may be considered for targeted delivery of important therapeutic agents like mRNA, DNA, and microRNA (Khan et al., 2015).

10.4 LIST OF PULMONARY DELIVERY SYSTEMS LAUNCHED IN THE MARKET (TABLE 10.1)

TABLE 10.1 List of Pulmonary Delivery Systems Launched in the Market

Sl. No.	Drug	Brand/System	Manufacturer
1.	Azelastine	Astelin	Meda Pharmaceuticals
2.	Beclometasone	Beconase	GlaxoSmithKline
3.	Levocabastine	Livostin	Jansen-Cilag
4.	Olapatadine	Patanase	Alcon Laboratories
5.	Sodium cromoglicate	Nasalcrom	Sanofi-Aventis
6.	Mupirocin	Bactroban	GlaxoSmithKline
7.	Triamcinolone acetonide	Nasacort	Sanofi Aventis
8.	Terbutaline 0.25 mg	Bricanyl	AstraZeneca
9.	Beclometasone dipropionate 250 mcg	Becloforte	Cipla Limited
10.	Fluticasone propionate	Flixotide	GlaxoSmith Kline

TABLE 10.1 *(Continued)*

Sl. No.	Drug	Brand/System	Manufacturer
11.	Salbutamol	Salbutamol dry powder capsules	Cipla Limited
12.	Salbutamol pressurized inhalation (100 μg)	Asthalin	Cipla
13.	Ipratropium Bromide 20 mcg	ATEM	Chiesi Farmaceutici
14.	Xinafoate	Seretide Evohaler	GlaxoSmithKline
15.	Albuterol	Ventolin	GlaxoSmithKline
16.	Levalbuterol HCl	Xopenex	3M Pharmaceuticals
17.	Fluticasone 50 μg	Flixotide	GlaxoSmithKline
18.	Formoterol Fumarate 12 mcg	–	Ultratech
19.	Amikacin	ARIKAYCE	Insmed
20.	Tobramycin	TOBI PODHALER	Novartis
21.	Poractant alfa	Curosurf	Chiesi USA
22.	Ciprofloxacin	Apulmiq	Aradigm
23.	Paclitaxel	Paclitaxel-based DPI	InhaTarget
24.	Cisplatin	Cisplatin-based DPI	InhaTarget
25.	Phospholipid-containing formulation with drug suspension	Pulmosphere	Novartis
26.	Fluticasone furoate	ELLIPTA DPI	GlaxoSmithKline
27.	Budesonide	Pulmicort Flexhaler DPI	AstraZeneca
28.	Albuterol sulfate	PROAIR RESPICLICK DPI	Teva Pharmaceuticals
29.	Salbutamol/salmeterol/fluticasone propionate/salmeterol+fluticasone propionate	Diskus/Accuhaler DPI	GlaxoSmithKline
30.	Budesonide	AKITA Jet nebulizer	Vectura
31.	Albuterol sulfate	ProAir HFA pMDI	Teva Pharmaceutical Industries
32.	Salbutamol	Ventolin HFA pMDI	GlaxoSmithKline
33.	Tiotropium bromide monohydrate	Respimat soft mist liquid inhaler	Boehringer Ingelheim Limited
34.	Tobramycin	PARI eFlow rapid nebulizer PARI International	PARI International
35.	Fluticasone propionate Salmeterol xinafoate	Conix DPI	3M Drug Delivery Systems

10.5 PATENTED NOVEL PULMONARY SYSTEMS (TABLE 10.2)

TABLE 10.2 Patented Novel Pulmonary System

Patent Number	Name of the Inventor(s)	Description of the Invention	Device or Drug System Developed	Disease Targeted	Patent Taken From	Year	References
US20180086798A1	Emilio Hirsch and Alessandra Ghigo	Novel pi3k gamma inhibitor peptide for treatment of respiratory system diseases.	Drug system	Asthma, cystic fibrosis, or chronic obstructive pulmonary disease.	United State	2019	Hirsch & Ghigo (2019)
CN102186458A	T. Hoffman	Method for treatment of COPD and other pulmonary diseases.	Device and drug system.	COPD and other pulmonary diseases.	Chinese	2011	Hoffman (2011)
JP2018519117A	Lanz, Jean-Sebastian Misalal, Adrian	Device for stimulating tracheal/bronchial air.	Device	Respiratory diseases	Japanese	2021	Lanz et al. (2021)
CA2601532C	Paolo Licciulli	Apparatus for the administration of pharmaceutical products in aerosol form by controlling pressure.	Device and drug system	Respiratory diseases	French	2014	Licciulli (2014)
EP1824544B1	Rajiv Doshi, Robert A. Howard, Motohide Hatanaka	Respiratory devices and methods of use. A nasal respiratory device has designed for delivery of drug to the lungs.	Device	Respiratory diseases	European Patent Office	2013	Doshi et al. (2013)
WO2005025540A3	David Morton, David Ganderton, John Staniforth, Yorick Kamlag	Mucoactive agents for treating a pulmonary disease.	Drug system	Cystic fibrosis and chronic obstructive pulmonary disease.	WIPO (PCT)	2005	Morton et al. (2005)
US20200268655A1	Chi-Shuo Chen, Rong-Shing Chang, Jong-Shinn Wu, Yun-Chien Cheng, Hui-Ling Lin	Plasma aerosol inhalation device and method for thinning the sputum.	Device and drug system.	Respiratory diseases.	United States	2020	Chen et al. (2020)

TABLE 10.2 *(Continued)*

Patent Number	Name of the Inventor(s)	Description of the Invention	Device or Drug System Developed	Disease Targeted	Patent Taken From	Year	References
US20210338718A1	Kenneth Liechty, Carlos Zgheib, Sarah Ashley Hilton	Methods for preventing and treating pulmonary inflammation and fibrosis.	NPs drug system	Pulmonary disease	United States	2021	Liechty et al. (2021)
US20220008666A1	Bernhard Muellinger, Susan Snape	Pharmaceutical administration to neonates, infants, and children.	–	Pulmonary disease such as asthma.	United States	2022	Muelinger et al. (2022)
WO2021183903A1	Stephen Douglas Galbraith, Robert M. Rauker	Portable ventilator.	Device	Pulmonary disease	WIPO (PCT)	2021	Galbraith & Rauker (2021)
CN110664819A	Sun Hongbin	Oleanolic acid preparation for inhalation, preparation method and medical application thereof.	Drug system	Respiratory diseases	Chinese	2020	Hongbin (2020)

10.6 CONCLUSION, CHALLENGES, AND FUTURE PROSPECTS

The pulmonary route has been considered as a potential means to achieve delivery of important therapeutics used against various diseases like asthma, COPD, TB, LC, acute respiratory tract infections, COVID-19 amongst others, which offers its local action and also facilitates quick systemic effects. Numbers of researchers have focused on the use of Novel pulmonary DDSs for several diseases owing to their unique properties and many modification possibilities as discussed earlier including surface tenability, rapid onset of delivery, chemical stability, noninvasiveness, self-administrability, better usefulness, localized effect, reduced side effects and electrical properties. Henceforth, over the last twenty years comprehensive studies of novel pulmonary DDSs have highlighted their possible uses in diverse fields including drug delivery to achieve site specific action along with to overcome problem with pulmonary DDS. Irrespective of countless applications of novel pulmonary DDSs, their use has been restricted owing to their extreme clearance mechanism by the alveolar macrophages, site of deposition of inhaled formulations, rapid drug degradation, and limited control over the deposition of the drug particles. Conversely, there is tremendous scope for the researchers to explore this arena to minimize possible side effect(s) associated with the use of novel pulmonary DDSs.

KEYWORDS

- **acute respiratory tract infections**
- **asthma**
- **chronic obstructive pulmonary disorder**
- **COVID-19**
- **lung cancer**
- **particle-based pulmonary delivery systems**
- **tuberculosis**

REFERENCES

Ahire, E. D., & Kshirsagar, S. J., (2021). Immune responses induced by different vaccine platforms against coronavirus disease-19. *Journal: Exploration of Immunology*, 243–257.

Ahire, E. D., Sonawane, V. N., & Surana, K. R., (2020). Role of drug repurposing in current treatment strategies against COVID-19; systemic review. *Pharm. Reson.*, 24–29.

Ahire, E. D., Sonawane, V. N., Surana, K. R., Jadhav, K. R., Sonawane, D. D., & Shah, A. A., (2020). Convalescent plasma therapy: A promising approach in the treatment of COVID-19. *Int. J. Pharm. Sci. Res.*, *11*, 4078–4086.

Ahire, E. D., Talele, S. G., & Shah, H. S., (2020). Nanoparticles as a promising technology in microbial pharmaceutics. In: *Applied Pharmaceutical Science and Microbiology* (pp. 133–158). Apple Academic Press.

Ahire, E., Thakkar, S., Borade, Y., & Misra, M., (2020). Nanocrystal based orally disintegrating tablets as a tool to improve dissolution rate of vortioxetine. *Bulletin of Faculty of Pharmacy, Cairo University*, *58*(1, 2), 11–20.

Ahire, E., Thakkar, S., Darshanwad, M., & Misra, M., (2018). Parenteral nanosuspensions: A brief review from solubility enhancement to more novel and specific applications. *Acta Pharmaceutica Sinica B*, *8*(5), 733–755.

Ahmad, J., Akhter, S., Rizwanullah, M. D., Amin, S., Rahman, M., Ahmad, M. Z., Rizvi, M. A., et al., (2015). Nanotechnology-based inhalation treatments for lung cancer: State of the art. *Nanotechnology Science Application, 8*, 55-66.

Anderson, C. F., Grimmett, M. E., Domalewski, C. J., & Cui, H., (2020). Inhalable nanotherapeutics to improve treatment efficacy for common lung diseases. *Wiley Interdisciplinary Reviews, Nanomedicine and Nanobiotechnology, 12*(1), e1586.

Asuntha, A., Brindha, A., Indirani, S., & Srinivasan, A., (2016). Lung cancer detection using SVM algorithm and optimization techniques. *Journal of Chemical and Pharmaceutical Sciences, 9*(4), 3198–3203.

Bailey, M. M., & Berkland, C. J., (2009). Nanoparticle formulations in pulmonary drug delivery. *Medicinal Research Reviews*, *29*(1), 196–212.

Beasley, R., Semprini, A., & Mitchell, E. A., (2015). Risk factors for asthma: Is prevention possible? *The Lancet*, *386*(9998), 1075–1085.

Bellini, R. G., Guimarães, A. P., Pacheco, M. A., Dias, D. M., Furtado, V. R., De Alencastro, R. B., & Horta, B. A. C., (2015). Association of the anti-tuberculosis drug rifampicin with a PAMAM dendrimer. *Journal of Molecular Graphics and Modeling, 60*, 34-42.

Bivas-Benita, M., Zwier, R. I., Junginger, H. E., & Borchard, G., (2005). Non-invasive pulmonary aerosol delivery in mice by the endotracheal route. *European Journal of Pharmaceutics and Biopharmaceutics, 61*(3), 214–218.

Boehm, A., Pizzini, A., Sonnweber, T., Loeffler-Ragg, J., Lamina, C., Weiss, G., & Tancevski, I., (2019). Assessing global COPD awareness with Google Trends. *European Respiratory Journal, 53*, 1900351.

Broaddus, V. C., Ernst, J., King, T. E., Lazarus, J. S., Sarmiento, K. F., Schnapp, L. M., Stapleton, R., & Gotway, M. B., (2022). Murray & Nadel's textbook of respiratory medicine. *Aerosols and Drug Delivery* (7th edn.).

Burks, A. W., Holgate, S. T., O'Hehir, R. E., Broide, D. H., Bacharier, L. B., Hershey, G. K., Peebles, R. S., & Middleton, E., (2020). *Middleton's Allergy: Principles and Practice* (9th edn.). Amsterdam: Elsevier.

Busse, W. W., (2019). Biological treatments for severe asthma: A major advance in asthma care. *Allergology International, 68*(2), 158–166.
Bussi, C., & Gutierrez, M. G., (2019). Mycobacterium tuberculosis infection of host cells in space and time. *FEMS Microbiology Reviews, 43*(4), 341–361.
Caimmi, D., Martocq, N., Trioleyre, D., Guinet, C., Godreuil, S., Daniel, T., & Chiron, R., (2018). Positive effect of liposomal amikacin for inhalation on *Mycobacterium* abscesses in cystic fibrosis patients. *Open Forum of Infectious Diseases*, *5*(3), ofy034.
Chen, C. S., Chang, R. S., Wu, J. S., Cheng, Y. C., & Lin, H. L., (2020). *Plasma Aerosol Inhalation Device and Method for Thinning the Sputum.* US Patent US20200268655A1.
Conti, D. S., Brewer, D., Grashik, J., Avasarala, S., & Da Rocha, S. R., (2014). Poly(amidoamine) dendrimer nanocarriers and their aerosol formulations for siRNA delivery to the lung epithelium. *Molecular Pharmaceutics, 11*, 1808-1822.
Costa-Gouveia, J., Pancani, E., Jouny, S., Machelart, A., Delorme, V., Salzano, G., Iantomasi, R., et al., (2017). Combination therapy for tuberculosis treatment: Pulmonary administration of ethionamide and booster co-loaded nanoparticles. *Scientific Reports, 7*(1), 1–14.
Dabbagh, A., Kasim, N. H., Yeong, C. H., Wong, T. W., & Rahman, N. A., (2018). Critical parameters for particle-based pulmonary delivery of chemotherapeutics. *Journal of Aerosol Medicine and Pulmonary Drug Delivery, 31*(3), 139–154.
Deacon, J., Abdelghany, S. M., Quinn, D. J., Schmid, D., Megaw, J., Donnelly, R. F., Jones, D. S., et al., (2015). Antimicrobial efficacy of tobramycin polymeric nanoparticles for *Pseudomonas aeruginosa* infections in cystic fibrosis: Formulation, characterization, and functionalization with dornase alfa (DNase). *Journal of Controlled Release: Official Journal of Controlled Release, 198*, 55–61.
Deb, P. K., Abed, S. N., Jaber, A. M., & Tekade, R. K., (2018). Chapter 5 – particulate level properties and its implications on product performance and processing. *Dosage Form Design Parameters, Advances in Pharmaceutical Product Development and Research, II*, 155–220.
Decramer, M., Rennard, S., Troosters, T., Mapel, D. W., Giardino, N., Mannino, D., Wouters, E., et al., (2008). COPD as a lung disease with systemic consequences – clinical impact, mechanisms, and potential for early intervention. COPD: *Journal of Chronic Obstructive Pulmonary Disease, 5*(3), 235–256.
Doshi, R., Howard, R. A., & Hatanaka, M., (2013). Respiratory devices and methods of use. *A Nasal Respiratory Device Has Designed for Delivery of Drug to the Lungs.* European Patent EP1824544B1.
European Respiratory Society. The economic burden of lung disease. In: *European Lung White Book.* https://www.erswhitebook.org/chapters/the-economic-burden-oflung-disease/ (accessed on 6 December 2023).
Frohlich, E., & Sharareh, S. B., (2021). Oral inhalation for delivery of proteins and peptides to the lungs. *European Journal of Pharmaceutics and Biopharmaceutics, 163*, 198–211.
Galbraith, S. D., & Rauker, R. M., (2021). *Portable Ventilator.* WIPO (PCT) Patent WO2021183903A1.
Gaspar, D. P., Gaspar, M. M., Eleutério, C. V., Grenha, A., Blanco, M., Gonçalves, L. M. D., Taboada, P., et al., (2017). Microencapsulated solid lipid nanoparticles as a hybrid platform for pulmonary antibiotic delivery. *Molecule Pharmaceutics, 14*(9), 2977–2990.
Grassin-Delyle, S., Buenestado, A., Naline, E., Faisy, C., Blouquit-Laye, S., Louis-Jean, C., Guen, M. L., et al., (2012). Intranasal drug delivery: An efficient and non-invasive route

for systemic administration: Focus on opioids. *Pharmacology and Therapeutics, 134*(3), 366–379.

Gross, C. P., Anderson, G. F., & Powe, N. R., (1999). The relation between funding by the National Institutes of Health and the burden of disease. *New England Journal of Medicine, 340*, 1881–1887.

Gunjal, S. S., Ahire, E. D., Surana, K. R., Sonawane. V. N., Pagar, K. R., Vijayraj, N. S., & Khushali, R. P. (2021). Influence of allopathy, homeopathy, ayurveda, and unani systems of medicine in the management of COVID-19. *COVID-19 Pandemic: Questions, Answers and Hypotheses, 1.*

Hadinoto, K., Phanapavudhikul, P., Kewu, Z., & Tan, R. B., (2007). Dry powder aerosol delivery of large hollow nanoparticulate aggregates as prospective carriers of nanoparticulate drugs: Effects of phospholipids. *International Journal of Pharmaceutics, 333*(1, 2), 187–198.

Hargreave, F. E., & Nair, P., (2009). The definition and diagnosis of asthma. *Clinical & Experimental Allergy, 39*(11), 1652–1658.

Haworth, C. S., Bilton, D., Chalmers, J. D., Davis, A. M., Froehlich, J., Gonda, I., Thompson, B., et al., (2019). Inhaled liposomal ciprofloxacin in patients with noncystic fibrosis bronchiectasis and chronic lung infection with *Pseudomonas aeruginosa* (ORBIT-3 and ORBIT-4): Two phase 3, randomized controlled trials. *Lancet Respiratory Medicine, 7*(3), 213–226.

Hirsch, E., & Ghigo, A., (2019). *Novel pi3k Gamma Inhibitor Peptide for Treatment of Respiratory System Diseases.* US Patent US20180086798A1.

Hitzman, C. J., Wattenberg, L. W., & Wiedmann, T. S., (2006). Pharmacokinetics of 5-fluorouracil in the hamster following inhalation delivery of lipid-coated nanoparticles. *Journal of Pharmaceutical Sciences, 95*(6), 1196–1211.

Hoffman, T., (2009). *Method for Treatment of COPD and other Pulmonary Diseases.* Chinese Patent CN102186458A.

Holgate, S. T., (2013). Mechanisms of asthma and implications for its prevention and treatment: A personal journey. *Allergy, Asthma & Immunology Research, 5*(6), 343–347.

Hongbin, S., (2019). *Oleanolic Acid Preparation for Inhalation, Preparation Method and Medical Application Thereof.* Chinese Patent CN110664819A.

Hu, X., Yang, F. F., Quan, L. H., Liu, C. Y., Liu, X. M., Ehrhardt, C., & Liao, Y. H., (2014). Pulmonary delivered polymeric micelles—Pharmacokinetic evaluation and biodistribution studies. *European Journal of Pharmaceutics and Biopharmaceutics, 88*(3), 1064–1075.

Huang, Z., Kłodzińska, S. N., Wan, F., & Nielsen, H. M., (2021). Nanoparticle-mediated pulmonary drug delivery: State of the art towards efficient treatment of recalcitrant respiratory tract bacterial infections. *Drug Delivery and Translational Research, 11*, 1634–1654.

Inapagolla, R., Guru, B. R., Kurtoglu, Y. E., Gao, X., Lieh-Lai, M., Bassett, D. J. P., & Kannan, R. M., (2010). *In vivo* efficacy of dendrimer-methylprednisolone conjugate formulation for the treatment of lung inflammation. *International Journal of Pharmaceutics, 399*, 140-147.

Jaafar-Maalej, C., Véronique, A., Abdelhamid, E., & Hatem, F., (2011). Beclomethasone-loaded lipidic nanocarriers for pulmonary drug delivery: Preparation, characterization, and *in vitro* drug release. *Journal of Nanoscience and Nanotechnology, 11*, 1841–1851.

James, S. L., Abate, D., & Abate, K. H., (2018). Global, regional, and national incidence, prevalence, and years lived with disability for 354 diseases and injuries for 195 countries and territories, 1990–2017: A systematic analysis for the Global Burden of Disease Study 2017. *Lancet, 392*, 1789–1858.

Javadzadeh, Y., & Shadi, Y., (2017). Chapter 20 – therapeutic nanostructures for pulmonary drug delivery. *Nanostructures for Drug Delivery Micro and Nano Technologies*, 619–638.

Jean-Sebastian, L., & Adrian, M., (2021). *Device for Stimulating Tracheal / Bronchial Air*. Japan Patent CN102186458A.

Ji, P., Yu, T., Liu, Y., Jiang, J., Xu, J., Zhao, Y., Hao, Y., et al., (2016). Naringenin-loaded solid lipid nanoparticles: Preparation, controlled delivery, cellular uptake, and pulmonary pharmacokinetics. *Drug Design Development and Therapy, 10*, 911–925.

Karlinsky, J. B., & Snider, G. L., (1978). Animal models of emphysema. *American Review of Respiratory Disease, 117*, 1109–1133.

Khan, O. F., Zaia, E. W., Jhunjhunwala, S., Xue, W., Cai, W., Yun, D. S., Barnes, C. M., et al., (2015). Dendrimer-inspired nanomaterials for the *in vivo* delivery of siRNA to lung vasculature. *Nano Letters, 15*(5), 3008–3016.

Kim, G., Piao, C., Oh, J., & Lee, M., (2018). Self-assembled polymeric micelles for combined delivery of anti-inflammatory gene and drug to the lungs by inhalation. *Nanoscale, 10*(18), 8503–8514.

Labiris, N. R., & Dolovich, M. B., (2003). Pulmonary drug delivery. Part II: The role of inhalant delivery devices and drug formulations in therapeutic effectiveness of aerosolized medications. *British Journal of Clinical Pharmacology, 56*(6), 600–612.

Licciulli, P., (2014). *Apparatus for the Administration of Pharmaceutical Products in Aerosol Form by Controlling Pressure*. Canada Patent CA2601532C.

Liechty, K., Zgheib, C., & Hilton, S. A., (2019). *Methods for Preventing and Treating Pulmonary Inflammation and Fibrosis*. US Patent US20210338718A1.

Liu, Y., Pleasants, R. A., Croft, J. B., Wheaton, A. G., Heidari, K., Malarcher, A. M., Ohar, J. A., et al., (2015). Smoking duration, respiratory symptoms, and COPD in adults aged ≥45 years with a smoking history. *International Journal of COPD, 10*, 1409–1416.

Makled, S., Nafee, N., & Boraie, N., (2017). Nebulized solid lipid nanoparticles for the potential treatment of pulmonary hypertension via targeted delivery of phosphodiesterase-5-inhibitor. *International Journal of Pharmaceutics, 517*(1), 312–321.

McShane, H., (2011). Tuberculosis vaccines: Beyond bacille Calmette–Guérin. *Philosophical Transactions of the Royal Society B: Biological Sciences, 366*(1579), 2782–2789.

Mehta, M., Deeksha, Tewari, D., Gupta, G., Awasthi, R., Singh, H., Pandey, P., et al., (2019). Oligonucleotide therapy: An emerging focus area for drug delivery in chronic inflammatory respiratory diseases. *Chemico-Biological Interaction, 308*, 206-215.

Mendes, L. P., Pan, J., & Torchilin, V. P., (2017). Dendrimers as nanocarriers for nucleic acid and drug delivery in cancer therapy. *Molecules, 22*, 1401–1422.

Miller, C. J., (2009). DVM, DACVIM, Chapter 192 – aerosolized medications. *Small Animal Critical Care Medicine*, 814–817.

Mishra, B., & Singh, J., (2020). Novel drug delivery systems and significance in respiratory diseases. *Targeting Chronic Inflammatory Lung Diseases Using Advanced Drug Delivery Systems, 57*–95.

Mitsopoulos, P., Omri, A., Alipour, M., Vermeulen, N., Smith, M. G., & Suntres, Z. E., (2008). Effectiveness of liposomal-N-acetylcysteine against LPS-induced lung injuries in rodents. *International Journal of Pharmaceutics, 363*, 106–111.

Morton, D., Ganderton, D., Staniforth, J., & Kamlag, Y., (2004). *Mucoactive Agents for Treating a Pulmonary Disease*. WIPO (PCT) Patent WO2005025540A3.

Muellinger, B., & Snape, S., (2019). *Pharmaceutical Administration to Neonates, Infants, and Children*. US Patent US20220008666A1.

Mukherjee, B., Paul, P., Dutta, L., Chakraborty, S., Dhara, M., Mondal, L., & Sengupta, S., (2017). Chapter 14 – pulmonary administration of biodegradable drug nanocarriers for more efficacious treatment of fungal infections in lungs: Insights based on recent findings. *Multifunctional Systems for Combined Delivery, Biosensing and Diagnostics*, 261–280.

Mussi, S. V., Silva, R. C., Oliveira, M. C., Lucci, C. M., Azevedo, R. B., & Ferreira, L. A. M., (2013). New approach to improve encapsulation and antitumor activity of doxorubicin loaded in solid lipid nanoparticles. *European Journal of Pharmaceutical Sciences, 48*, 282–290.

Myrdal, P. B., Sheth, P., & Stein, S. W., (2014). Advances in metered dose inhaler technology: Formulation development. *AAPS PharmSciTech, 15*(2), 434–455. doi: 10.1208/s12249-013-0063-x.

Narasimhan, P., Wood, J., MacIntyre, C. R., & Mathai, D., (2013). Risk factors for tuberculosis. *Pulmonary Medicine*, 1–11.

Nassimi, M., Schleh, C., Lauenstein, H. D., Hussein, R., Lübbers, K., Pohlmann, G., Switalla, S., et al., (2009). Low cytotoxicity of solid lipid nanoparticles in *in vitro* and ex vivo lung models. *Inhalation Toxicology, 21*(S1), 104–109.

Nassimi, M., Schleh, C., Lauenstein, H., Hussein, R., Hoymann, H., Koch, W., Pohlmann, G., et al., (2010). A toxicological evaluation of inhaled solid lipid nanoparticles used as a potential drug delivery system for the lung. *European Journal of Pharmaceutics and Biopharmaceutics, 75*, 107–116.

National Heart, Lung, and Blood Institute. *Guidelines for the Diagnosis and Management of Asthma (EPR-3)*. https://www.nhlbi.nih.gov/files/docs/guidelines/asthma_qrg.pdf (accessed on 6 December 2023).

Pai, M., (2020). Tuberculosis: The story after the primer. *Nature Reviews Disease Primers, 6*(1), 1–23.

Passi, M., Shahid, S., Chockalingam, S., Sundar, I. K., & Packirisamy, G., (2020). Conventional and nanotechnology-based approaches to combat chronic obstructive pulmonary disease: Implications for chronic airway diseases. *International Journal of Nanomedicine, 15*, 3803–3826.

Patil, J. S., & Sarasija, S., (2012). Pulmonary drug delivery strategies: A concise, systematic review. *Lung India, 29*(1), 44–49. doi: 10.4103/0970-2113.92361.

Peter, J. B., (2000). Mechanisms in COPD: Differences from asthma. *Chest, 117*(2), 105–145.

Preedy, E. C., & Prokopovich, P., (2013). 2 – History of inhaler devices, inhaler devices, fundamentals, design and drug delivery. *Woodhead Publishing Series in Biomaterials, 13*–28.

Rajabnezhad, S., Casettari, L., Lam, J. K., Nomani, A., Torkamani, M. R., Palmieri, G. F., Rajabnejad, M. R., & Darbandi, M. A., (2016). Pulmonary delivery of rifampicin microspheres using lower generation polyamidoamine dendrimers as a carrier. *Powder Technology, 291*, 366-374.

Ramalingam, V., Varunkumar, K., Ravikumar, V., & Rajaram, R., (2018). Target delivery of doxorubicin tethered with PVP stabilized gold nanoparticles for effective treatment of lung cancer. *Science Report, 8*(1), 3815–3827.

Ridge, C. A., McErlean, A. M., & Ginsberg, M. S., (2013). Epidemiology of lung cancer. *Seminars in Interventional Radiology, 30*(2), 93–98.

Rojanarat, W., Changsan, N., Tawithong, E., Pinsuwan, S., Chan, H. K., & Srichana, T., (2011). Isoniazid proliposome powders for inhalation-preparation, characterization, and cell culture studies. *International Journal of Molecular Sciences, 12*, 4414–4434.

Rosière, R., Amighi, K., & Wauthoz, N., (2019). Chapter 10 – nanomedicine-based inhalation treatments for lung cancer. *Nanotechnology-Based Targeted Drug Delivery Systems for Lung Cancer*, 249–268.

Rosière, R., Van, W., M., Gelbcke, M., Mathieu, V., Hecq, J., Mathivet, T., Vermeersch, M., et al., (2018). New folate-grafted chitosan derivative to improve delivery of paclitaxel-loaded solid lipid nanoparticles for lung tumor therapy by inhalation. *Molecule Pharmaceutics, 15*(3), 899–910.

Salvati, A., Pitek, A. S., Monopoli, M. P., Prapainop, K., Bombelli, F. B., Hristov, D. R., Kelly, P. M., et al., (2013). Transferrin-functionalized nanoparticles lose their targeting capabilities when a biomolecule corona adsorbs on the surface. *Nature Nanotechnology, 8*, 137–143.

Serisier, D. J., Bilton, D., De Soyza, A., Thompson, P. J., Kolbe, J., Greville, H. W., Cipolla, D., et al., (2013). ORBIT-2 Investigators. Inhaled, dual release liposomal ciprofloxacin in non-cystic fibrosis bronchiectasis (ORBIT-2): A randomized, double-blind, placebo-controlled trial. *Thorax, 68*(9), 812–817.

Silva, L. F. C., Kasten, G., De Campos, C. E. M., Chinelatto, A. L., & Lemos-Senna, E., (2013). Preparation and characterization of quercetin-loaded solid lipid microparticles for pulmonary delivery. *Powder Technology, 239*, 183–192.

Sing, R., Nawale, L., Arkile, M., Wadhwani, S., Shedbalkar, U., Chopade, S., Sarkar, D., & Chopade, B. A., (2016). Phytogenic silver, gold, and bimetallic nanoparticles as novel antitubercular agents. *International Journal of Nanomedicine, 11*, 1889–1897.

Sorino, C., Negri, S., Spanevello, A., Visca, D., & Scichilone, N., (2020). Inhalation therapy devices for the treatment of obstructive lung diseases: The history of inhalers towards the ideal inhaler. *European Journal of Internal Medicine, 75*, 15–18.

Sung, J. C., Pulliam, B. L., & Edwards, D. A., (2007). Nanoparticles for drug delivery to the lungs *Trends Biotechnology, 25*, 563–570.

Talele, S. G., Ahire, E. D., Surana, K. R., Sonawane, V. N., & Talele, G. S., (2022). Corona virus disease (COVID-19): A past and present prospective. *Asian Journal of Pharmaceutical Research, 12*(1), 45–53.

Tan, Y., Yang, Z., Pan, X., Chen, M., Feng, M., Wang, L., Liud, H., et al., (2012). Stability and aerosolization of pressurized metered dose inhalers containing thymopentin nanoparticles produced using a bottom-up process. *International Journal of Pharmaceutics, 427*(2), 385–392.

Tanuja, B., (2020). COVID: A basic review. *International Join of Alternative and Complementary Medicine, 1*(2), 46–48.

Thombre, N. A., Niphade, P. S., Ahire, E. D., & Kshirsagar, S. J., (2022). Formulation development and evaluation of microemulsion based lornoxicam gel. *Biosciences Biotechnology Research Asia, 19*(1), 69–80.

Tortorici, A. M., & Veesler, D., (2019). Structural insights coronavirus entry. *Advance in Virus Res., 105*, 94–116.

Trapani, A., Gioia, S. D., Ditaranto, N., Cioffi, N., Goycoolea, F. M., Carbone, A., Garcia-Fuentes, M., et al., (2013). Systemic heparin delivery by the pulmonary route using chitosan and glycol chitosan nanoparticles. *International Journal of Pharmaceutics, 447*, 115–123.

Trotta, V., Lee, W. H., Loo, C. Y., Young, P. M., Traini, D., & Scalia, S., (2016). Co-spray dried resveratrol and budesonide inhalation formulation for reducing inflammation and oxidative stress in rat alveolar macrophages. *European Journal of Pharmaceutical Sciences, 30*(86), 1–43.

Uhlig, S., Marco, R., & Slutsky, S. A., (2004). Biotrauma hypothesis of ventilator-induced lung injury. *American Journal of Respiratory and Critical Care Medicine, 169*(2), 314–320.

Varshosaz, J., Ghaffari, S., Mirshojaei, S. F., Jafarian, A., Atyabi, F., Kobarfard, F., & Azarmi, S., (2013). Biodistribution of amikacin solid lipid nanoparticles after pulmonary delivery. *BioMed. Research International, 2013*, 1–8.

Ventola, C. L., (2017). Progress in nanomedicine: Approved and investigational nanodrugs. *PT., 42*(12), 742–755.

Verschraegen, C. F., Gilbert, B. E., Loyer, E., Huaringa, A., Walsh, G., Newman, R. A., & Knight, V., (2004). Clinical evaluation of the delivery and safety of aerosolized liposomal 9-nitro-20 (s)-camptothecin in patients with advanced pulmonary malignancies. *Clinical Cancer Research, 10*(7), 2319–2326.

Vestbo, J., Prescott, E., & Lange, P., (1996). Association of chronic mucus hypersecretion with FEV1 decline and chronic obstructive pulmonary disease morbidity. Copenhagen City Heart Study Group. *American Journal of Respiratory and Critical Care Medicine, 153*, 1530–1535.

Vladimir, K., Jan, C., Vladimir, Z., Katerina, N., Jakub, Z., Jaroslav, Z., Vratislav, S., et al., (2013). Chronic obstructive pulmonary disease: Official diagnosis and treatment guidelines of the Czech pneumological and phthisiological society; a novel phenotypic approach to COPD with patient-oriented care. *Biomedical Papers, 157*(2), 189–201.

Wang, W., Zhu, R., Xie, Q., Li, A., Xiao, Y., Liu, H., Wang, S., & Cui, D., (2012). Enhanced bioavailability and efficiency of curcumin for the treatment of asthma by its formulation in solid lipid nanoparticles. *International Journal of Nanotechnology, 7*, 3667–3677.

Wang, X., Chen, Q., Zhang, X., Ren, X., Zhang, X., Meng, L., Liang, H., et al., (2018). Matrix metalloproteinase 2/9-triggered-release micelles for inhaled drug delivery to treat lung cancer: Preparation and *in vitro/in vivo* studies. *International Journal of Nanomedicine, 13*, 4641–4659.

Wu, C., Xu, J., Hao, Y., Zhao, Y., Qiu, Y., Jiang, J., Yu, T., et al., (2017). Application of a lipid-coated hollow calcium phosphate nanoparticle in synergistic co-delivery of doxorubicin and paclitaxel for the treatment of human lung cancer A549 cells. *International Journal of Nanomedicine, 12*, 7979–7992.

Wu, Z., & McGoogan, J. M., (2020). Characteristics of and important lessons from the corona virus disease 2019 (COVID-19) outbreak in China summary of a report of 72 314 cases from the Chinese center for disease control and prevention. *Journal of the American Medical Association, 323*(1), 1239–1242.

Xu, J., Jiangang, T., & Jidong, W., (2020). Design and application in the delivery system of intranasal antidepressants. *Frontiers in Bioengineering and Biotechnology, 8*(626882), 1–13.

Yurdasiper, A., Arıcı, M., & Ozyazici, M., (2018). Chapter 7 – nanopharmaceuticals: Application in inhaler systems. *Emerging Nanotechnologies in Immunology, The Design, Applications and Toxicology of Nanopharmaceuticals and Nanovaccines Micro and Nano Technologies, 165*–201.

Zappa, C., & Mousa S. A., (2016). Non-small cell lung cancer current treatment and future advances. *Translational Lung Cancer Research, 5*(3), 288–300.

Zhong, Q., Bielski, E. R., Rodrigues, L. S., Brown, M. R., Reineke, J. J., & Da Rocha, S. R., (2016). Conjugation to poly(amidoamine) dendrimers and pulmonary delivery reduce cardiac accumulation and enhance antitumor activity of doxorubicin in lung metastasis. *Molecular Pharmaceutics, 13*, 2363–2375.

CHAPTER 11

New Frontiers for the Management of Diabetic Retinopathy

UMESH DILIP LADDHA,[1] KAILAS K. MORAVKAR,[2] EKNATH D. AHIRE,[1] and SANJAY J. KSHIRSAGAR[1]

[1]*MET's Institute of Pharmacy, Bhujbal Knowledge City, Adgaon, Nasik, Maharashtra, India*

[2]*Principal Research Engineer, Regeron Inc., Chuncheon, Gangwon-Do, South Korea*

ABSTRACT

Diabetes mellitus (DM) is known as a global epidemic which has affected both developed and developing countries equally. The major concern with diabetes is its adverse impact on almost all parts of the body. Diabetes significantly affects the eyes and leads to the development of a microvascular complication known as diabetic retinopathy (DR). It is associated with the development of new but weak blood vessels at the retinal site, which get ruptured due to blood flow and result in blood leakage. This leads to blindness in patients aged 25-65. Currently, DR is treated by invasive methods such as vitrectomy, intravitreal injection, and laser treatment. These methods are associated with pain, secondary infections, high cost, and poor patient compliance. Recent advancements in research have tried to address the limitations of the current treatment of retinopathy. This chapter provides insight into the development of DR, the limitations of current treatment, and the applications of novel delivery systems for better care of retinopathy.

Novel Drug Delivery Systems in the Management of Chronic Diseases.
Raj K. Keservani, Umesh D. Laddha, Eknath D. Ahire, and Sanjay J. Kshirsagar (Eds.)

11.1 DIABETES

Diabetes mellitus (DM) or simply Diabetes is the most common chronic metabolic disorder associated with elevated blood glucose levels (hyperglycemia) (Kharroubi et al., 2015). The fasting blood glucose level > 126 mg/dl and random plasma glucose level > 200 mg/dl, HbA1c > 6.5% is the alarming condition of hyperglycemia (Khawandanah, 2019). It has affected both developed and developing countries equally. DM is the result of impaired insulin secretion or flawed action of insulin or a combination of both (Punthakee et al., 2018). Many times, affected people are unaware of the existence of DM and unfortunately, only half of the affected population receive proper treatment (Adapa & Sarangi, 2015). The major complication associated with DM is that it simultaneously starts to damage other essential systems of the body. Chronic hyperglycemia significantly affects the eyes, kidneys, and nerves, as well as increases the risk of cardiovascular complications (Deshmukh & Jain, 2015). Uncontrolled diabetes may lead to stupor and coma. Many times, uncontrolled DM also results in the death of the patient due to ketoacidosis (Galtier, 2010).

11.2 CLASSIFICATION OF DM

When the body gets a lesser amount of insulin or too much food or little exercise, it predominantly results in hyperglycemia, i.e., DM (Ali, 2011; Shanker et al., 2011). Although the categorization of diabetes is important because it directly affects treatment strategies, it is not easy to classify patients in a single class, and many times it demands a revision of the classified person (Thunander et al., 2012; Cakan, 2012).

The most preferable classification is as given by the American Diabetes Association (ADA) in 1997. ADA has classified DM into three types as below (Figure 11).

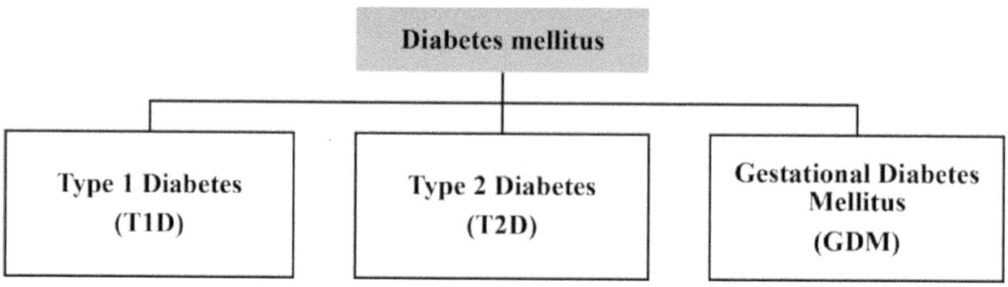

FIGURE 11.1 Types of diabetes mellitus.

1. **Type 1 Diabetes:** It is also referred as the insulin dependent diabetes mellitus (IDDM). T1D is develops due to the autoimmune damage of the β-cells which further suppressed and finally completely inhibits the insulin production. It is more responsible for ketoacidosis. T1D is alternatively well-known as the juvenile diabetes (Punthakee, 2018; Uppu & Parinandi, 2011).
2. **Type 2 Diabetes:** It also known as NIDDM, i.e., non-insulin dependent diabetes mellitus (NIDDM) as well as the adult-onset diabetes. Resistance to insulin, deficiency of insulin or insulin absence are the causative factors for development of T2D. It is no likely responsible for ketoacidosis (Punthakee, 2018; Uppu & Parinandi, 2011). Unhealthy diet, obesity, genetic factor and deskbound routine are the key factors in development of T2D.
3. **Gestational Diabetes Mellitus (DM):** It is observed during pregnancy where it shows varying degree of glucose intolerance. However; it may or may not further continue after pregnancy (Lemos et al., 2011; Vilar, 2006).

11.3 DIABETES COMPLICATIONS

Complications of DM are progressive which affects to almost all parts of body (Sayin et al., 2015). DM and associated complications are known as the most considerable cause of morbidity and mortality worldwide (Forbes et al., 2005; Jang et al., 2011). DM is associated with various complications including eye disorders, cardiovascular disease, renal disorders, cerebrovascular diseases and many more. Complications of DM include retinopathy, hypertension, renal disease, erectile dysfunction, neuropathy, electrolyte imbalance and complications in pregnancy.

11.4 DIABETES AND EYES

Anatomy, protective, and physiological factors made ocular site as complex and unique (Kumar et al., 2013). It is known as the most imperative organ and its defense mechanism renders ocular drug delivery as the most challenging task to ophthalmologist (Almeida et al., 2013; Kaul et al., 2012; Rajoria & Gupta, 2012). Diabetes has adverse impact almost in all parts of body (Lutty, 2013). DM affects progressively to eye and if not treated may results into poor vison to blindness. Cornea and retina are the major sufferer of impact of hyperglycemia. It has been reported that in comparison to normal cornea, the diabetic cornea experiences four times more glucose in tear film. Moreover; patients with diabetes are more prone for visual loss due to impact of hyperglycemia on retina. Eyes complications due to DM are referred as diabetic eye disease (DED). DM can lead to develops various complications in context with eye viz. retinopathy, macular edema, increased ocular pressure, i.e., glaucoma, cataract, and many other ocular surface diseases (Threatt et al., 2013). Furthermore; diabetic patients are more prone for any ocular disease in comparison to normal person (Victoria et al., 2016).

Eye structure is broadly divided into two parts, viz. frontal part and posterior part (Cholkar et al., 2013). Hence; ocular diseases can also be classified as anterior part disease and posterior part disease as per site of occurrence (Waugh & Grant, 2010). Diabetic retinopathy (DR) is the major complication of diabetes which causes significant damage to retina. Retina located to back of the eye where delivery of drug is most crucial and challenging aspect (Loftsson, 2017). Anatomy and protective barriers of eye restricts the movement of active moiety in deep of eye. The BAB, i.e., blood-aqueous-barrier followed to BRB, i.e., blood-retinal-barrier control the uptake of drug by eye (Weng et al., 2017).

Ocular absorption of drug is restricted by various barriers which can be sort out into three parts viz. static, dynamic, and metabolic barriers which are summarized in Figure 11.2.

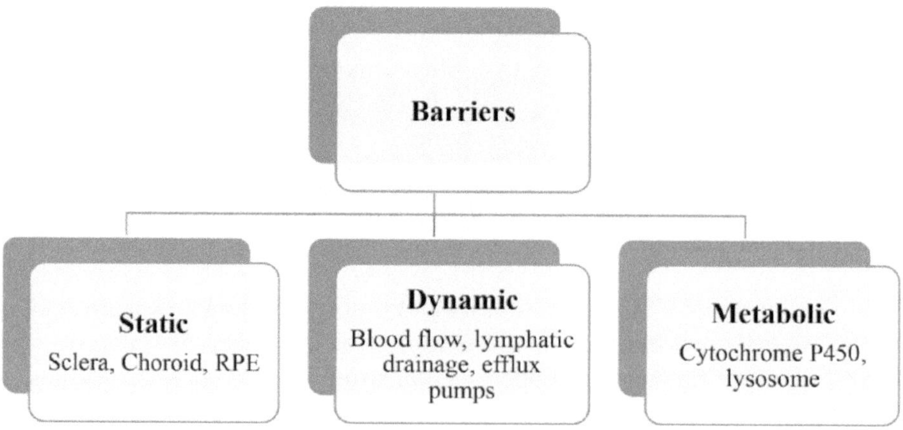

FIGURE 11.2 Barriers for ocular absorption of drug.

The ocular barriers along with the traveling distance between the cornea and retina followed to defense mechanism of eye affects the delivery of drug in desired concentration in posterior segment (Loftsson, 2008; Weng et al., 2017). Furthermore; rapid ocular clearance along with penetration barriers results into poor ocular bioavailability of drug (~5%) (Laddha & Mahajan, 2017). To target the retina, one has to consider the barriers associated with it. Therefore; if someone wish to target drug to retinal site by topical administration then the one of the possible ways is to administer high dose of drug which may leads to systemic toxicity. Another possibility is systemic administration of therapeutic dose, in which, again the bioavailability will be restricted due to inner blood retinal barrier (Dubald et al., 2018). Henceforth; there is need to find out some strategies to treat back of eye, i.e., posterior segment without any tissue damage (Patil et al., 2012).

11.5 DIABETIC RETINOPATHY (DR)

Diabetes mellitus (DM) is now considered as a global epidemic, which has already affected 382 million of citizens and it is predicted that by year 2035 this number will be double (Das et al., 2015). Along with diabetes, management of its associated complications is the prime challenge in front of all health care provider as well as researchers (Daniel et al., 2012).

Patients with diabetes are more prone for development of DR which is known to be the most devastating microvascular complication (Renu et al., 2014). Risk of DR is at alarming level. WHO has defined DR as 'characteristic group of lesions found in the retina of individuals having DM.' DR can lead to legal blindness to a patient of wide age range from 20–65 years (Abdulrahman & Alghadya, 2011). Furthermore; prevalence of DR is 100% in patient with Type 1 diabetes and 60% in patients with Type 2 DM. DR can occur at any stage of incidence of diabetes (Joanna et al., 2013).

11.5.1 RISK FACTORS FOR DR

WHO has classified factors that increase the risk of DR into two types vi,z. modifiable risk factors and non-modifiable risk factors (Figure 11.3).

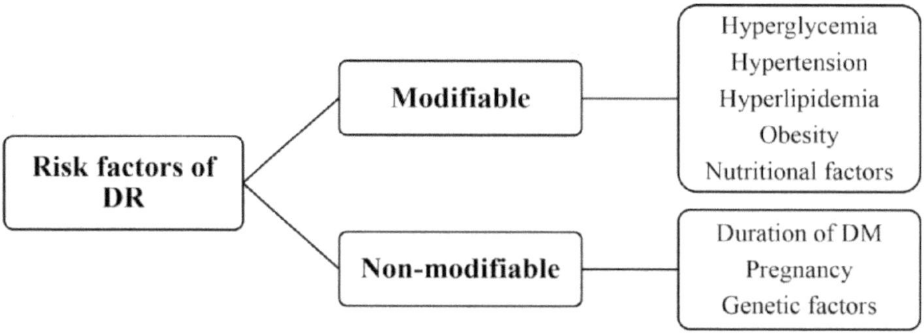

FIGURE 11.3 Risk factors of diabetic retinopathy.

11.5.2 PREVALENCE OF DR

According to WHO, DR is contributed for 5% of global blindness which accounts around 37 million of people (Daniel et al., 2012). Moreover; one survey was performed from 1980 to 2008 in different 35 countries from U.S., Australia, Europe, and Asia showed that the overall prevalence of DR was 34.6% and 7% of vision threatening DR (VTDR) (Daniel et al., 2016). Approximately 24,000 become blind each year due to DR in the United States. It has been projected that the number of DR and VTDR to be increase by 191 million and 57 million, respectively up to year 2030. As per survey of National eye Institute, at present in India 7.7 million of people are suffering from DR, which may reach to 11 million by 2030 and 15 million by 2050 (Figure 11.4).

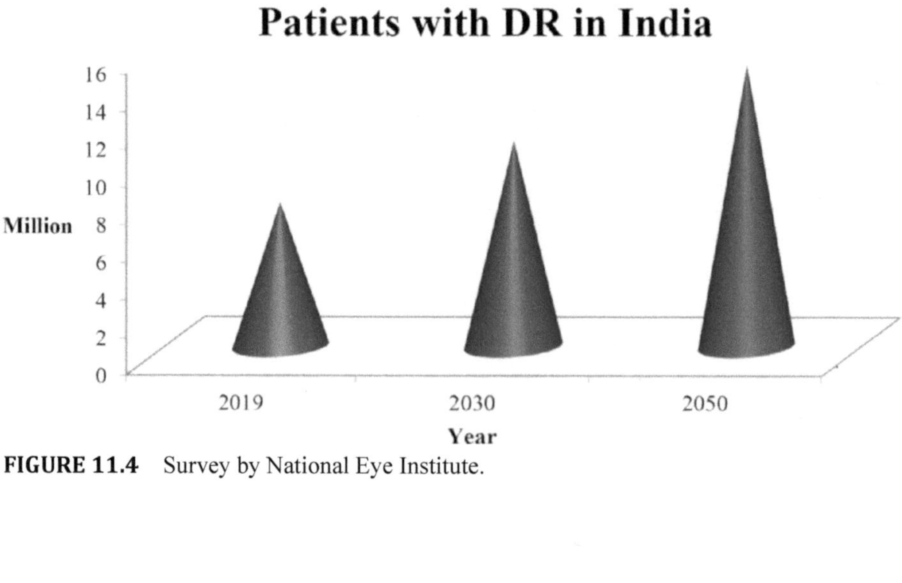

FIGURE 11.4 Survey by National Eye Institute.

11.5.3 *SYMPTOMS OF DR*

The early stage of DR does not show any symptoms. However, as time passes few of the symptoms start to persist depending on the stage of DR (niddk.nih.gov) (Figure 11.5):

- Blur vision;
- Double vision;
- Blank spots in vision;
- Spotty vision;
- Difficulty reading;
- Difficulty in color interpretation;
- Difficult to predict the object from eye corner site.

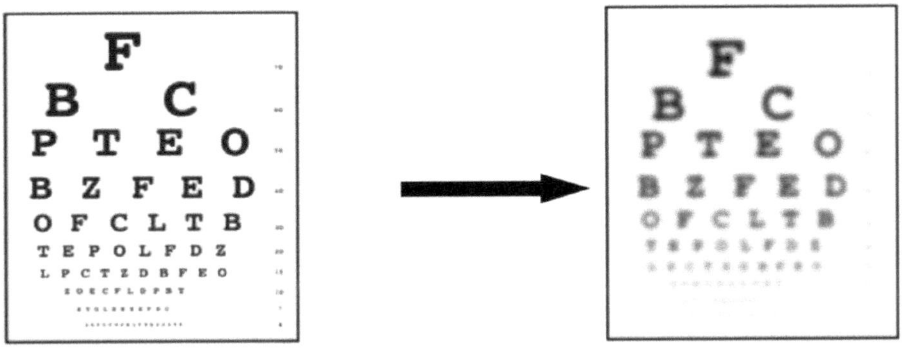

FIGURE 11.5 Normal vision vs. blurry vision.

11.5.4 *CAUSE AND DEVELOPMENT OF DR*

The main cause of DR is diabetes itself. Hyperglycemia damages the small blood vessels of retina which provides nourishment to retina. Moreover; increased glucose level can also block these vessels completely. At this stage, damaged blood vessels leak the fluid into retina which causes swelling of macula. This condition is called diabetic macular edema (DME). As when there is swelling to macula, it affects the reading ability and face reorganization capacity of person. If this damage is continued, then vestibular epithelial growth factor secretion increases which leads to develop new abnormal blood vessels at retina and vitreous site. This process is known as neovascularization and this stage is called proliferative diabetic retinopathy (PDR). These new vessels are formed in response to poor nourishment to retina. However; these newly form capillaries are weak and hence; breaks easily which leak blood into the vitreous gel. The leaking blood prevent light movement towards the retina. It results into development of spotty vision. Amount of damage may be same or different in both eyes. This bleeding may cause scar tissue formation and leads to retinal detachment. It may also lead to high pressure and pain in case when the blood vessels grow on the iris. All this ends with loss of vision (American society of retina specialist). Difference in normal and with DR is shown in Figure 11.6.

FIGURE 11.6 Normal vs. diabetic retinopathy vision.

Source: Modified from National Eye Institute, National Institute of Health

11.5.5 TYPES OF DR

Diabetic retinopathy (DR) can be classified into two types, as shown in Figure 11.7.

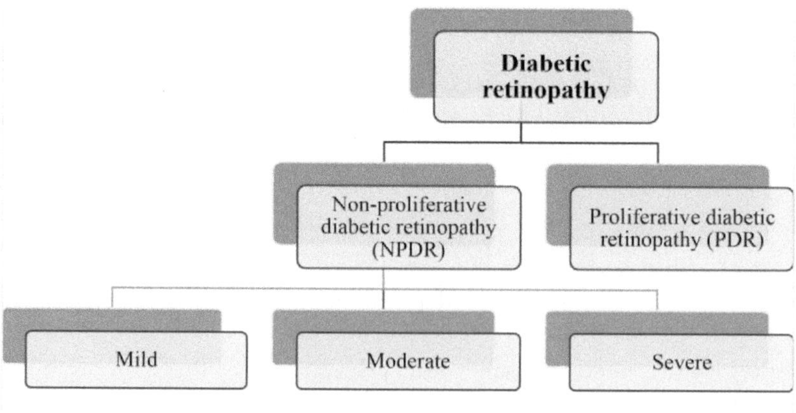

FIGURE 11.7 Types of diabetic retinopathy.

Source: Adapted from Jonathan 2

1. **Non-Proliferative Diabetic Retinopathy (NPDR): It is characterized by:**
   - Microaneurysms;
   - Exudates;
   - Hemorrhages;
   - Microinfarcts.

   Depending on changes taketakes place, NPDR is classified into three types:
   - Mild;
   - Moderate; and
   - Severe.

   Characteristics of each type are summarized in Table 11.1.

TABLE 11.1 Characteristics of Types of Non-Proliferative Diabetic Retinopathy

Type of NPDR	Characteristics
Mild to moderate	Microaneurysms.
	Intra-retinal hemorrhages.
	Hard exudate with or without macular edema.
Moderate to severe	Extensive intra-retinal hemorrhages.
	And/or cotton wool spots.
	And/or venous beading.
	And/or intra-retinal microvascular abnormalities (IRMA).
Severe to very severe	Cotton wool spots + venous beading + IRMA in 4:2:1 quadrant.

   Structural difference between normal eye and eye with NPDR is shown in Figure 11.8.

2. **Proliferative DR:** It is the complicated case of DR. This stage is associated with formation of new but weak blood vessels (neovascularization) as shown in Figure 11.9. It is associated with the new vessel formation from two pathways as:

   i. **New Vessels on the Disk (NVD):** where new blood vessels arise on the optic disk.

   ii. **New Vessel Elsewhere (NVE):** New blood vessels develop at any place of retina.

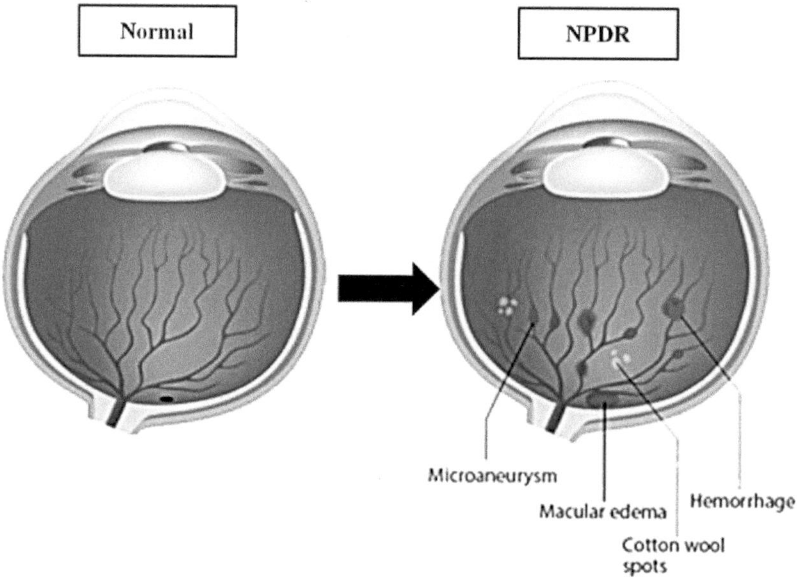

FIGURE 11.8 Characteristic changes in eye with NPDR.

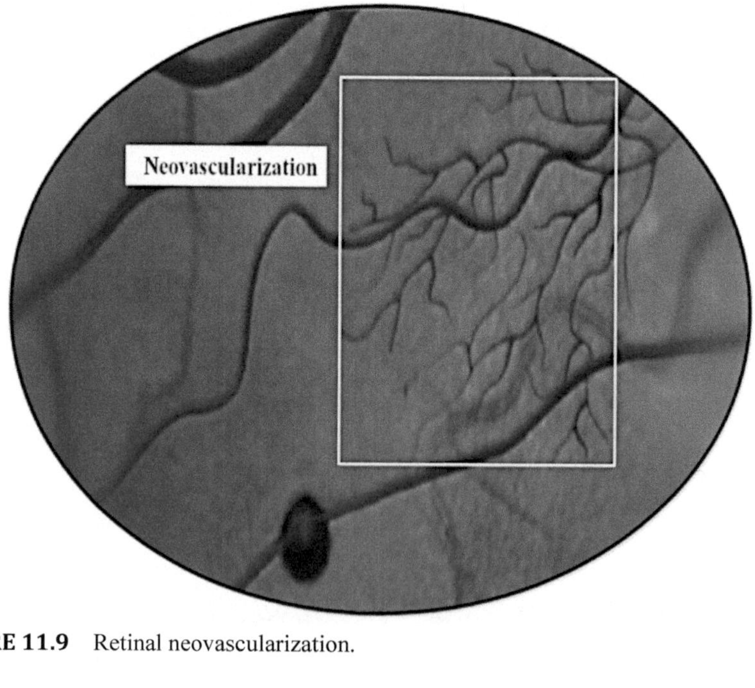

FIGURE 11.9 Retinal neovascularization.

PDR can be categorized into three parts as early PDR, PDR with high risk and PDR with advanced eye disease (Table 11.2).

TABLE 11.2 Types and Characteristics of PDR

Type of PDR	Characteristics
EPDR	Retinal hemorrhage.
PDR with high-risk criteria	Vitreous hemorrhage.
	NVD >1/3-disc diameter.
	NVE > 1/2-disc diameter.
PDR with advanced eye disease	Retinal detachment.
	Neovascularization of iris.

11.5.5.1 DIABETIC MACULAR EDEMA (DME)

It occurs at any stage of DR and result in loss of vision. The most crucial area of retina is macula which is responsible for central as well as color vision. It possesses the highest density of photoreceptors.

In DME, damage to blood vessels leads to leakage of blood or fluid into the macula, which results in swelling and thickening of the retina as shown in Figure 11.10 (Borja et al., 2017; Xu et al., 2009; Nguyen et al., 2009; Abdulrahman & Alghadyan, 2011).

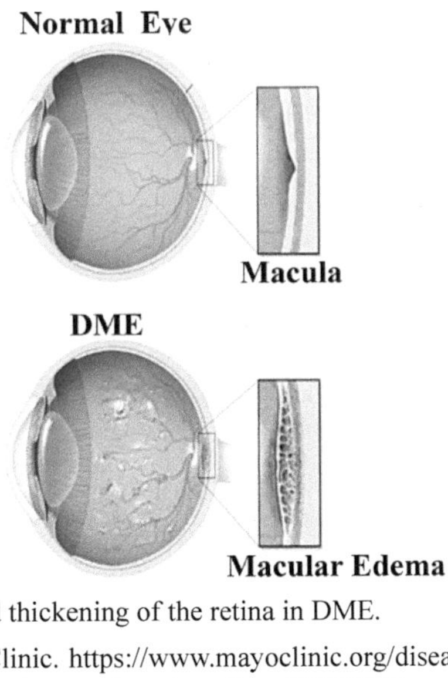

FIGURE 11.10 Swelling and thickening of the retina in DME.

Source: Adapted from Mayo Clinic. https://www.mayoclinic.org/diseases-conditions/diabetic-retinopathy/multimedia/diabetic-macular-edema/img-20124558

11.5.6 DIABETIC RETINOPATHY (DR): PATHOLOGY

Hyperglycemia leads to damaging effect on the vascular as well as pigment epithelial cells of retina (Hammes et al., 1995; Mizutani et al., 1996; Decanini et al., 2008). With progress of diabetes, basement membrane goes on thickening along with retinal capillary non-perfusion which alters the production pattern of many of mediators which includes growth factors, coagulation molecules, vasoactive agents, etc. Together; it results into drastic change in cell turnover (apoptosis, proliferation), amplified blood flow, modified capillary diameter and finally proaggrgant pattern and angiogenesis. These pathological changes lead to breakdown of BRB, retinal hemorrhage and retinal detachment (Gardner et al., 2000; Ciulla et al., 2003; Yam & Kwok, 2007).

Development of DR is associated with increased formation of advanced glycation end (AGE) products, i.e., AGEs and more oxidative stress. Furthermore; these biochemical modifications surge several other activities like promotion of apoptosis, inflammation, angiogenesis which together damages the retina and results into worst known complication of diabetes as DR (Ciulla et al., 2003; Abu El-Asrar et al., 2009). The cascade of DR is shown in Figure 11.11.

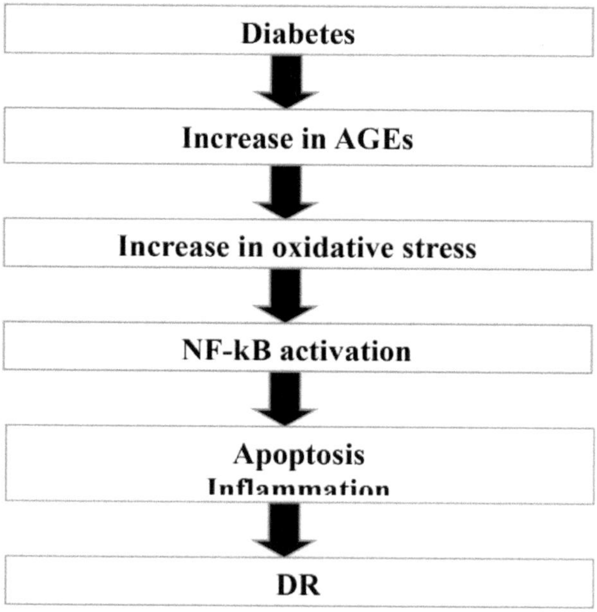

FIGURE 11.11 Development of DR.

11.5.7 CURRENT MANAGEMENT OF DR

Current treatment of DR involves use of any or combination of laser treatment, intravetrous injection and vitrectomy. Still, the complete cure is not available for DR, for a sake the loss of vision can be prevented or extended by either of above-mentioned therapy (Ciudin et al., 2013; Boyd, 2019).

1. **Laser Photocoagulation Treatment:** Laser treatment has been considered as the evidence-based practice for DME and PDR. This treatment uses laser to produce small areas of heat on the retinal site. It creates small laser burns and leads to shrinkage of blood vessels and ultimately those weak and small vessels get disappear. The main purpose of this therapy is not to get better vision acuity; rather its ,aim is to prevent further proliferation either at NVD or NVE and, so far, prevent bleeding at the retinal and vitreous sites.

   Laser therapy has been divided into two types:

   i. **Macular Laser Therapy (MLT):** In this case there is not any need of hospitalization of patient. Typically, it takes 10 minutes for actual treatment. Patient sits at slit lamp as usual ophthalmic examination with slight modification to accept a laser fiberoptic cable. Prior to treatment, cornea is got anesthetized by drops of proxymetacaine or benoxinate. During treatment patient do not experience any pain but may get sensation of bright flashes. During treatment it is most important that patient must keep eye in proper position, otherwise it may affect to fovea.

      After treatment, patient may be dazzled, however this effect fades with time. No after treatment or special precautions are required after MLT.

      ➢ **Limitations:**

      - Some reduction in color vision;
      - Development of tiny dark points at central vision;
      - Accidental visual loss.

   ii. **Panretinal Laser Therapy (PLT):** This treatment is similar to MLT, however; it involves a greater number of larger, intense burns which are applied very rapidly in comparison to MLT. Therefore; depending on patient it may or may not require hospitalization. In general, it involves 1,500–2,000 burns of size 500 μm with speed of 0.1–0.2 per seconds (Figure 11.12).

It is associated with pains after repeated treatment. PLT takes around 30 minutes for completion. After treatment doctor may prescribes oral NSAIDs like ibuprofen, flurbiprofen, etc. (Jonathan, 2003).

➢ Limitations:

- Reduces visual acuity;
- Impair peripheral retinal function;
- Reduces light sensitivity;
- Reduces visual field;
- During treatment, any movement of patient can permanently damage to retinal vessels.

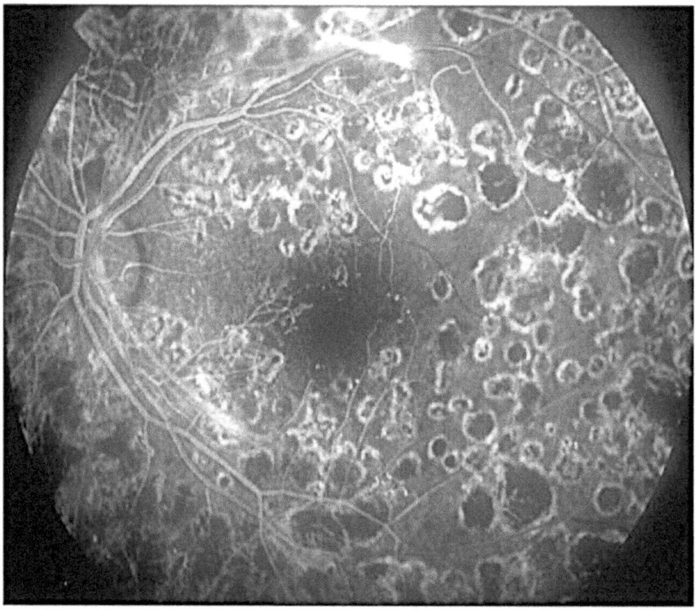

FIGURE 11.12 Panretinal laser therapy (PLT).

Source: Adapted from Jonathan (2003)

2. **Vitrectomy:** It involves the replacement of affected vitreous fluid with mild anesthesia. In this method temporary water-tight openings called ports are placed in the eye which helps to insert and remove instruments called a Vitrector (Figure 11.13). Through port a clear salt solution is smoothly pushed into the eye. The same instruments can also be used to take away scar tissue or to restore a detached retina.

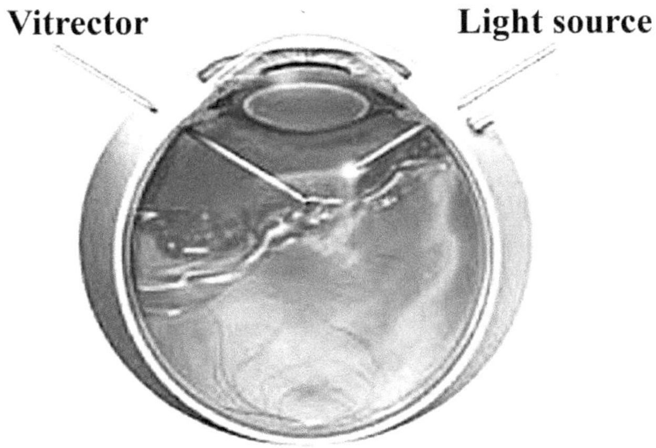

FIGURE 11.13 Vitrectomy.
Source: Adapted from VRMNY. https://www.vrmny.com/conditions/vitrectomy/

3. **Intravetrous Injection:** Several studies have shown the effectiveness of various therapeutic agents administered intravetrous way in management of DR (Figure 11.14). Steroids and anti-VEGF agents are the main leading molecules for treatment of DR. However; use of steroids is more effective in DME than DR.

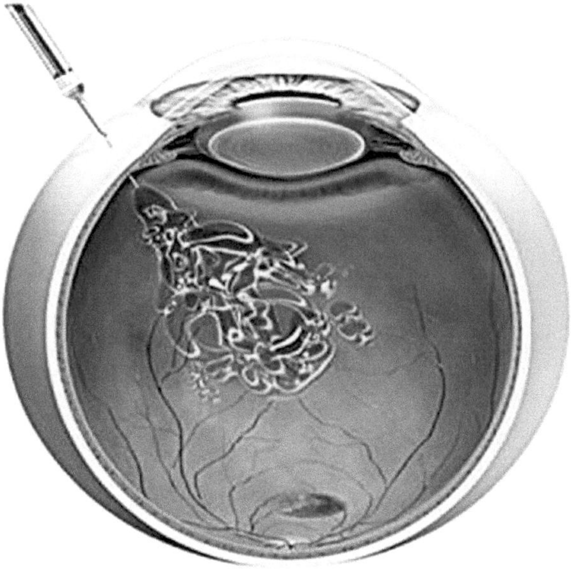

FIGURE 11.14 Intravetrous injection.

Use of intravetrous steroids is proffered to manage DME as it shows anti-inflammatory and antiangiogenic effects that stabilize of the inner blood-retina-barrier. Few of cases also showed the down regulation of VEGF in PDR after steroids administration. However; effect of intravetrous steroids is temporary and may revert within three months (Edelman et al., 2005; Brooks et al., 2004).

Recently FDA has approved few drugs for treatment of DR in the form of intravetrous injection. Approved drugs are summarized in Table 11.3.

TABLE 11.3 Approved Drug for Intravetrous Injection in Management of DR

Drug Name	Dose	Route	Brand Name	Manufacturer	FDA Approval	References
Ranibizumab	0.3–0.5 mg	Intravitreal injection	*Lucentis*	Novartis Pharmaceuticals UK Ltd.	FDA approved (06/02/2015)	Drugs.com, The electronic Medicines Compendium (eMC)
Aflibercept	2 mg	Intravitreal injection	*Eylea*	Bayer AG, Germany	FDA approved (25/03/2015)	Drugs.com European Medicines Agency
Bevacizumab	–	Intravitreal injection	Avastin	Genentech/Roche, United States	Non-approved for DR	Eyewiki UW Health

Although intravetrous injection helps to deliver a therapeutic amount of drug at the same time, it is associated with some major limitations as:

- Tremendous pain;
- Increased risk of bleeding;
- Chances to have secondary infection;
- High cost.

Overall, this current treatment is associated with several limitation and poor patient compliance. Therefore; there is surged to develop alternative method for the DR treatment which will be painless and ultimately more patient friendly.

11.5.8 PEROXISOME PROLIFERATOR ACTIVATED RECEPTOR-G AND DIABETIC RETINOPATHY (DR)

Various clinical studies showed the evidence that proved the advantageous role of peroxisome proliferator-activated receptor-γ, i.e., PPAR-γ in management of DR. PPAR-γ is located at RPE, photoreceptors, outer segment, retina, and even on cornea (Herzlich et al., 2008). It has been reported that PPAR-γ agonist can be able to inhibit cascade of DR at different stage of progression. PPAR-γ has showed ability to control or inhibit the process of angiogenesis and neovascularization (Muranaka et al., 2006). Therefore; PPAR- can be consider a new hope in management of DR.

PPAR-γ belongs to peroxisome proliferator-activator receptors (PPARs) which are well known nuclear hormone receptor. PPARs were known as a peroxisome proliferator 'binding-protei' able to cause hepatocyte peroxisome proliferation. Peroxisome's main function is the elimination of molecular oxygen and breaking down hydrogen peroxide. Retinoid X Receptor, i.e., RXR along with PPARs results in glucose and lipid metabolism.

PPARs are observed in three forms:

- PPARα-468 amino acids;
- PPARβ-441 amino acids; and
- PPARγ-479 amino acids.

Many researchers have reported the key role of PPAR-γ in management DR. Till date seven isoforms of PPAR-γ have been identified. Thiazolidinedione (TZDs) are known to specifically activate PPARγ. TZDs reported to hamper retinal endothelial cell proliferation and tube formation in response to VEGF treatment (Ciudin et al., 2013; Treacy & Hurst, 2012).

11.5.9 PPAR-γ: A NEW HOPE IN MANAGEMENT OF DR

PPAR-γ has proved its involvement in control of DR by inhibition of progress of DR at different stage. PPAR-γ agonist possesses antioxidant activity that can control and reduce the formation of AGEs formation which plays crucial role in DR and associated microvascular complications (Watson et al., 2005; Sulistio et al., 2008). Study also revels the role of PPAR-γ in angiogenesis through in streptozotocin induced DR via inhibition of NF-KB (Kim et al., 2007). Inhibition of NF-kB can be correlated with inflammation and retinal leakage as well as leukostasis (Simó et al., 2010). Furthermore; PPAR-γ agonist can also control DR by up regulation of members of Bcl-2 family which prevents the complex process of apoptosis. PPAR-γ also reported to play a crucial role in management of DR by prevention or control of neovascularization due to down regulation of VEGF (Murata et al., 2001). Additionally; reduction in VEGF formation due to PPAR-gamma agonist reported the minimization of expression of COX-2 (Scoditti et al.,

2009; Toda & Nakanishi-Toda, 2007; Ergul, 2011). Overall involvement of PPAR-γ in control of DR with different mechanism is shown in Figure 11.15.

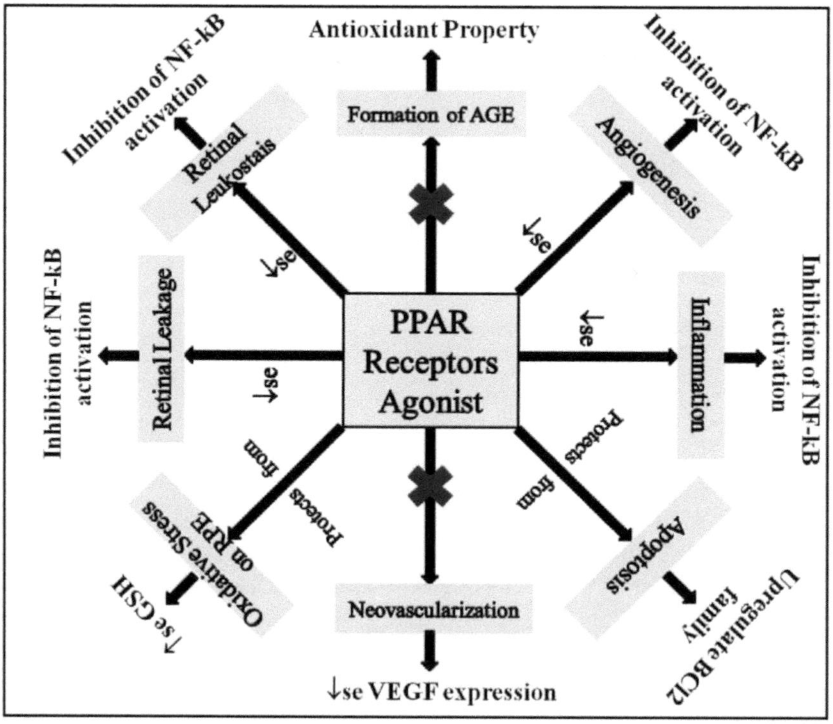

FIGURE 11.15 Involvement of PPAR-gamma receptor in control of DR.

11.6 NOVEL APPROACHES TO TARGET POSTERIOR SEGMENT OF EYE

By considering the involvement of PPAR-gamma receptors in DR, there is need to try PPAR-gamma agonist through topical administration. However; as discussed earlier defense mechanism eye restricts the entry of many drugs towards the posterior site and hence; one has to look at different novel approaches which may help to deliver the drugs at desired site. Most feasible approach is to fabricate nanotechnology-based formulations with small possible size as less than 200 nm which will help to overcome various barriers of ocular site and assist in delivery of desired amount of drug at posterior site (Cholkar et al., 2013). Nanotechnology based approach involves the formulation of nanoparticles, nanosuspension, nanoemulsion (NE), liposomes, etc., those can be easily administered through ocular site with better patient compliance. Many researchers have tried nanotechnology-based formulation for treatment of both frontal as well as back part of the eye and they found it as most promising approach (Patel et al., 2013).

11.6.1 NANOPARTICLES (NPS)

NPs are the colloidal particles with size 10–100 nm made up of various polymers viz. PLGA, PLA, Chitosan, Na-alginate, etc. Nowadays, many researchers working on eyesight have preferred PLGA and PLA as those are biodegradable and FDA approved polymers (Cholkar et al., 2013; Diebold & Calonge, 2010). Nanoparticles as a drug carrier can be prepared in the form of spheres or capsules. In the case of spherical particles, the drug is dispersed in the polymer into the polymeric matrix. On the other hand, in the case of capsules, the drug is present in the center which is then surrounded by polymer (Cholkar et al., 2013; Janoria et al., 2007; Kompella et al., 2013). Furthermore, cellular association of these nanoparticles has been improved by surface modification with surfactant like polysorbate 80. Ocular retention of these NPs can be improved by the addition of mucoadhesive material such as gellan gum, Carbopol, HPMC, Poloxamer, etc., which also acts as a structured vehicle and stabilizes the formulation. Much evidence has proved the effectiveness of NPs to target the back of the eye. Tahara et al. (2017) investigated the delivery of drugs to the retina by topical administration of surface-modified PLGA nanoparticles. Surface modification was tried with the help of chitosan, glycol chitosan, and polysorbate 80. Preclinical evidence showed better penetration of NPs to the retina after surface modification with P80. During the study, they used coumarin-6 as a model drug and fluorescent marker, and after ocular administration of PLGA NPs eye drops, the fluorescent intensity of coumarin-6 was observed in the retina. Mitra et al. (2016) administered nanoparticles of miR200-b in the form of intravitreal injection to check its impact on the progression of DR. Preclinical evidence showed the downregulation of VEGF. A low level of VEGF avoided the formation of new but weak capillaries at the retinal site, which further can be helpful to control the cascade of DR. Pandit et al. (2016) developed, optimized, and evaluated chitosan-PLGA nanoparticles of bevacizumab to target the retina. Bevacizumab, an anti-VEGF drug, was used to target the retina via SC route. They found good permeability for nanoparticles in comparison to a simple drug solution. Therefore, the use of NPs is the new hope for the treatment and management of DR (Ahire et al., 2020a, Ahirrao et al., 2022; Talele et al., 2021).

11.6.2 NANOSUSPENSION

It is a colloidal dispersion of non-polar drug in suitable medium. It is considered as the one the promising approach for crossing barriers of eyes. It is associated with several advantages in context with ocular site like ease of sterilization, less or no irritation, good retention time, etc. (Cholkar et al., 2013; Shid et al., 2013). Particle size is usually less than 500 nm helps in better uptake of particles (Shende & Godbole, 2016). Laddha & Kshirsagar (2020) developed surface modified nanoparticles of Pioglitazone by using PLGA. Surface modification was achieved by using polysorbate 80. These nanoparticles were dispersed in sterile saline solution and used for treatment of DR in rats. Study showed significant control over DR in comparison to control group (Ahire et al., 2018, 2020b).

11.6.3 NANOEMULSION (NE)

It possesses droplet size below 200 nm. It can be used same as classical type as O/W or W/O and it is considered as valuable formulation for back targeting of eye (Mahboobian et al., 2017; Achouri et al., 2012). NE is associated with several advantages like improved uptake and bioavailability, suitability for both hydrophobic and hydrophilic drugs, etc. Small droplet size together with surfactant (which also acts as penetration enhancer) significantly improves ocular permeability of drug (Wang et al., 2018). Various preclinical studies have proved the ability of NE to target posterior segment. NE showed improved absorption along with stabilization of drug. Mohammad M. Mahboobian et al. found dose reduction of brinzolamide from 1.0% to 0.4% due to formation of NE (Mahboobian et al., 2017; Thombre et al., 2022).

11.6.4 LIPOSOME

It includes the formation of lipid vesicles with phospholipid bilayers encircling a polar core. It possesses a size in the range of 10–1,000 nm.

Furthermore, they are classified as unilamellar and multilamellar liposomes. Since liposomes possess the same structure as that of the cell membrane, they are considered an ideal delivery system for eye sight. Several research studies have proven the effectiveness of liposomes in anterior as well as posterior ocular site drug delivery. However, they are associated with limitations such as chemical instability, oxidative degradation, lack of reproducibility, and high cost, which act as hurdles for the scalability of the formulation (Patel et al., 2013).

11.6.5 NANO-MICELLES

It is one of the trending approaches to deliver the drug at the targeted site. Micelles of surfactants (sodium dodecyl sulfate, Vit E TPGS, octoxynol-40, and C8 lecithin) or polymers (N-isopropyl acrylamide, vinyl pyrrolidone, acrylic acid, polyhydroxyethyl aspartamide) can be used to target the retina (Wang et al., 2018; Cholkar et al., 2013). It showed improvement in stability, permeability, and bioavailability of the drug. In recent years, nanomicelles loaded with active moiety such as dexamethasone, voclosporin, and prodrugs showed promising results in ocular complications (Wang et al., 2018).

11.7 CONCLUSION

Diabetic retinopathy (DR) is the worst microvascular complication which may result in blindness. At present, it is treated by means of invasive methods such as vitrectomy, laser surgery, and intravitreal injections. All these approaches are associated with pain, fear of loss of vision, secondary infections, no self-medications, and ultimately poor patient compliance. By considering numerous limitations associated with present treatment, in the last five years, researchers have investigated some novel approaches to target the posterior segment of the eye to target retinopathy and associated complications. The use of nano-systems has been found to be effective during preclinical investigation. However, there is a strong need to check the efficiency of prepared systems through clinical studies.

KEYWORDS

- **diabetes**
- **diabetic retinopathy**
- **liposome**
- **nanoparticles**
- **nanosuspension**
- **PPAR-gamma**
- **risk factors**

REFERENCES

Abdulrahman, A., & Alghadyan, M. D., (2011). Diabetic retinopathy: An update. *Saudi Journal of Ophthalmology, 25*, 99–111.

Abu El-Asrar, A. M., Al-Mezaine, H. S., & Ola, M. S., (2009). Pathophysiology and management of diabetic retinopathy. *Expert Review of Ophthalmology, 4*(6), 627–647.

Achouri, D., Alhanout, K., Piccerelle, P., & Andrieu, V., (2012). Recent advances in ocular drug delivery. *Drug Development and Industrial Pharmacy, 39*(11), 1–19.

Adapa, D., & Sarangi, T. K., (2015). A review on diabetes mellitus: Complications, management, and treatment modalities. *J. Med. Health. Sci., 4*(3), 1–18.

Ahire, E. D., Talele, S. G., & Shah, H. S., (2020). Nanoparticles as a promising technology in microbial pharmaceutics. In: *Applied Pharmaceutical Science and Microbiology* (pp. 133–158). Apple Academic Press.

Ahire, E., Thakkar, S., Borade, Y., & Misra, M., (2020). Nanocrystal based orally disintegrating tablets as a tool to improve dissolution rate of Vortioxetine. *Bulletin of Faculty of Pharmacy, Cairo University, 58*(1, 2), 11–20.

Ahire, E., Thakkar, S., Darshanwad, M., & Misra, M., (2018). Parenteral nanosuspensions: A brief review from solubility enhancement to more novel and specific applications. *Acta Pharmaceutica Sinica B, 8*(5), 733–755.

Ahirrao, S. P., Sonawane, M. H., Bhambere, D. S., Udavant, P. B., Ahire, E. D., & Kanade, R., (2022). Cocrystal formulation: A novel approach to enhance solubility and dissolution of etodolac. *Biosciences Biotechnology Research Asia, 19*(1), 111.

Ali, Z. H., (2011). Health and knowledge progress among diabetic patients after implementation of a nursing care program based on their profile. *Journal of Diabetes and Metabolism, 2*(2), 121.

Almeida, H., Amaral, M. H., Loba, P., & Lobo, J. M. S., (2013). *In Situ* gelling systems: A strategy to improve the bioavailability of ophthalmic pharmaceutical formulations. *Drug Discovery Today,* 1–13.

American Society of Retina Specialist. (2013). *Retina Health Series: Fact Sheets, 312*, 578–8760.

Borja, C., Santiago, D., & Maria, O. G. A., (2017). Update on diagnosis and treatment of DR: A consensus guidelines of the working group of ocular health. *Journal of Ophthalmology*, 1–10.

Boyd, K., (2019). *Retinopathy Treatment*. Published by American Academy of ophthalmology. https://www.aao.org/eye-health/diseases/diabetic-retinopathy-treatment (accessed on 6 December 2023).

Brooks, H. L., Caballero, S., Newell, C. K., Steinmetz, R. L., Watson, D., Segal, M. S., Harrison, J. K., et al., (2004). Vitreous levels of vascular endothelial growth factor and stromal derived factor 1 in patients with diabetic retinopathy and cystoid macular edema before and after intraocular injection of triamcinolone. *Archives of Ophthalmology, 122*(12), 1801–1807.

Cakan, N., Kizilbash, S., & Kamat, D., (2012). Changing spectrum of diabetes mellitus in children: Challenges with initial classification. *Clinical Pediatrics, 51*(10), 939–944.

Cholkar, K., Dasari, S., Pal, D., & Mitra, A., (2013). Eye: Anatomy, physiology, and barriers to drug delivery. *Ocular Transporters and Receptors*, 1–36.

Cholkar, K., Patel, A., Agrahari, V., & Mitra, A., (2013). Ocular drug delivery systems: An overview. *World Journal of Pharmacology, 2*(2), 47–64.

Ciudin, A., Hernandez, C., & Simo, R., (2013). Molecular implications of the PPARs in diabetic eye. *PPAR Research*, 1–11.

Ciulla, T. A., Amador, A. G., & Zinman, B., (2003). Diabetic retinopathy and diabetic macular edema: Pathophysiology, screening, and novel therapies. *Diabetes Care, 26*(9), 2653–2664.

Daniel, G., Aristomenis, T., & Demetrios, V., (2012). Therapeutic interventions against inflammatory and angiogenic mediators in proliferative diabetic retinopathy. *Mediators of Inflammation*, 1–10.

Daniel, S. W. T., Cheung, G. C. M., & Wong, T. Y., (2016). Diabetic retinopathy: Global prevalence, major risk factors, screening practices and public health challenges: A review. *Clinical & Experimental Ophthalmology, 44*(4), 260–277.

Das, A., Stroud, S., Mehta, A., & Rangasamy, S., (2015). New treatments for diabetic retinopathy. *Diabetes, Obesity and Metabolism, 17*(3), 219–230.

Decanini, A., Karunadharma, P. R., Nordgaard, C. L., Feng, X., Olsen, T. W., & Ferrington, D. A., (2008). Human retinal pigment epithelium proteome changes in early diabetes. *Diabetologia, 51*(6), 1051–1061.

Deshmukh, C. D., & Jain, A., (2015). Diabetes mellitus: A review. *Int. J. Pure App. Biosci, 3*(3), 224–230.

Diebold, Y., & Calonge, M., (2010). Applications of nanoparticles in ophthalmology. *Progress in Retinal and Eye Research, 29*(6), 586–609.

Dubald, M., Bourgeois, S., Andrieu, V., & Fessi, H., (2018). Ophthalmic drug delivery systems for anti-biotherapy: A review. *Pharmaceutics, 10*(1), 1–31.

Edelman, J. L., Lutz, D., & Castro, M. R., (2005). Corticosteroids inhibit VEGF-induced vascular leakage in a rabbit model of blood–retinal and blood–aqueous barrier breakdown. *Experimental Eye Research, 80*(2), 249–258.

Ergul, A., (2011). Endothelin-1 and diabetic complications: Focus on the vasculature. *Pharmacological Research, 63*(6), 477–482.

Forbes, J. M., Soldatos, G., & Thomas, M. C., (2005). Below the radar: Advanced glycation end products that detour around the side is HbA1c not an accurate enough predictor of long-term progression and glycaemic control in diabetes? *Clin. Biochem. Rev., 26*, 123–134.

Galtier, F., (2010). Definition, epidemiology, risk factors. *Diabetes & Metabolism, 36*, 628–651.

Gardner, T. W., Antonetti, D. A., Barber, A. J., Lanoue, K. F., & Nakamura, M., (2000). New insights into the pathophysiology of diabetic retinopathy: Potential cell-specific therapeutic targets. *Diabetes Technology & Therapeutics, 2*(4), 601–608.

Hammes, H. P., Strödter, D., Weiss, A., Bretzel, R. G., Federlin, K., & Brownlee, M., (1995). Secondary intervention with aminoguanidine retards the progression of diabetic retinopathy in the rat model. *Diabetologia, 38*(6), 656–660.

Herzlich, A. A., Tuo, J., & Chan, C. C., (2008). Peroxisome proliferator-activated receptor and age-related macular degeneration. *PPAR Research*, 389–407.

https://www.niddk.nih.gov/health-information/diabetes/overview/preventing-problems/diabetic-eye-disease (Niddk) (accessed on 6 December 2023).

https://eyewiki.aao.org/Bevacizumab (accessed on 6 December 2023).

https://pubchem.ncbi.nlm.nih.gov (accessed on 6 December 2023).

https://www.accessdata.fda.gov/scripts/cder/iig/index.cfm (accessed on 6 December 2023).

https://www.drugs.com/lucentis.html (Drugs.com) (accessed on 6 December 2023).

https://www.drugs.com/mtm/aflibercept-ophthalmic.html (Drugs.com) (accessed on 6 December 2023).

https://www.ema.europa.eu/en/medicines/human/EPAR/eylea (European Medicines Agency) (accessed on 6 December 2023).

https://www.medicines.org.uk/emc/product/307/smpc (The electronic Medicines Compendium (eMC)) (accessed on 6 December 2023).

Jang, C., Lim, J. H., Park, C. W., & Cho, Y. J., (2011). Regulator of calcineurin 1 isoform 4 (RCAN1. 4) Is overexpressed in the glomeruli of diabetic mice. *The Korean Journal of Physiology & Pharmacology, 15*(5), 299–305.

Janoria, K. G., Gunda, S., Boddu, S. H., & Mitra, A. K., (2007). Novel approaches to retinal drug delivery. *Expert Opinion on Drug Delivery, 4*(4), 371–388.

Joanna, M., Tarr, K. K., Mohit, C., Eva, M. K., & Rakesh, C., (2013). Pathophysiology of diabetic retinopathy. *ISRN Ophthalmology*, 1–13.

Jonathan, G. F., (2003). Laser management of diabetic retinopathy. *Journal of the Royal Society of Medicine, 96*(6), 277–279.

Kaul, S., Kumar, G., & Kothiyal, P., (2012). An insight into ocular insert. *Int. J. Pharm. and Pharm. Sci. Res., 3*, 1905–1912.

Kharroubi, A. T., & Darwish, H. M., (2015). Diabetes mellitus: The epidemic of the century. *World Journal of Diabetes, 6*(6), 850.

Khawandanah, J., (2019). Double or hybrid diabetes: A systematic review on disease prevalence, characteristics, and risk factors. *Nutrition & Diabetes, 9*(1), 1–9.

Kim, E. K., Kwon, K. B., Koo, B. S., Han, M. J., Song, M. Y., Song, E. K., Han, M. K., et al., (2007). Activation of peroxisome proliferator-activated receptor-γ protects pancreatic β-cells from cytokine-induced cytotoxicity via NFκB pathway. *The International Journal of Biochemistry & Cell Biology, 39*(6), 1260–1275.

Kompella, U. B., Kadam, R. S., & Lee, V. H., (2010). Recent advances in ophthalmic drug delivery. *Therapeutic Delivery, 1*(3), 435–456.

Kumar, B. P., Harish, G., Bhowmik, D., Duraivel, S., & Kumar, B. P., (2013). Ocular inserts: A novel controlled drug delivery system. *The Pharma Innovation, 1*(12), 1–16.

Laddha, U. D., & Mahajan, H. S., (2017). An insight to ocular in situ gelling systems. *International Journal of Advances in Pharmaceutics, 6*(2), 31–38.

Laddha, U., & Kshirsagar, S., (2020). Formulation of PPAR-gamma agonist as surface modified PLGA nanoparticles for non-invasive treatment of diabetic retinopathy: *In vitro* and *in vivo* evidences. *Heliyon, 6*(8), 1–10.

Lemos, C. T., Detsch, J., Pimazoni-Netto, A., De Almeida, A., & Sztal-Mazer, S., (2011). Glycemic variability and mean weekly glucose in the evaluation and treatment of blood glucose in gestational diabetes mellitus; evidence for lower neonatal complications. *J. Diabetes Metab., 2*(137), 2–5.

Loftsson, T., & Stefansson, E., (2017). Cyclodextrins and topical drug delivery to the anterior and posterior segments of the eye. *International Journal of Pharmaceutics, 531*(2), 413–423.

Loftsson, T., Sigurdsson, H. H., Konradsdottir, F., Gisladottir, S., Jansook, P., & Stefansson, E., (2008). Topical drug delivery to the posterior segment of the eye: Anatomical and physiological considerations. *Die Pharmazie- An International Journal of Pharmaceutical* Sciences*, 63*(3), 171–179.

Lutty, G. A., (2013). Effects of diabetes on the eye. *Investigative Ophthalmology and Visual Science, 54*(14), ORSF81–ORSF87.

Mahboobian, M., Seyfoddin, A., Rupenthal, I., Aboofazeli, R., & Foroutan, M., (2017). Formulation development and evaluation of the therapeutic efficacy of brinzolamide containing nanoemulsions. *Iranian Journal of Pharmaceutical Research, 16*(3), 847–857.

Mitra, R. N., Nichols, C. A., Guo, J., Makkia, R., Cooper, M. J., Naash, M. I., & Han, Z., (2016). Nanoparticle-mediated miR200-b delivery for the treatment of diabetic retinopathy. *Journal of Controlled Release, 236*, 31–37.

Mizutani, M., Kern, T. S., & Lorenzi, M., (1996). Accelerated death of retinal microvascular cells in human and experimental diabetic retinopathy. *The Journal of Clinical Investigation, 97*(12), 2883–2890.

Muranaka, K., Yanagi, Y., Tamaki, Y., Usui, T., Kubota, N., Iriyama, A., Terauchi, Y., et al., (2006). Effects of peroxisome proliferator-activated receptor γ and its ligand on blood–retinal barrier in a streptozotocin-induced diabetic model. *Investigative Ophthalmology & Visual Science, 47*(10), 4547–4552.

Murata, T., Hata, Y., Ishibashi, T., Kim, S., Hsueh, W. A., Law, R. E., & Hinton, D. R., (2001). Response of experimental retinal neovascularization to thiazolidinediones. *Archives of Ophthalmology, 119*(5), 709–717.

Pandit, J., Sultana, Y., & Aqil, M., (2016). Chitosan coated PLGA nanoparticles of bevacizumab as novel drug delivery to target retina: Optimization, characterization, and *in-vitro* toxicity evaluation. Artificial cells, *Nanomedicine and Biotechnology*, 1–11.

Patel, A., Cholkar, K., Agrahari, V., & Mitra, A. K., (2013). Ocular drug delivery systems: An overview. *World Journal of Pharmacology, 2*(2), 47–64.

Patil, A. P., Tagalpallewar, A. A., Rasve, G. M., Bendre, A. V., & Khapekar, P. G., (2012). A novel ophthalmic drug delivery systems: *In situ* gel. *Int. J. Pharm. Sci. and Res., 3*, 2938–2946.

Punthakee, Z., Goldenberg, R., & Katz, P., (2018). Definition, classification, and diagnosis of diabetes, prediabetes. and metabolic syndrome. *Canadian Journal of Diabetes, 42*, S10–S15.

Rajoria, G., & Gupta, A., (2012). *In Situ* gelling system: A novel approach for ocular drug delivery. *Am. J. PharmTech Res., 2*, 2249–3387.

Renu, A. K., R. A., Zhong, Q., Santos, J. M., Thandampallayam, M., Putt, D., & Gierhart, D. L., (2014). Beneficial effects of the nutritional supplements on the development of diabetic retinopathy. *Nutrition & Metabolism, 11*(1), 1–10.

Sayin, N., Kara, N., & Pekel, G., (2015). Ocular complications of diabetes mellitus. *World Journal of Diabetes, 6*(1), 92–108.

Scoditti, E., Massaro, M., Carluccio, M. A., Distante, A., Storelli, C., & De Caterina, R., (2010). PPARγ agonists inhibit angiogenesis by suppressing PKCα-and CREB-mediated COX-2 expression in the human endothelium. *Cardiovascular Research., 86*(2), 302–310.

Shanker, J. H., Mahmood, S. E., Joshi, M. C., & Shaifali, I., (2011). Obesity indices amongst diabetics in an urban population of Western Nepal. *J. Diabetes Metab., 2*, 134.

Shende, P., & Godbole, R., (2016). Current and novel techniques in the ophthalmic drug delivery systems. *International Journal of Pharmaceutical Sciences and Research, 7*(9), 3557–3566.

Shid, R., Dhole, S., Kulkarni, N., & Shid, S., (2013). Nanosuspension: A review. *International Journal of Pharmaceutical Science Revision and Research, 22*(1), 98–106.

Simó, R., Villarroel, M., Corraliza, L., Hernández, C., & Garcia-Ramírez, M., (2010). The retinal pigment epithelium: Something more than a constituent of the blood-retinal barrier—Implications for the pathogenesis of diabetic retinopathy. *Journal of Biomedicine and Biotechnology*, 1–15.

Sulistio, M. S., Zion, A., Thukral, N., & Chilton, R., (2008). PPARγ agonists and coronary atherosclerosis. *Current Atherosclerosis Reports, 10*(2), 134–141.

Tahara, K., Karasawa, K., Onodera, R., & Takeuchi, H., (2017). Feasibility of drug delivery to the eye's posterior segment by topical instillation of PLGA nanoparticles. *Asian Journal of Pharmaceutical Sciences, 12*(4), 394–399.

Talele, S. G., Ahire, E. D., Talele, G. S., & Derle, D. V., (2021). An innovative approach as self-emulsifying drug delivery system for phytoconstituents. In: *Enhancing the Therapeutic Efficacy of Herbal Formulations* (pp. 69–84). IGI Global.

Thombre, N. A., Niphade, P. S., Ahire, E. D., & Kshirsagar, S. J., (2022). Formulation development and evaluation of microemulsion based lornoxicam gel. *Biosciences Biotechnology Research Asia, 19*(1), 69–80.

Threatt, J., Williamson, J. F., Huynh, K., Davis, R. M., & Hermayer, K., (2013). Ocular disease, knowledge, and technology applications in patients with diabetes. *The American Journal of the Medical Sciences, 345*(4), 266–270.

Thunander, M., Törn, C., Petersson, C., Ossiansson, B., Fornander, J., & Landin-Olsson, M., (2012). Levels of C-peptide, body mass index and age, and their usefulness in classification of diabetes in relation to autoimmunity, in adults with newly diagnosed diabetes in Kronoberg, Sweden. *European Journal of Endocrinology, 166*(6), 1021–1029.

Toda, N., & Nakanishi-Toda, M., (2007). Nitric oxide: Ocular blood flow, glaucoma, and diabetic retinopathy. *Progress in Retinal and Eye Research, 26*(3), 205–238.

Treacy, M., & Hurst, T., (2012). The case for intraocular delivery of PPAR agonists in the treatment of diabetic retinopathy. *BMC Ophthalmology, 12*(46), 1–5.

Uppu, R. M., & Parinandi, N. L., (2011). Insulin sensitization and resistance interrelationship revisited with a quantitative molecular model approach. *J. Diabetes Metab., 2*(106), 2.

Victoria, J., Vieira, P., Dimitrios, K., & Darren, J. L., (2016). Ocular complications of diabetes and therapeutic approaches. *BioMed Research International*. Article ID 3801570.

Vilar, L., (2006). *Endocrinologia Clínica* (3rd edn., pp. 630–642). Rio de Janeiro: Guanabara Koogan.

Wang, Y., Xu, X., Gu, Y., Cheng, Y., & Cao, F., (2018). Recent advance of nanoparticle-based topical drug delivery to the posterior segment of the eye. *Expert Opinion on Drug Delivery, 15*(7), 687–701.

Watson, G. S., Cholerton, B. A., Reger, M. A., Baker, L. D., Plymate, S. R., Asthana, S., Fishel, M. A., et al., (2005). Preserved cognition in patients with early Alzheimer disease and amnestic mild cognitive impairment during treatment with rosiglitazone: A preliminary study. *The American Journal of Geriatric Psychiatry, 13*(11), 950–958.

Waugh, A., & Grant, A., (2010). *Ross and Wilson, Anatomy and Physiology in Health and Illness* (11th edn., pp. 190–205). Churchill Livingstone, Elsevier.

Weng, Y., Liu, J., Jin, S., Guo, W., Liang, X., & Hu, Z., (2017). Nanotechnology based strategies for treatment of ocular disease. *Acta Pharmaceutica Sinica, 7*(3), 281–291.

Xu, H., Chen, M., & Forrester, J. V., (2009). Para-inflammation in the aging retina. *Progress in Retinal and Eye Research, 28*(5), 348–368.

Yam, J. C., & Kwok, A. K., (2007). Update on the treatment of diabetic retinopathy. *Hong Kong Med., J., 13*, 46–60.

PART IV

Novel Drug Delivery System and Future Aspects

CHAPTER 12

Artificial Intelligence: A New Era of Novel Drug Delivery System

ROHAN K. BARSE

Department of Pharmaceutics, Yashwantrao Bhonsale College of Pharmacy, Sawantwadi, Sindhudurg, Maharashtra, India

ABSTRACT

AI and machine learning are both gradually developing and contributing to the overall growth of the pharma sector. In the last two decades, advancements in AI have led to a dramatic change in the industrial scenario due to highly automated and predictive capabilities. Drug discovery and the development of dosage forms are at the heart of pharmaceutical research, which is a significantly lengthy and risky operation. AI has shown potential in handling drug discovery and the development of dosage forms through a highly automated system. AI includes steps like problem definition, data preparation, design of the AI architecture, model training and evaluation, and understanding and explaining the results. ANFIS (ANNs), GRNN, RSM, and DNN are different model approaches utilized in AI for the novel drug delivery system (NDDS). Artificial intelligence (AI) has incredible intensification potential in research and development.

12.1 DEVELOPMENT OF ARTIFICIAL INTELLIGENCE (AI)

Artificial neurons as artificial intelligence (AI) were proposed by Walter Pitts and Warren McCulloch in the year 1943. Furthermore, in the year 1949, the Hebbian rule is the learning rule for updating the interrelation strength connecting neurons. "Computing machinery and intelligence started innovating in the year 1950. John McCarthy at the Dartmouth conference

Novel Drug Delivery Systems in the Management of Chronic Diseases.
Raj K. Keservani, Umesh D. Laddha, Eknath D. Ahire, and Sanjay J. Kshirsagar (Eds.)

adopted the word AI in the year 1956. That was considered the introduction of AI in academia. Furthermore, AI considered innovations such as advancing algorithm programs that can resolve mathematical problems (1966), humanoid robot (1972), XCON (1987), a vacuum cleaner (2002). AI came into the business world until the year 2006. AI implementation also started by companies like Netflix, Twitter, and Facebook." Google now, as an Android app, has been put in place by Google, which is able to provide information to the user as a prediction. Now, AI has taken over to an incredible level. The views of Deep learning, big data, and data science are now moving like a boom (Vyas et al., 2018; Mijwel, 2015).

AI entered the health section in 1950. Computers used in medicine with the research were developed in 1975. AIM workshop and DeepQA in 2007 increased the attention of the health sector on the usefulness of AI. The first FDA-approved AI-DL application was initiated by 2017, and to date, many AI trials have been accomplished (Gupta et al., 2021; Kaul et al., 2020) (Table 12.1).

TABLE 12.1 History of AI

Year	AI
1950	Development of turning tests.
1952	Machine learning.
1956	Appear the term AI.
1961	First industrial robot related to AI.
1966	First electronic person.
1971–9172	Research assests on computers in biomedicines and mycin developed.
1973	Sumex AIM created.
1975	NIH sponsored AIM workshop.
1976	Starting of AI winter.
1980	Development of Emycin.
2010	First Pharmaboat released.
2017	First FDA approved cloud build DL application in healthcare.

12.2 INTRODUCTION

AI is learning and investigating mainly intelligent computer programs, which provide results in a similar way to human attention process. AI is used for analyzing machine learning to imitate the cognitive tasks of individuals (Mak et al., 2019; Hassanzadeh et al., 2019). AI is a computer-based system that assists with problem-solving through symbolized programming.

AI is an expansion of computer systems in such a way that they are able to perform programs similar to those requiring human intelligence (Sessa et al., 2020). AI is "the preparing computer-based system which can carry work as that of human intelligence" (Oxford, 2019) (Figure 12.1).

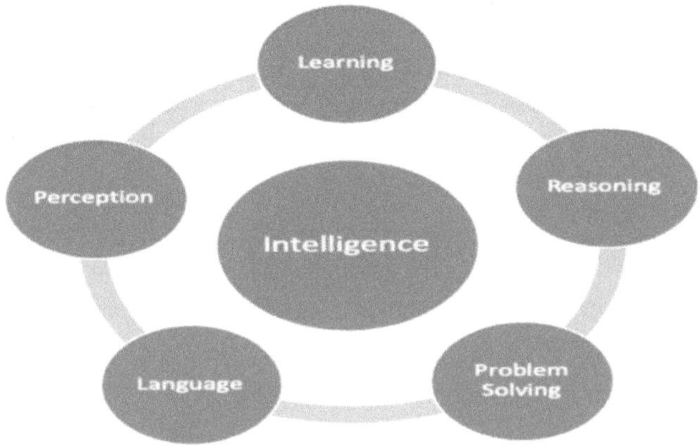

FIGURE 12.1 Abilities of artificial intelligence.

Source: Reprinted from Sessa et al., 2020. https://creativecommons.org/licenses/by/4.0/

The AI has different abilities like learning, reasoning, insight, problem solving and language.

12.3 ARTIFICIAL INTELLIGENCE (AI) APPLICATION IN PHARMACY

- Drug discovery and development;
- Clinical research;
- Disease diagnosis;
- Novel medication;
- Data analytic.

12.4 ARTIFICIAL INTELLIGENCE (AI): NDDS DEVELOPMENT

1. **Drug Discovery and Development:** As per USFDA, approximately 33% of drugs move from Phase II to III, while around 25 to 30% move from Phase III to the next phase.

   Drug development has time-consuming, low success, costly and gradual process that initiating with recognized a successful molecule. The foremost intension of the drug discovery is to detect an active particle whose ultimate objective is to affect the human body and prove its aspect and purity. The drug discovery motivated when existing drugs have limited efficacy and/or severe toxicity.

Main barrier in drug development have become a concern, leading to the high failure rate in selecting new drug candidates.

In last two decades AI developed significantly in context with pharm sector and contributed to development of sophisticated and automated systems. It contributes to development of new drugs and dosage form with maximum prediction on computer-based system which performs same on platform.

AI benefits in drug discovery and drug developments are creating novel ideas for the novel compounds; predict the desired properties of the entirely novel compound, predicting the properties of potential molecule.

ML terminology used for parse data with algorithm, neural network (NN) is one of the best models, which used DL, i.e., deep learning for instruct algorithms that classify data which also overcomes the clustering issues. Even though ML addresses the major dimensional data problem, but still the increased number of variables not of responses are still problematic. The number of unrelated variables or diviners increases, the capacity of the noise becomes notable, resulting in the diminished predictive performance of most ML algorithms (Figures 12.2 and 12.3).

		Traditional Statistical Methods	ML	DL/AI
Clinical Trials Data Analysis	Ph1 – 4 Trials Focused on Inference or Estimation of Trt Effect	✓		
	Ph 0 Trials	✓	✓	
	Small – Medium Dimensional Data for Evidence Generation & Translational Research	✓	✓	
	High-dimensional Translational Data Focused on Prediction		✓	
Translational Research or Drug Discovery Data Analysis	Small – Medium Dimensional Data	✓	✓	
	High-dimensional data		✓	✓
Development of Systems	Systems with human-like reasoning to optimize drug development process (e.g., in manufacturing or trial operations)			✓

FIGURE 12.2 Significance of AI and machine learning in drug development (Kolluri et al., 2022).

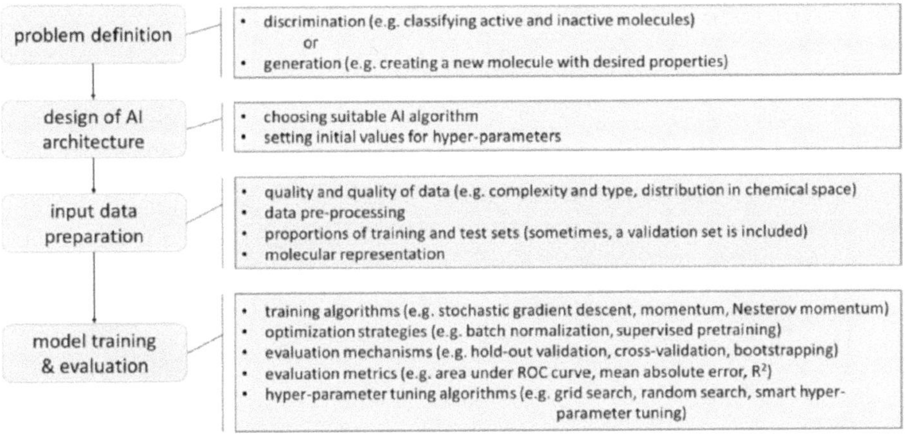

FIGURE 12.3 Workflow for AI (Yang et al., 2019).

Source: Reprinted from Yang et al. 2019. CC-BY-NC. https://pubs.acs.org/doi/10.1021/acs.chemrev.8b00728

AI includes steps like problem definition, data preparation, design of the AI architecture, model training and evaluation, and understanding and explaining the results. First, one needs to know whether a specific issue belongs to the domain of generative tasks. In next step, selection of specific algorithm and use of initial values for hyper parameters. Commonest algorithm programs are SVM, RF, and ANNs. The provisional architecture is determined in this second step, further in next further third place is to begin data set. Once the provisional architecture and data set is finalized then model training and evaluation is necessary. Developed final AI model must be able to express the molecular representations.

2. **Design of NDDS by Artificial Neuronal Network:** In NDDS, associating networks have been functionalized. AI system with ANN was used to correlate the drug release and various formulation variables. AI and ANN also showed involvement in optimization. ANN trained by data of research papers also showed very promising results after stringent validation and testing which may help in-silico models for dosage form development (Lima et al., 2016; Yichun et al., 2014). ANN is computational tools that emulate the interconnected neurological anatomical human brain make up and the human brain capacity to learn and solve problems through pattern recognition. ANN simulate the investigating human brain conduct by modeling data facts and perceiving patterns for complicated multi-dimensional

connection that exist between input and output or choose goal sets of data. Once trained an ANN can be used to anticipate and speculate outputs for a given a set of input conditions and may therefore be used to develop optimize both process variables and formulation in order to engineer and construct put together high quality, safe, and effective dosage forms (Figure 12.4).

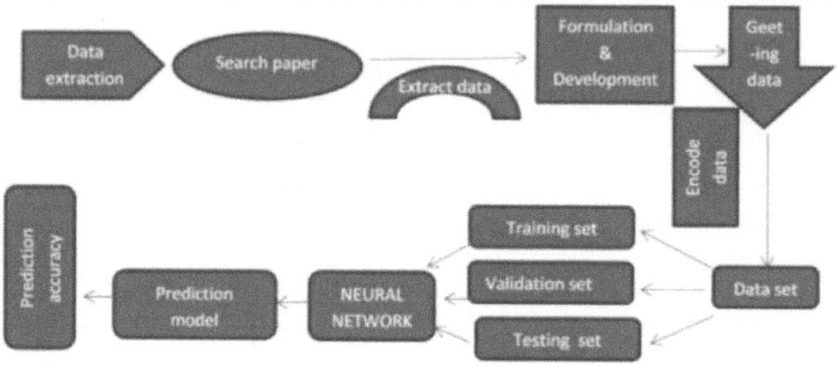

FIGURE 12.4 Establish model process (Chaudhari et al., 2020).
Source: Reprinted from Chaudhari et al. 2020. Open Access.

Rebouh looked into the drug release ways from cellulose derivative (CD) matrices in the oral tablets. The adaptive neural-fuzzy inference system (ANFIS) was used to conjecture the greatest formulation variables to acquire the ideal ibuprofen (IB) as a model drug for sustained drug delivery. ANFIS is a type of artificial neural network (ANN) that is based on Takagi–Sugeno fuzzy inference system. The ANFIS model exactly models the link between IB release behavior and tablet formulation variables. The preferred model can be contemplated as an efficient tool with anticipating potential that is applicable for the scheming, testing, and try out of novel dosage forms based on polymers (Rebouh et al., 2019).

Hasanjani et al. utilized Adaptive neuro-fuzzy inference system is a neuro fuzzy approach where the merging is made in the NN and the fuzzy inference system (FIS) that is a computational practice for simultaneous resolution of fluoxetine (FLX) and sertraline (SRT) in pharmaceutical formulations (Hasanjani et al., 2014). The collaboration of FL systems and NN has manufactured a functional system capable of adapting its structure based on complete information; in other words, these systems may be set side by side to biological systems, in terms of reasoning learning and high-level thinking. It is a boosting tool for determining the behavior of indistinct described complex systems. The objective of a neuro-fuzzy system is to perceive the variables and/or structure of neuro-fuzzy systems. These neuro-fuzzy systems can merge the benefits of the two significant paradigms into a single capsule. Determination of FLX and SRT in pharmaceutical formulations and biological fluid. Several features, which construct them acceptable for a spacious range of scientific applications. The system was found to be more accurate and reliable. FIS is a knowledge rendition where each fuzzy rule expresses a local behavior of the system. The network structure that executes FIS is mentioned as ANFIS and appoints hybrid learning rules and regulations to train a Sugeno-style FIS with linear rule outputs. ANFIS is applicable to train a FIS that is applied to simultaneous determination of SRT and FLX hydrochloride in various synthetic mixtures (Cundari et al., 2001).

Hyang prepared produced granulated pellet-containing tablets and further contrast differentiation of novel granulated pellet and traditional pellet-containing tablets by ANNs (Huang et al., 2015). MATLAB version 7.9.0 (Mathworks Inc., Natick, MA) was used in the progress of ANNs. The pellet size was the median diameter dignified by laser scattering, using scirocco dry dispersion unit at a feed pressure of 3.0 bars (Mastersizer, 2000; Malvern Instruments, Malvern, UK). It was observed by study that new formulations ANNs have been used to constructively predict the similarity factor values of drug release profiles for PCT and GPCT in comparison with the corresponding original pellets, and ANNs model appreciable to GPCT.

In recent years, ANNs based on the working of the human brain have been successfully initiated in pharmaceutics for the forecasting of solid-water distribution coefficients, optimization of formulation and process parameters, etc. This is because not only do ANNs have generalization capabilities, precision, fitting degree in dealing with multiple factors, high adaptability, and multiple responses, but also the predictive ability of these open systems can be upgraded by selecting different training functions, algorithms, and training samples (Debunne et al., 2004; Koppen et al., 2009; Barron et al., 2009).

Chaibva et al. tried the ANN to develop salbutamol sulfate matrix system. ANN was trained by using central composite design with input variables viz. xanthan gum, Carbopol974P, HPMC K100M, and Surelease. ANN with nine nodes showed significant predictive ability. Furthermore, such an ANN system showed reliable prediction of dissolution rate (Chaibva et al., 2010).

Jayaram used ANN to prepare O/W emulsion of fatty alcohol. Optimization predicted by the system was found to be more practical (Jayaram Kumar et al., 2011).

Ghaffari investigated the impact of coating weight load and concentration of the polymeric system on the theophylline release. ANN was trained by using five different classes like gradient descent, quasi-Newton (Levenberg–Marquardt: LM), and genetic algorithm (GA). ANN was of four nodes with a single hidden layer. The predictive ability was determined for the developed model. IBP, BBP > LM > QP (quick propagation) > GA was the performance order. The ANN system showed that increasing coating and decreasing polymeric concentration slow down the drug release. Furthermore, the concentration of the polymer affects the drug release behavior crucially (Ghaffari et al., 2006).

ANNs show supremacy as a modeling methodology for datasets exhibiting nonlinear relationships, and thus for both data fitting and prediction capabilities. ANN is a computer-based learning system that simulates the human thinking ability and can be used to quantify a nonlinear relationship between pharmaceutical responses and causal factors (Achanta et al., 1995; Bourquin et al., 1997a). An extended-release tablet of aspirin was prepared by using Ibricused, which is a GRNN model. Tablets prepared with the help of AI, when investigated for drug release, showed maximum reliability with results predicted by AI. The study showed the ability of AI in the development of NDDS (Svetlana et al., 2002).

ANNs are biologically motivated computer algorithms that act as universal function approximators, having the ability to model thoroughly complex relationships where the response variables are non-linearly related to the independent variables. Two of the leading advantages of the ANN are – (i) there is no need to take for granted an underlying data distribution; and (ii) ANNs are significant for highly non-linear multivariate problems. However, a critical evaluation of the ANN can reveal several disadvantages including – (a) minimizing overfitting requires a great deal of computational effort; and (b) it is not so complicated to discover the individual relations between the input variables and the output variables (Hussain et al., 1991).

ANN system was also used in optimization of Nimodipine zero order release matrix tablet. Independent variables were concentration of PEG 4000, PVP K30, two grades of HPMC and drug release time at which 90% of the drug was dissolved, 90% of nimodipine released in 2 and 8 h, Y2h, and Y8h, respectively were responses. It was observed that a feed-forward backpropagation ANN with eight hidden units exhibited better fit for all responses compared to the MLR models. The ANN was further checked by pruning, which preserved only PEG 4000 and one of the HPMC as inputs.

The estimation of the similarity factor showed the potential of ANNs to predict the results (Hussain et al., 1991).

Yuksel et al. prepared acrylic microspheres by ANNs. Polymer ratio, stirring rate and dispersion agent concentration were the preparative variables tried at different levels by using both ANN and MLR systems. Results showed better prediction by ANN than MLR method (Yüksel et al., 2001).

Mihajlovic tried to use AI system to identify the impact of changing spray-drying parameters on the development of a naratriptan/maltodextrin/lactose composite, as well as to characterize the application of design of trials and multilayer perceptron (MLP) NN in learning the spray-drying process (Mihajlovic et al., 2011). The selected MLP structure had three layers: the first layer had three input units, the second layer had three hidden units, and the third layer had four output units. The selected MLP was trained through 10,000 epochs. Design of experiments using response surface methodology permits insight into the linkup between variables, and an ANN comes up with superior prediction potential, authorizing simultaneous diagnosis of several outputs. Design of experiments and ANN proved to be useful tools for optimization of the spray-drying process, where a design space for attaining the best process yields and ideal optimum particle characteristics was established. There is a possibility that this conclusion can be used for other similar systems with different APIs.

Neural computing technology is competent in solving problems involving complex pattern recognition. This technology is put here to pharmaceutical product development. Delta back-propagation network is the frequent algorithm used to study the relation between drug release and formulation variables in capsule preparation. ANN showed response values for the number of validation experiments more accurately than the conventional system (Kandimalla et al., 1999). Similarly, Svetlana et al. studied ANN in the development of ER Tablet and found drastic reliable prediction than the already available system (Svetlana et al., 2012).

Han et al. investigated the usefulness of ANN and DNN in the prediction of oral disintegrating tablets obtained by direct compression. Data was loaded using 145 formulation data from Web of Science. Data prediction by ANN and DNN were contrasted for disintegration time. ANN and DNN showed together better prediction for ODT (Han et al., 2018) (Table 12.2).

TABLE 12.2 AI Approached in NDDS Development

AI Approached	Dosage Form	References
Adaptive neural-fuzzy inference system (ANFIS)	SR ibuprofen	Rebouh et al. (2019)
ANFIS	Estimation of fluoxetine (FLX) and sertraline (SRT)	Hasanjani et al. (2014)
ANN	Pellet-containing tablets	Huang et al. (2015)
ANN	Salbutamol controlled release tablet	Chaibva et al. (2010)
ANNs	O/W emulsion	Jayaram Kumar et al. (2011)
Generalized regression neural network (GRNN)	ER aspirin tablets	Svetlana et al. (2002)
ANNs	Nimodipine matrix tablet	Hussain et al. (1991)
Multilayer perceptron (MLP) neural network.	Naratriptan/maltodextrin/lactose composite production	Mihajlovic et al. (2011)
ANN	ODT	Han et al. (2018)
ANN	Bupropion nanospheres	Zaki et al. (2015)
RSM	liposomal gel	Bhattacharjee et al. (2020)
RSM	Starch-alginate	Nayak et al. (2013)

12.5 CONCLUSION

AI and machine learning are both gradually growing and helping the pharmaceutical industry as a whole. Because of its highly automated and predictive capabilities, advancements in AI have profoundly altered the industrial landscape during the last two decades. The core of pharmaceutical research, which is a very time-consuming and dangerous process, is drug discovery and dosage form development. Through the use of highly automated systems, AI has demonstrated its potential in the management of drug discovery and dosage form development. An issue must be defined, the data must be prepared, the AI architecture must be designed, the model must be trained and evaluated, and the findings must be understood and explained. Different model approaches are used in the AI for the innovative medication delivery system, including ANFIS, ANNs, GRNN, RSM, and DNN. The field of AI has enormous growth potential.

KEYWORDS

- **artificial intelligence**
- **artificial neural network**
- **automation**
- **data prediction**
- **drug discovery**
- **machine learning**
- **pharmaceuticals**

REFERENCES

Achanta, A. S., Kowalski, J. G., & Rhodes, C. T., (1995). Artificial neural networks: Implications for pharmaceutical sciences. *Drug Dev. Ind. Pharm., 21*, 119–155.

Barron, L., Havel, J., Purcell, M., et al., (2009). Predicting sorption of pharmaceuticals and personal care products onto soil and digested sludge using artificial neural networks. *Analyst, 134*, 663–670.

Bhattacharjee, A., Das, P. J., Dey, S., Nayak, A. K., Roy, P. K., Chakrabarti, S., et al., (2020). Development and optimization of besifloxacin hydrochloride loaded liposomal gel prepared by thin film hydration method using 3^2 full factorial design. *Colloids Surf, A: Physicochem. Eng. Asp., 585*, 124071.

Bourquin, J., Schmidli, H., Hoogevest, P., & Leuenberger, H., (1997a). Application of artificial neural networks (ANNs) in the development of solid dosage forms. *Pharm. Dev. Technol., 2*, 111–121.

Cundari, T. R., & Russo, M., (2001). Database mining using soft computing techniques. an integral neural network-fuzzy logic-genetic algorithm approach. *J. Chem. Inf. Sci., 41*, 281–287.

Debunne, A., Vervaet, C., Mangelings, D., & Remon, J. P., (2004). Compaction of enteric-coated pellets: Influence of formulation and process parameters on tablet properties and *in vivo* evaluation. *Eur. J. Pharm. Sci., 22*, 305–314.

Ebube, N. K., Owusu-Ababio, G., & Adeyeye, C. M., (2000). Preformulation studies and characterization of the physicochemical properties of amorphous polymers using artificial neural networks. *International Journal of Pharmaceutics, 196*(1), 27–35.

Faith, C., Michael, B., & Roderick, B. W., (2010). Optimization of salbutamol sulfate dissolution from sustained release matrix formulations using an artificial neural network. *Pharmaceutics, 2*(2), 182–198.

Ghaffari, A., Abdollahi, H., Khoshayand, M. R., Soltani, B. I., Dadgar A., & Rafiee-Tehrani M., (2006). Performance comparison of neural network training algorithms in modeling of bimodal drug delivery. *International Journal of Pharmaceutics, 327*, 126–138.

Hamid, R. A. H., Mahmoud, R. S., & Parviz, A., (2014). Adaptive neuro-fuzzy inference system (ANFIS) applied for spectrophotometric determination of fluoxetine and sertraline in pharmaceutical formulations and biological fluid. *Iranian Journal of Pharmaceutical Sciences, 10*(3), 19–34.

Hassanzadeh, P., Atyabi, F., & Dinarvand, R., (2019). The significance of artificial intelligence in drug delivery system design. *Adv. Drug Deliv. Rev., 151*, 169–190.

Huang, Y., Yao, Q., Zhu, C., Zhang, X., Qin, L., & Wang, Q., (2015). Comparison of novel granulated pellet-containing tablets and traditional pellet-containing tablets by artificial neural networks. *Pharm Dev Technol., 20*(6), 670–675.

Hussain, A. S., Yu, X., & Johnson, D., (1991). Application of neural computing in pharmaceutical product development. *Pharm. Res., 8*(10), 1248–1252.

Jayaram, K. K., Gopal, M. P., & Surabhi, P., (2011). Application of artificial neural networks in optimizing the fatty alcohol concentration in the formulation of an O/W emulsion. *Acta Pharma., 61*(2), 249–256.

Kandimalla, K. K., Kanikkannan, N., & Singh, M., (1999). Optimization of a vehicle mixture for the transdermal delivery of melatonin using artificial neural networks and response surface method. *J. Control Release, 61*.

Kaul, V., Enslin, S., & Gross, S. A., (2020). History of artificial intelligence in medicine. *Gastrointest Endosc. 92*(4), 807–812. doi: 10.1016/j.gie.2020.06.040. Epub 2020 Jun 18. PMID: 32565184.

Koppen, M., Kasabov, N., & Coghill, G., (2009). *Advances in Neuro-Information Processing: 15th International Conference, ICONIP 2008, Auckland, New Zealand*. Springer.

Lima, A. N., Philot, E. A., Trossini, G. G., Scott, L. B., Maltarollo, V. G., & Honorio, K. M., (2016). Use of machine learning approaches for novel drug discovery. *Expert Opin. Drug Discov., 11*, 225–239.

Mak, K. K., & Pichika, M. R., (2019). Artificial intelligence in drug development: Present status and future prospects. *Drug Discov. Today, 24*(3), 773–780.

Manoj, K. C., & Vipul, P. P., (2020). A review article on artificial intelligence; change in modern technique of pharmaceutical formulation and development. *JETIR, 7* (9), 1466–1473. https://www.jetir.org/papers/JETIR2009198.pdf .

Maurizio, S., Abdul, R. K., David, L., Morten, A., & Murat, K., (2020). Artificial intelligence in pharmacoepidemiology: A systematic review. Part 1 – Overview of knowledge discovery techniques in artificial intelligence. *Frontiers in Pharmacology, 11*, 1928, 1–28.

Nayak, A. K., Pal, D., & Hasnain, M. S., (2013). Development, optimization and *in vitro-in vivo* evaluation of pioglitazone-loaded jackfruit seed starch-alginate beads. *Curr. Drug Deliv., 10*(5), 608–619.

Oxford, (2019). *Artificial Intelligence*. Oxford Dict. Available at: https://www.lexico.com/en/definition/artificial_intelligence (accessed on 6 December 2023).

Rebouh, S., Lefnaoui, S., Bouhedda, M., Yahoum, M. M., & Hanini, S., (2019). Neuro-fuzzy modeling of ibuprofen-sustained release from tablets based on different cellulose derivatives. *Drug Deliv. Transl. Res., 9*(1), 162–177.

Rohan, G., Devesh, S., Mehar, S., Swati, T., Rashmi, A., & Pravir, K., (2021). Artificial intelligence to deep learning: Machine intelligence approach for drug discovery. *Molecular Diversity, 25*, 1315–1360.

Run, H., Yilong, Y., Xiaoshan, L., & Defang, O., (2018). Predicting oral disintegrating tablet formulations by neural network techniques. *Asian J. Pharm. Sci., 13*(4), 336–342.

Sessa M, Khan AR, Liang D, Andersen M and Kulahci M (2020) Artificial Intelligence in Pharmacoepidemiology: A Systematic Review. Part 1—Overview of Knowledge Discovery Techniques in Artificial Intelligence. Front. Pharmacol. 11:1028. doi: 10.3389/fphar.2020.01028

Sheela, K., Jianchang, L., Rachael, L., Yanwei, Z., & Wenwen, Z., (2022). Machine learning and artificial intelligence in pharmaceutical research and development: A review. *The AAPS Journal, 24*, 19.

Svetlana, I., Jelena, D., Jelena, P., & Zorica, D., (2012). Artificial neural networks in evaluation and optimization of modified release solid dosage forms. *Pharmaceutics, 4*(4), 531–550.

Svetlana, I., Milica, J., Jelena, P., & Ljiljana, S., (2002). The application of generalized regression neural network in the modeling and optimization of aspirin extended-release tablets with Eudragit RS PO as matrix substance. *J. Control Release, 82*(2, 3), 213–222.

Tijana, M., Svetlana, I., & Aleksandar, M., (2011). Application of design of experiments and multilayer perceptron neural network in optimization of the spray-drying process. *Drying Technology, 29*(14), 1638–1647.

Vyas, M., Thakur, S., Riyaz, B., et al., (2018). *Asian J. Pharmaceutics, 12*(2), 72–76.

Xin, Y., Yifei, W., Ryan, B., Gisbert, S., & Shengyong, Y., (2019). Concepts of artificial intelligence for computer-assisted drug discovery. *Chem. Rev., 119*, 10520–10594.

Yichun, S., Yingxu, P., & Yixin, C., (2003). Application of artificial neural networks in the design of controlled release drug delivery systems. *Adv. Drug Deliv. Rev., 55*(9), 1201–1215.

Yüksel, N., Türkoglu, M., Baykara, T., Yüksel, N., Türkoglu, M., & Baykara, T., (2000). Modelling of solvent evaporation method for the preparation of controlled release acrylic microsphere using neural networks. *Journal of Microencapsulation, 5*(17), 541–551.

Zaki, M. R., Varshosaz, J., & Fathi, M., (2015). Preparation of agar nanospheres: Comparison of response surface and artificial neural network modeling by a genetic algorithm approach. *Carbohydr. Polym., 122*, 314–320.

Index

B

C

D

J

K

L

M

O

Q

R

X

Z